Essential Systems Analysis

Essential Systems Analysis

by Stephen M. McMenamin & John F. Palmer

foreword by Tom DeMarco

Yourdon Press
1133 Avenue of the Americas
New York, New York 10036

Library of Congress Cataloging in Publication Data

McMenamin, Stephen M., 1955-
    Essential systems analysis.

    Bibliography: p.
    Includes index.
    1. System analysis. I. Palmer, John F. II. Title.
T57.6.M374 1984   001.64   84-11913
ISBN 0-917072-30-8

This book was set in Times Roman by Yourdon Press, 1133 Avenue of the Americas, New York, N.Y., using a PDP-11/70 running under the UNIX® operating system.*

---

*UNIX is a registered trademark of Bell Laboratories.

To my mother,
and in memory of my father

— S.M.McM.

To my wife, Barbara

— J.F.P.

Foreword

When Associate Justice Potter Stewart stepped down from the Supreme Court a few years ago, he called a press conference to make some parting remarks. Reporters asked him if he had any regrets about his time on the high court, and he replied with a characteristic gentle humor, "Well . . . perhaps that quote, 'I may not be able to define *obscenity,* but I know it when I see it.' I worry they're going to carve that on my grave-stone."

In the terms that Steve McMenamin and John Palmer introduce in *Essential Systems Analysis,* the problem that Justice Stewart had wrestled with was how to determine the *essence* of a complicated phenomenon. In the crucial systems analysis activity, this problem is huge: Discovering the true essence of a system is probably the most important thing analysts need to do, and the most difficult. The essence of a system is too easily confused with one particular method selected to implement that essence, what the authors call the system's *incarnation.* The inability to separate essence from incarnation leads analysts to write system specifications that are overly restrictive, and developers to build systems that carry over meaningless features from past incarnations.

Essence and incarnation correspond roughly to what I have referred to in the past as the logical and physical perceptions of a system. In my own work, I took a Stewart-like approach to separating the two: I may not have been able to define *logical* to anyone's satisfaction, but I knew it when I saw it. Although I went to some lengths to help my readers attain the same "I know it when I see it" facility, that wasn't enough for many analysts. Worse still, those who did learn to distinguish between logical and physical had little success imparting that skill to others. Analysts clearly needed a more rigorous distinction between what a system *has to be* and what one particular implementation of the system *happens to be.* McMenamin and Palmer have made this distinction clearly and persuasively. With it, they have concocted an entirely new approach to system modeling.

An astonishing number of books are published about system development that merely rehash other people's ideas. Some years ago, I coined a metric of this phenomenon, the Same Old Stuff Index. Books with a high SOSI offer the reader no new insight and so no real value. The highest compliment I can pay McMenamin and Palmer is that their SOSI is nearly zero — their book is simply full of good new ideas. Because of this, *Essential Systems Analysis* has impact. It tends to change the way you go about the work of analysis. It has changed me; I regularly apply its event partitioning techniques, for instance, and its synthesis of function models and information models, and many of its useful terms. I believe it will have the same kind of effect on you.

Tom DeMarco
The Atlantic Systems Guild
New York, New York

April 1984

Acknowledgments

While developing the concepts and procedures that make up *Essential Systems Analysis,* we enjoyed the generous assistance of our clients, colleagues, editors, and friends. We cannot hope to thank them all individually, but we can mention at least these, our principal sources of advice and assistance.

The methods we propose have been proved in the field, on real-life system development efforts. So, our first thanks must go to those of our clients who accepted the challenges and risks of pioneering an experimental system development technique. Much of this work was accomplished by employees of the Southern California Edison Company, under the leadership of William H. Bentley and David R. Tommela.

The first of these pioneer efforts was the Materials Management System (MMS) project, begun in 1978 under the direction of Michael L. Mushet. Through MMS, we learned about the anatomy of large systems and the organization of large analysis efforts, and we were able to try out our first procedure for deriving the logical equivalent of a current physical model. James R. Horstman served as the internal structured analysis expert on the MMS team.

Much of this knowledge was employed on the Continuing Property Records (CPR) project, which began in 1979 and which was the first to employ many techniques that are rapidly becoming standard practice. These included event and object partitioning to avoid excessive physical modeling, iterative modeling at multiple levels of detail, the blitzing technique for rapid development of project planning models, and the gallery approach to product reviews. The CPR project was led by Larry L. Proctor, Janet Halliwell, and Glenn W. Collins; Nina C. Snett was responsible for many of the improvements in structured analysis methods achieved by this project team.

During the same time, the Boeing Company's Seattle Service Division and Boeing Computer Services undertook a large and ambitious analysis effort known as the Integrated Employee Records System (IERS). The IERS project was also among the first to make use of our event/object approach for partitioning a logical system model. The IERS experience helped us to refine our techniques, particularly those for modeling stored data requirements. We thank all the members of this project, particularly Ruth Bertilson, Les Fauver, John Hill, Frank Jones, and Art Miller.

Also in 1979, Salomon Bros. used an embryonic form of essential systems analysis on their Bank Link project, where a blitzed event/object-partitioned essential model was produced. New ideas on the development of new physical models came out of our joint efforts on this project. Barrie Mitchell led this project to an almost flawless installation.

An early version of our technique was used at the J. Aron & Co. on a project to develop a foreign exchange (FOREX) system. Lou Gutentag led this effort and, in the process, helped us develop standards for minimal models of the current implementation of a system.

After the pioneers helped us take our first steps with essential systems analysis, a second wave of clients helped us standardize our approach. From 1980 to 1983, these clients helped us work out a party line that would communicate essential systems analysis even when we weren't around to point out how everything is "intuitively obvious to a casual observer."

The assistance of Rob Herson at Texaco allowed us to firm our blitzing approach to the point that it could be made part of Texaco's Analysis, Design, and Management Standard. Commander Brian Sonner and Lt. Commander Rod Smith at the Office of the Comptroller, U.S. Coast Guard, also helped us polish our blitzing approach through its use on their Standard Automated Accounting System project. David Shapiro, Eric Joyall, and Rakesh Jain at Bell Canada and Jan Lauridson and Keld Jorgensen at Sparekassernes Datacenter provided us with numerous projects on which to refine our data modeling ideas.

Thanks again are due to Eric Joyall and the Structured Analysis and Design Project at GUIDE, IBM's large-system user group. In 1982, they provided us the opportunity to consolidate our ideas on the transition between analysis and design and gave us a forum to present those ideas at GUIDE 54. Ken DeLavigne at IBM also deserves kudos for helping us see that the analysis-design transition/new physical phase is a lot more than preparing data flow diagrams to be turned into structure charts through transform analysis.

For their support and encouragement, we thank our associates at the Atlantic Systems Guild: Tom DeMarco, Tim Lister, James Robertson, and Suzanne Robertson.

Other colleagues also deserve our gratitude, particularly Truitt Allen, Wayne E. Bissell, Brian Dickinson, Pat Duran, Mike Fife, Richard K. Fisher, Matt Flavin, Dorothy K. Kushner, Lou Mazzucchelli, Ira Morrow, Meilir Page-Jones, Sandra Rapps, Gary Schuldt, Michael Silves, III, Charles A. Tryon, Mark Wallace, Paul T. Ward, and Carroll T. Zahn.

Tim Lister and Wendy Eakin provided us with the opportunities to develop our ideas. While head of seminars at Yourdon, Tim gave us several course development projects that helped us evolve our techniques. As Director of Yourdon Press, Wendy gave us the opportunity to write this book, and hung in with us throughout its writing.

Finally, we want to thank the staff of Yourdon Press. Susan Moran was absolutely indispensable not only as our editor, but also as an astute technical advisor, helping us clarify the presentation of many of the book's complex ideas. Both she and Janice Wormington did a masterful job of helping us tighten our prose. We would also like to thank Lorie Mayorga for her review of the initial manuscript, Dan Murray for his artwork, and George Armstrong for the cover design.

Contents

Part One: FOUNDATION   1

Chapter 1: IDENTIFYING TRUE REQUIREMENTS   3
1.1 Distinguishing between true and false requirements   3
1.2 The danger of false requirements   4
1.3 Defining requirements with structured analysis   5
1.4 Distinguishing between logical and physical   5
1.5 The scope of this book   8
1.6 Summary   9

Chapter 2: PLANNED SYSTEMS   10
2.1 Interactive systems   10
2.2 Planned response systems   12
2.3 Summary   14

Chapter 3: THE ESSENCE OF A SYSTEM: LOGICAL REQUIREMENTS   16
3.1 The concept of perfect technology   16
3.2 The components of a system's essence   17
    3.2.1 Fundamental activities   17
    3.2.2 Essential memory   18
    3.2.3 Custodial activities   20
    3.2.4 Compound essential activities   21
3.3 Summary   22

Chapter 4: THE INCARNATION OF A SYSTEM: PHYSICAL CHARACTERISTICS   23
4.1 The incarnation of essential features   25
4.2 The planned response system within the implementation   25
4.3 The impact of imperfect technology   29
    4.3.1 Fragmentation   29
    4.3.2 Redundancy   30
    4.3.3 Extraneousness   30
    4.3.4 Convolution   32
    4.3.5 Conglomeration   32
    4.3.6 Vastness   32
4.4 Summary   33

Chapter 5: THE ESSENCE OF SYSTEM DEVELOPMENT   34
5.1 Imperfect system development technology   35
    5.1.1 Imperfect system development processors   35
    5.1.2 Multiple system development processors   36
    5.1.3 Planning and control activities   36
    5.1.4 Quality control activities   37
5.2 Essential system development activities   38
5.3 Summary   38

Chapter 6: SYSTEM MODELING   39

6.1 The benefits of modeling   40
6.2 The choice of a modeling approach   40
6.3 The principles of essential modeling   42
     6.3.1 The budget for complexity   42
     6.3.2 Technological neutrality   43
     6.3.3 Perfect internal technology   44
     6.3.4 The minimal essential model   45
     6.3.5 The benefits of the modeling principles   46
6.4 Summary   46

Part Two: ESSENTIAL MODELING TOOLS AND STRATEGIES   47

Chapter 7: PARTITIONING THEMES FOR ESSENTIAL MODELS   49

7.1 Partitioning the system into essential activities   49
     7.1.1 Recognizing events   49
     7.1.2 The event-partitioned system   51
     7.1.3 The advantages of event partitioning   56
7.2 Partitioning essential memory   57
     7.2.1 Objects   57
     7.2.2 Object partitioning   59
7.3 Problems with other partitioning themes   61
     7.3.1 The single data store partitioning solution   61
     7.3.2 The data store for each element partitioning solution 61
     7.3.3 The private component file partitioning solution 64
     7.3.4 The benefits of object partitioning   66
7.4 Leveling   66
7.5 Summary   68

Chapter 8: MODELING ESSENTIAL ACTIVITIES   69

8.1 Modeling the stimulus   69
     8.1.1 A stimulus from an external event   69
     8.1.2 A stimulus from a temporal event   71
8.2 Modeling the planned response   73
     8.2.1 Controlling the complexity of minispecifications 74
     8.2.2 Avoiding false requirements   74
     8.2.3 Redefining structured English   75
     8.2.4 Using a nonsequential modeling tool   75
8.3 Modeling planned response results   76
     8.3.1 Modeling external results   77
     8.3.2 Modeling essential memory updates   78
8.4 Modeling essential memory accesses   78
     8.4.1 Defining an essential memory access   79
     8.4.2 Modeling essential memory accesses in minispecifications 80
     8.4.3 Modeling essential memory accesses on the DFD   80
     8.4.4 Modeling accesses with the data dictionary   81
8.5 Summary   84

Chapter 9: MODELING ESSENTIAL MEMORY   85

9.1 Modeling data elements   85
9.2 Modeling objects   86
     9.2.1 Modeling large objects   87
     9.2.2 Intra-object relationships   92
9.3 Modeling interobject relationships   93
     9.3.1 Data structure diagrams   94
     9.3.2 Entity-relationship diagrams   95
     9.3.3 Minispecifications   96

9.4 Choosing the right approach   96
    9.4.1 Interobject relationship modeling vs. essential access modeling   97
    9.4.2 Global models vs. local models   97
    9.4.3 Specialized models vs. general models   98
    9.4.4 Requirement models vs. semantic models   98
    9.4.5 Graphic models vs. narrative models   98
9.5 The need for a better convention   99
9.6 Summary   100

Chapter 10: STRATEGIES FOR MODELING THE ESSENCE OF A SYSTEM   101

10.1 Creating the new essence   102
    10.1.1 First decision: identifying the system's purpose   102
    10.1.2 Second decision: identifying fundamental activities   103
    10.1.3 Third decision: identifying required information   104
    10.1.4 Fourth decision: identifying custodial activities   105
10.2 Modeling the new essence   106
10.3 Deriving the new essence   108
    10.3.1 Why derive the essence?   108
    10.3.2 Why model the current incarnation?   109
10.4 Choosing a modeling strategy   112
    10.4.1 Factors in choosing a modeling strategy   112
    10.4.2 The reasons for our choice of strategy   115
10.5 Summary   115

Part Three: THE ANATOMY OF EXISTING SYSTEMS   **117**

Chapter 11: THE ANATOMY OF SINGLE PROCESSOR SYSTEMS   119

11.1 Processors   119
11.2 Single processor/single essential activity systems   121
    11.2.1 Essential activities   123
    11.2.2 Internal quality control activities   124
    11.2.3 Internal transportation activities   124
11.3 Optimizing processor performance   125
11.4 Single processor/multiple essential activity systems   127
11.5 Summary   127

Chapter 12: THE ANATOMY OF MULTIPLE PROCESSOR SYSTEMS   128

12.1 The fragmentation of essential activities among processors   128
    12.1.1 Fragmentation according to processor skill and cost   129
    12.1.2 Fragmentation according to processor capacity and cost   131
12.2 The interprocessor infrastructure   133
    12.2.1 Sharing stored data   133
    12.2.2 Communicating intermediate products   145
12.3 The interprocessor administration   148
12.4 Summary   150

Chapter 13: THE CONSOLIDATION OF SYSTEM ACTIVITIES   151

13.1 Consolidation of essential activity fragments   151
13.2 Consolidation of nonessential system activities   152
    13.2.1 Consolidation of the infrastructure   154
    13.2.2 Consolidation of data stores   157
    13.2.3 Consolidation of administrative activities   158
    13.2.4 Consolidation of activities in a multiple processor system 160
13.3 Nested processors   160
13.4 Superprocessor incarnations   165
13.5 Summary   167

Chapter 14: DISCOVERING THE ESSENTIAL SYSTEM THROUGH SYSTEM ARCHAEOLOGY   168
14.1 The discovery strategy   169
14.2 Finding and classifying the essential activity fragments   171
14.3 Modeling an essential activity   171
14.4 Integrating the essential activity models   171
14.5 Summary   172

Part Four: FINDING THE ESSENTIAL ACTIVITIES   **173**
Chapter 15: EXPANDING THE CURRENT PHYSICAL MODEL   175
15.1 Expanding the data flow diagrams   177
15.2 Expanding the data dictionary   181
15.3 Expanding the minispecifications   184
15.4 Feedback effects   186
15.5 A tranquilizer for expandophobes   187
15.6 Summary   187

Chapter 16: REDUCING THE EXPANDED PHYSICAL MODEL   188
16.1 Removing the interprocessor infrastructure   188
16.2 Removing the administration   197
16.3 Reconnecting the essential fragments   199
16.4 The results of the reduction process   201
16.5 Summary   201

Chapter 17: CLASSIFYING THE ESSENTIAL FRAGMENTS   202
17.1 Event partitioning revisited   204
17.2 Identifying external events   205
     17.2.1 Naming external events   208
17.3 Identifying essential fragments for each external event   209
     17.3.1 Identifying the system's complete response   215
17.4 Identifying temporal events   221
     17.4.1 Naming temporal events   223
17.5 Identifying essential fragments for each temporal event   224
17.6 Building the initial essential model   226
17.7 Summary   229

Part Five: DEFINING AN ESSENTIAL ACTIVITY   **231**
Chapter 18: COMPLETING THE DERIVATION PROCESS: AN OVERVIEW   233
18.1 Partitioning the remaining work   233
     18.1.1 The two-pass approach   234
18.2 The difference between creating and deriving   237
18.3 Summary   237

Chapter 19: RECOGNIZING REMAINING PHYSICAL CHARACTERISTICS   238
19.1 Physical characteristics of essential activity fragments   238
     19.1.1 Fragmentation   238
     19.1.2 Extraneousness   240
     19.1.3 Redundancy   241
     19.1.4 Convolution   241
     19.1.5 Nonessential sequence   244
19.2 Physical characteristics of essential memory fragments   251
     19.2.1 Fragmentation   252
     19.2.2 Extraneousness   253
     19.2.3 Redundancy and convolution   253
19.3 Physical names   254
19.4 Summary   254

Chapter 20: DERIVING AN ESSENTIAL ACTIVITY   255

20.1   Response-based derivation   255
20.2   Selecting an essential activity   257
20.3   Uncovering the core of the essential activity   257
20.4   Removing extraneous physical features   260
20.5   Consolidating the remaining fragments   265
20.6   Partitioning the essential memory   266
       20.6.1 Identifying objects   267
       20.6.2 Attributing data elements to object data stores   269
       20.6.3 Resolving attribution anomalies   270
20.7   Minimizing the essential accesses   270
20.8   Establishing the essential order among the activities   276
20.9   Establishing the physical ring   276
20.10 Summary   278

Chapter 21: MODELING AN ESSENTIAL ACTIVITY   279

21.1   Creating the high-level essential activity model   280
21.2   Modeling the details of a small essential activity   281
21.3   Modeling the details of a large essential activity   283
21.4   Modeling interobject relationships   287
21.5   Summary   287

Part Six: INTEGRATING THE ESSENTIAL ACTIVITY MODELS   289
Chapter 22: INTEGRATING THE ESSENTIAL ACTIVITIES   291

22.1   Cross-checking essential accesses   291
22.2   Resolving missing custodial activities   293
22.3   Removing unnecessary custodial activities   295
22.4   Summary   298

Chapter 23: BUILDING A GLOBAL ESSENTIAL MODEL   299

23.1   Consolidating object data stores   300
23.2   Drawing a global data flow diagram   304
23.3   Remodeling derived essential memory   304
23.4   Factoring out common functions   307
23.5   The completely essential model   309

Chapter 24: REVIEWING THE QUALITY OF THE MODEL   310

24.1   Repartitioning data stores   311
24.2   Creating subminispecs   312
24.3   Building upper levels of data flow diagrams   314
24.4   Summary   317

Part Seven: MODELING THE NEW SYSTEM   319
Chapter 25: ADDING NEW ESSENTIAL FEATURES   321

25.1   Building mini-models   322
25.2   Defining new essential features   323
       25.2.1 When the users or analysts understand essential systems analysis   323
       25.2.2 When the users or analysts don't know essential systems analysis   325
       25.2.3 Integrating new essential requirements   325
25.3   Understanding the psychology of new essential requirements   327
25.4   Summary   327

Chapter 26: SELECTING AN INCARNATION   329

26.1   An incarnation's general goals   329
26.2   Strategy for deriving an incarnation   329
       26.2.1 Specifying the external interface   330

26.2.2 *Allocating essential activities to processors*   330
26.2.3 *Establishing the infrastructure*   331
26.2.4 *Establishing the administration*   336
26.2.5 *Optimizing the incarnation*   336
26.3 Summary   340

Chapter 27: CREATING A SOFTWARE DESIGN BLUEPRINT
FROM THE INCARNATION MODEL   341

27.1 Designing hierarchical software   341
27.2 Transforming the network model into a hierarchical model   342
27.3 Applying structured design   342
27.4 Summary   345

Part Eight: MANAGING ESSENTIAL SYSTEMS ANALYSIS   **347**

Chapter 28: TIME: THE CRITICAL DEVELOPMENT RESOURCE   349

28.1 Why time is critical   349
28.2 How project time is wasted   350
28.2.1 *The current physical tarpit*   351
28.3 Summary   354

Chapter 29: OPTIMIZING THE PROCESS OF DEFINING ESSENCE   355

29.1 How the essential modeling framework saves time   355
29.2 Deviating from the idealized derivation approach   356
29.3 The optimized procedure: blitzing an essential model   357
29.3.1 *Establishing the system's purpose*   358
29.3.2 *Creating an essential context diagram*   358
29.3.3 *Creating lists of objects and events*   359
29.3.4 *Drafting the preliminary essential model*   361
29.4 Mopping up   361
29.4.1 *Blitzing a current physical model*   363
29.4.2 *Establishing an essential modeling plan*   364
29.4.3 *Building individual essential activity models*   365
29.4.4 *Integrating the essential activity models and performing the final pass*   366
29.5 Summary   366

Chapter 30: EFFECTIVE PROJECT MANAGEMENT: A LEVELED APPROACH   367

30.1 Leveled system development   369
30.2 Problems with the unleveled approach   371
30.3 The benefits of the leveled approach   371
30.4 Plumbing the depths of a detailed activity   372
30.5 Project scouts   372
30.6 Intermediate levels of development   374
30.7 Summary   374

Chapter 31: PLANNING LONG PROJECTS: THE PROJECT AS MOVING TARGET   375

31.1 Destructive external factors   375
31.2 Sitting duck projects   376
31.3 Turning a sitting duck into a moving target   377
31.4 Model-based project planning   377
31.5 Summary   380

Afterword   **381**

Bibliography   **383**

Index   **387**

Part One

Foundation

We begin *Essential Systems Analysis* by establishing the foundation upon which the rest of the book rests. Chapter 1 sets our sights on the problem of identifying what we call the true requirements for a system. We point out the importance of discovering true requirements, the historical difficulty we have had in doing so, and the consequences of failure. We also trace the attempts of other systems analysis disciplines to make this determination, often described as the distinction between the physical and logical aspects of a system.

In Chapter 2, we focus on systems that carry out predefined actions as a result of anticipated events. Such planned response systems make up the vast majority of systems analysis assignments, and our modeling tools and strategies are well suited to this type of system.

Chapter 3 introduces the central abstraction of our work, the concept of the essence of a system. We define the essence as those aspects of the system that transcend all possible technological constraints. We use the notion of a system's essence to distinguish true requirements from the limitations of a particular form of system implementation technology.

In Chapter 4, we contrast the essential view of a system with its more familiar appearance in reality, something we call the system's incarnation. We identify the nature of imperfect implementation technology and describe how its limitations complicate a real system and obscure its essence.

Chapter 5 employs the concept of essence to arrive at a simplified view of the system development process, one that ignores the political and managerial differences among projects and concentrates on the essential activities that must be carried out on projects of all sizes and durations.

In Chapter 6, we focus on one of the essential system development activities and the primary topic of the book: modeling the essence of a system. We offer a set of principles that will guide your use of the modeling tools and the detailed strategies described in subsequent parts.

Chapter 1

Identifying
True Requirements

The tools and techniques for performing systems analysis have improved tremendously over the last decade. Imprecise narrative text, incomplete flowcharts, and faulty human memory have given way to formal techniques for modeling business systems and data. In addition, software packages now support at least some of the analysis tasks.

Yet, despite these improvements, many systems analysis projects fail. The reason is simply that the resulting specification does not express the true requirements of the system to be developed. A true requirement is a feature or capability that a system must possess in order to fulfill its purpose, *regardless* of how the system is implemented. We call the complete set of true requirements for a system the *essence* of the system or the *essential* requirements.

In this book, we propose a method for discovering and defining essential system requirements. The method is based upon a conceptual framework composed of three elements: a description of the characteristics of systems, a classification of the components of a system's essence, and a classification of implementation constraints. Structured analysis tools are used to represent these system requirements.

In this chapter, we introduce the topic by first focusing on the importance of discovering the true requirements and the dangers of false requirements, followed by a brief look at the reasons for analysts' difficulty in specifying true requirements and a review of other systems analysis disciplines.

1.1 Distinguishing between true and false requirements

The specification should contain all the true requirements and nothing but the true requirements. Otherwise, the specification will suffer from two deficiencies: It will not contain *all* of the true requirements for the system, and it will contain *false* requirements.

A system requirement is false if the system could fulfill its purpose without implementing the requirement. A feature that is clearly irrelevant to the purpose of the system or an activity that must be carried out solely to accommodate the technology used to implement the system are examples of a false requirement. Another type of false requirement is a statement that specifies an activity that is truly needed, but describes the activity in terms of the technology used to carry it out. This type of false requirement imposes a particular procedure on a required system activity, a procedure that may be sensible but is not the *only* sensible way to accomplish the true requirement.

There are two major categories of false requirements: technological requirements and arbitrary requirements. Technological false requirements originate either because the analysts carry the technology of an existing system into the requirements specification, or analysts anticipate technological characteristics of the new system and include them in the specification. Arbitrary false requirements also originate in one of two ways. First, the analysts may describe a system requirement so that it makes the system do more than is necessary to accomplish the system's purpose. Second, the modeling tools themselves, whether charts or narrative English, can impose their own bias upon the organization of system activities. Whatever the cause, false requirements are undesirable because they do not express the essence of the system as precisely and accurately as possible.

1.2 The danger of false requirements

In most development projects, analysts study an existing system both to identify its problems and to build a specification of a new system to solve the problems. With company profitability, careers, and sometimes even public safety riding on the outcome of the analysis effort, project members are strongly motivated to do a complete job. Disaster is likely when these highly motivated analysts start to analyze an existing system without the ability to tell a true requirement from one specifying the technology used to implement the system. This inability can lead to analysis efforts that become stalled, specifications that are incorrect and hard to understand, systems that are unreliable and difficult to maintain, and projects that finish late and over budget.

Anxious to produce a thorough specification and lacking the discrimination of a true-requirements connoisseur, the analysts study and document everything to avoid missing anything. Progress slows as analysts spend weeks or even months documenting irrelevant details. Meanwhile, political pressure builds to the bursting point. When the patience of users and management runs out, they react by shelving projects, firing unlucky project leaders and members, or arbitrarily ending the analysis to move onto other project tasks. Even if the project survives this kind of analysis paralysis, the specification produced is often wrong in two ways. One flaw in the typical specification is that it omits some true requirements entirely. So many false requirements clutter the thinking and documentation of the analysts that they easily overlook features needed by the new system. Obviously, a system will be deficient if it is built to a specification that is incomplete. So, when false requirements prevent analysts from noticing the omission of true requirements, the ultimate effect is a system that doesn't do what it ought to do.

Second, the specification often contains an irrational preference, or technological bias, for a particular kind of implementation technology. No matter what the source of the bias is, analysts may end up specifying unneeded batch files, edits to check for mistakes that are no longer made, unnecessarily repeated data and activities, overly distributed or overly centralized systems, and a whole host of similar false requirements. If such a bias passes undetected all the way through the development process, then that bias will result in a new system that is unnecessarily complex. This complexity will interfere with the reliable operation of the system and also make the system harder to maintain and enhance.

1.3 Defining requirements with structured analysis

In the mid-1970s, the problem of specifying true requirements began to attract serious attention. Several new approaches to system requirements definition emerged, helping analysts produce complete statements of true requirements that were free from technological bias.

In 1975, Douglas T. Ross and Kenneth E. Schoman, Jr., published the first paper on their Structured Analysis Design Technique or SADT® [40]. SADT was a major step forward in requirements definition technology, because it was the first approach to propose a practical set of *graphic* modeling tools that overcame the obvious deficiencies of narrative specification. Unlike a narrative specification, SADT produces a well-organized specification that is easy to read, verify, and maintain. The centerpiece of an SADT specification or system model is an "activity diagram," which is an early variant of the data flow diagram employed by Edward Yourdon and Larry L. Constantine [54], among others. Although the activity diagram is an improvement over narrative text, some users believe the diagram carries too much information.

The larger problem with SADT is that Ross and Schoman did not say much about the process of developing a requirements specification. The degree to which users of SADT succeeded with the approach depended a lot on their ability to develop a method and the quality of the home-grown approach.

In 1977, in *Structured Systems Analysis: Tools & Techniques* [17], Chris Gane and Trish Sarson improved upon SADT by offering a rudimentary strategy for building a dataflow-based system requirements model. They also proposed further specialization of the modeling tools of structured analysis, specifically use of the data flow diagram in combination with the data dictionary and the process description. In order to avoid overcrowding the data flow diagram, Gane and Sarson advised analysts to take some of the information out of the DFD and place it in the dictionary or process description. A year later, books by Tom DeMarco and Victor Weinberg were published, both advocating roughly the same model structure as Gane and Sarson, with a few minor changes in the terminology [11, 52].

All of these authors focused on the documentation tools and conventions of structured analysis, which can help an analyst in many cases tell if true requirements have been omitted or technological bias has been introduced. However, these tools and conventions addressed only part of the problem of how to identify true system requirements. What the authors failed to provide were conceptual guidelines for analysts to distinguish true requirements from technological influences. Rather, these structured analysis pioneers offered only vague notions of the difference between system requirements and physical features, as discussed below.

1.4 Distinguishing between logical and physical

Just as they pioneered structured analysis modeling tools, Ross and Schoman also introduced the notion that structured analysis should distinguish the logical system requirements from the physical aspects of the system. Gane and Sarson, DeMarco, and Weinberg all agreed in their respective works that distinguishing the logical from the physical is indeed important. In fact, Gane and Sarson as well as DeMarco advocated building a logical model of the existing system. This current logical model shows the aspects of the existing system that are independent of the implementation technology; it serves as the foundation of the requirements specification for the new system. What we call *essence* corresponds to what the structured analysis pioneers intended by *logical system,* although the two terms are defined differently.

Like most authors who tell us to separate form from function, the founders of structured analysis were not too specific about how to do it. Gane and Sarson, for example, did not offer a definition of *logical* or *physical.* Instead, they provided examples of several possible physical implementations of a dataflow (for example, a letter, a telephone call, or a satellite link). Readers must figure out for themselves what makes each of these implementations physical and what other physical aspects of systems they should look for. DeMarco used the word *policy* to illustrate the logical version of systems. He said that when you "logicalize" a system model, "a particular implementation of policy is replaced by a representation of the policy itself" [11]. Other authorities described the difference between the logical and physical systems as the difference between the "what" and the "how" of a system, or between the "strategic" and "tactical" views.

These definitions are not clear enough to allow analysts to distinguish between the system's essential purpose and one particular implementation. They seem adequate only if you already have a fairly strong intuition of your own about the difference between logical and physical that is compatible with the definition. However, intuition and vague definitions are not much help in the search for true requirements.

To understand what is vague about DeMarco's definition of logical, for example, ask, What is policy? Suppose you receive a memo from your boss stating that all requests for pencils should be written in ink. Is this policy? Should such a requirement become a part of a logical model of a system, or is it one of Gane and Sarson's "alternative physical implementations"? The problem is that the word *policy* has two unfortunate connotations. First, policy implies any political or managerial constraint that is beyond your control or dictated by a higher authority. However, a requirement that is policy to one person may seem like an implementation tactic to another, depending upon their positions within the organization and their relationships to the system. The second problem with *policy* is that it usually refers to the business world, so the word's meaning is not always clear when it is applied to a scientific, engineering, or process control system. Consequently, the system development team members can be confused about the distinction between the physical and logical aspects of a system.

At least some of this confusion has been caused by the choice of words. *Physical* and *logical* each have meanings in everyday English that are incompatible with their meanings in structured analysis. To say that something is logical means that it makes sense, or that it is consistent with some set of rules. But in systems analysis, if something is logical, that means it is independent of system implementation technology. Lots of things are consistent with a given set of rules, but nonetheless contain many characteristics of a certain technology. A programming language, for example, may be logical in that it makes sense, but it certainly reflects a particular data processing technology.

Physical is a problematical word as well. To say that something is physical is to imply that it has mass and other tangible features. Although the technological characteristics of a system are often physical, they are more often intangible. For example, the potential for a power failure dictates that certain backup and recovery procedures be built into your systems. These aspects of the system are considered physical even if the dreaded power failure never occurs. A characteristic of a system is physical if it arises from a particular technology.

On the other hand, a part of the system can be physical (in the sense of tangible) and have nothing to do with the technology that the system uses. Consider a company that sells books. Regardless of the sophistication of the current system implementation technology, the book selling business has always involved books. Of course, books are physical in the usual sense; they have mass and can be touched and carried around. But books are also logical by our definition, because no matter how you implement a book selling operation, you eventually have to sell books. Because of these confusing connotations, the terms *logical* and *physical* sometimes are more trouble than they are worth. Nevertheless, these words and an admonition to do good work in distinguishing logical from physical were all that most early structured analysis texts offered systems analysts.

DeMarco, however, offered a more fully developed strategy for building a require-ments specification. This strategy, which is summarized in Figure 1.1, consists of four steps: Model current operations, distill underlying policy, add new requirements, and select automation boundary. He offered an abstract set of typical physical characteris-tics, which became the basis for two procedures for deriving the current logical model: one for activities and one for stored data.

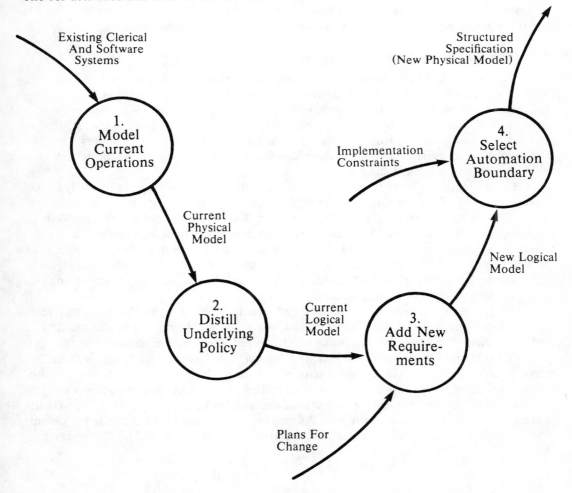

Figure 1.1. The process of structured analysis, according to DeMarco.

Despite the benefits that structured analysis offered, many systems analysis efforts that used the techniques in the late 1970s suffered. Project team members often fought long, acrimonious, and ultimately inconclusive battles over whether a particular form, activity, data element, or file belonged in the logical model. Unable to resolve these disputes, project managers usually chose to err on the side of caution, implementing the motto, "When in doubt, leave it in." The requirements specifications that these teams produced were therefore quite large, since so many of the existing and the new requirements seemed logical to at least some of the team members. Middle and upper management understandably showed little patience with projects that seemed forever mired in what we call the current physical tarpit. Some projects failed simply because management decided that they couldn't wait any longer for results. These problems continue today for all but a few projects that have advanced their techniques.

By the end of the 1970s, practitioners of structured analysis began to realize that there is more to identifying true system requirements than using the right modeling tools. The key to success is the thought process — the analysis strategy — that is carried out by the systems analyst during the construction of the requirements model. As consultants and educators, we turned our attention to the development and refinement of systems analysis strategies. The result is a new approach to discovering and modeling true system requirements, an approach we call *essential systems analysis.*

1.5 The scope of this book

Our first step in this book is to lay out the framework upon which our strategies are based. The first part of this framework is a definition of a class of systems that we call planned response systems, which includes software systems. We then provide a definition of the essence of a planned response system — its logical characteristics — and explain how it differs from a system's incarnation — its physical attributes. Finally, we describe the essence of the system development process and the principles that guide analysts in modeling the essence of a system.

Since identifying true requirements is the central problem of systems analysis, we devote most of our attention to a strategy for deriving the essence of an existing system. This is the purpose of Parts Two through Six. We also address the problem of creating essential requirements when there is no existing system or when there is a decision to scrap the existing system and start anew. Part Seven describes the process of modeling new essential requirements, and the transition from requirements analysis to software design. Finally, in Part Eight, we show how to use essential systems analysis to solve the problems of managing a real project.

To represent our ideas on requirements specification, we use the modeling tools of structured analysis and, to a lesser extent, the tools of information modeling. If you are unfamiliar with the modeling tools of structured analysis, we recommend that you consult DeMarco's *Structured Analysis and System Specification.* Although we use the tools of structured analysis and information modeling when we document our requirements definition work, this doesn't mean that they are the only documentation tools to be used with our techniques for identifying the system's essence. Indeed, our clients have applied our techniques with very different modeling tools. That is not a problem; the strategies of essential systems analysis transcend any one set of documentation tools.

1.6 Summary

The purpose of systems analysis is to produce a statement of the true requirements for the system to be built. A true requirement, also known as an essential or logical requirement, is a feature that the system must have no matter what technology is used to implement the system. System developers often accidentally omit true requirements from the specification and include false ones. False requirements can be totally irrelevant features, features needed by a certain technology, or features that describe a true requirement in terms of the technology used to carry it out.

When analysts confuse true and false requirements, the result is often disastrous. The analysts put far more requirements into the specification than necessary, not only making the specification confusing but also slowing down the project considerably. Furthermore, analysts may leave out true requirements as so many false requirements clutter their thinking.

Traditional structured analysis has tried to solve this problem by telling analysts to separate the logical requirements from the physical aspects of the system. Unfortunately, its proponents were unable to define the terms *logical* and *physical* adequately and did not give detailed procedures for distinguishing true requirements. To fill this need, we offer a strategy for discovering and defining essential requirements. We call this strategy essential systems analysis.

Chapter 2

Planned Systems

Part of the difficulty in modeling the essential requirements of a system arises because analysts do not understand what a system is. Without this understanding, how then can they know its true requirements? Analysts differ about what makes up a system, some falling back on a dictionary definition of *system:* "a group of interacting, interrelated, or interdependent elements forming, or regarded as forming, a collective entity."* Such a definition is too broad for our purposes, since it works equally well to define the metric system, a bridge, and an accounts payable system. It accounts both for systems that don't do anything (passive systems, like a bridge) and for systems that do accomplish things (active systems, like a turbine speed controller or a stock trading system).

EDP system developers usually build active systems. If we limited our definition to active systems only, it would still include all manner of active systems that exist in nature. However, EDP system developers aren't going to worry about spider web building, mitosis, photosynthesis, or the mating dance of the spiny lobster. If you're like the rest of us mortals, you deal with what Herbert Simon calls artificial systems [43], systems that are built by one small part of nature: the human community.

EDP system developers need a definition of system that applies only to the kind they build. To arrive at a working definition, we propose forgetting about using the word *system* as it is normally used in everyday speech, and instead, in the following paragraphs, we define the attributes of the systems that EDP developers build.

2.1 Interactive systems

EDP system developers build things belonging to a class of human-made systems whose most important feature is that they are interactive; that is, these systems act on things outside their control and these outside things act on the system. A token booth in the New York City subway system is a good example of an interactive system. The subway riders who interact with the token booth system are outside the control of token vendors. The riders may decide not to ride on a given day, or even if they do decide to ride, they may decide not to purchase tokens. The city worker in the token

---

The American Heritage Dictionary of the English Language. William Morris, ed. Boston: Houghton Mifflin, 1981.

booth is powerless to force riders to act in either of these ways. But if a passenger decides to purchase tokens, he or she must interact with the vendor by requesting a certain number of tokens and presenting money. In response, the token vendor acts in a way that affects the subway passenger: The token vendor hands the passenger tokens and possibly change.

Observing this interactive system, you see that the subway rider, who is a part of its environment, acts on the system, and the system reacts to the rider. Figure 2.1 depicts these interactions. Like the subway system, the systems that EDP system developers build are interactive; they have the general form shown in Figure 2.2.

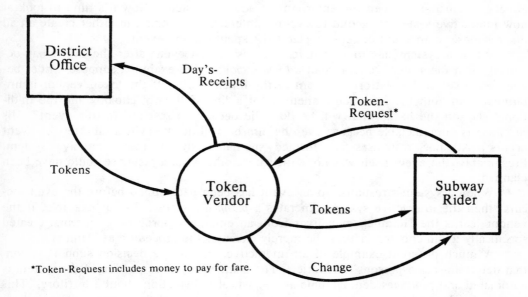

*Token-Request includes money to pay for fare.

Figure 2.1. A subway token vending system.

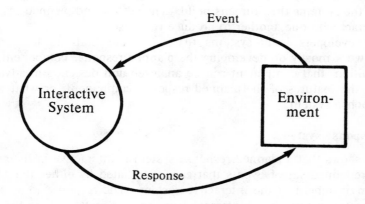

Figure 2.2. An interactive system.

In Figure 2.2, we assign the generic names *event* and *response* to the interactions between the system and its environment. An *event* is some change in the system's environment, and a *response* is the set of actions performed by the system whenever a certain event occurs. For example, the token vendor gives tokens and possibly change as a consequence of the event "subway rider requests tokens." The vendor tallies the day's receipts only after the event "end of selling day." We incorporate the event and response components into our definition of an interactive system by saying that an interactive system is an event-response mechanism.

So far, the discussion of interactive systems has concentrated on the cause and effect relationship between the environment and the system. Now it's time to look at how interactive systems respond to events. Interactive systems can either create an ad hoc response to an event or select a planned response to the event.

When a system has to invent its response to an event after the event has occurred, then the system generates an *ad hoc response*. For example, suppose a rider begins to choke on the platform in front of the token booth. If the token vendor is unfamiliar with both the Heimlich maneuver for aiding victims of choking and the traditional slap on the back, what will he do if he decides to respond to this event? His response is unpredictable because even he himself will not know it until after the event occurs. Ad hoc responses are formed spontaneously by the interactive system. Presumably, the event itself was not anticipated; otherwise, a response could have been planned.

When a system's response to an event has been determined before the event occurs, then the interactive system generates a *planned response*. In our example, if the vendor learns the Heimlich maneuver from an emergency procedures manual created specifically to aid choking victims, he merely executes the procedure as planned.

A much different example of an interactive system is a decision support system that determines salespersons' territories. The planned response portion of the system is automated and produces demographic and logistical information about a territory. This information is used by the human ad hoc response portion of the system to decide how to map out each territory. As this example suggests, ad hoc responses are almost always carried out by humans, because computers and other machines are designed to respond according to fully detailed instructions. This example brings out another characteristic of the systems that humans build: The ad hoc and planned response components can interact with one another to produce responses.

Although developers build systems that carry out both ad hoc and planned responses, they work mostly on developing the planned response component. After all, by its very definition, that component can be analyzed and designed in advance. Let's discuss some of the features of the planned response component, or what we now call the planned response system.

2.2 Planned response systems

Figure 2.3 shows that a planned response system is interactive in its own right. It responds to a predefined set of events that can be initiated by either the ad hoc component or the environment of the interactive system. The response to a certain event is the same no matter where the event came from and no matter where the response is directed to. The environment of a planned response system is then the union of the ad hoc component and the interactive system environment, as depicted in Figure 2.4.

Interactive System

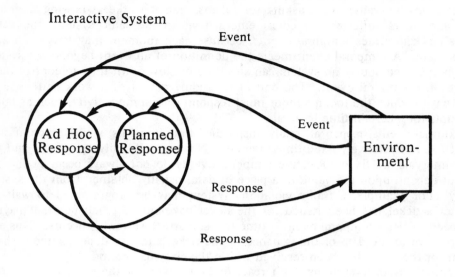

Figure 2.3. Two types of interactive response.

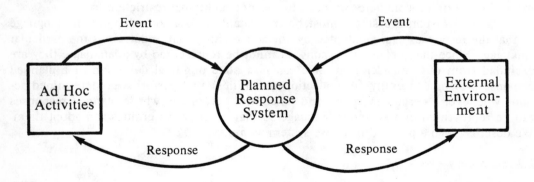

Figure 2.4. A planned response system.

In a planned response system, the responses are initiated by events that occur in the environment surrounding the system. In the token booth example, the token vendor does not execute the planned response of presenting tokens and change unless the initiating event of a subway rider requesting tokens takes place. The vendor also refrains from tallying the day's receipts until the exact moment arrives that defines the end of the day.

The events that a system responds to are of two types: *external events*, which are initiated by entities in the environment; and *temporal events*, which are initiated by the passing of time. In the token booth system, examples of external events are kids writing graffiti on the walls, riders asking for tokens, transit workers showing their entry passes, and riders suffering a medical emergency. In the same system, the subway station's clock indicates temporal events, because each moment may trigger an action by the system. A temporal event meets the definition of an event beyond the system's control because obviously no system can speed up or slow down time to control the occurrence of a temporal event. The end of the riding day or of a work shift is a temporal event to which the token vendor must respond with some activity, such as tallying the receipts for a certain time interval.

External events happen at totally unspecified time intervals, while temporal events occur at fixed or relatively fixed time intervals. Nanosecond, daily, quarterly, and yearly time intervals are fixed. Relative temporal events do not always occur at the same time, but depend upon the value of a piece of data, usually obtained from the system's memory. For example, a traffic violations system may be programmed to wait sixty days after a ticket has been issued for the ticket payment. At that point, if payment hasn't been made, a temporal event, "time to issue arrest warrant," occurs. This event occurs relative to the date of the issuance of the ticket and also depends upon whether payment for the ticket has been received within the sixty-day period.

Planned response systems don't react to every event in the environment. Many external and temporal events don't even raise a yawn from a particular system. For example, the token vendor may not care if a drunkard passes out in the station, just as he may not care if the Mets win the World Series or if the clock says 3:02:14 p.m.

Having described the characteristics of the planned response system, we can now try to define it: A planned response system is an entity (or collection of entities) that performs predetermined actions whenever a certain event beyond its control occurs. To make this definition complete, we need to add one additional restriction to it.

Planned response systems must be translatable into a formal, symbolic language so that the responses can be adopted by any active entity that understands the particular language. Since many planned responses cannot be represented by a language, they are excluded from our consideration. We can now state our final definition of a planned response system: an entity (or collection of entities) that performs predetermined actions whenever a certain event beyond its control occurs and whose planned responses can be represented in a symbolic language so that other active entities can adopt them. We describe such a planned response system as *transportable*.

2.3 Summary

In this chapter, we formally define the important features of the systems that EDP system developers are asked to build. These systems are interactive; they act on things beyond their control and these external things act on them. A change in a system's environment to which it reacts is called an event, which can be either external or temporal. The system's reaction is called a response. These responses are *ad hoc* if they are not determined in advance and *planned* if they are. The systems built by developers consist of both ad hoc and planned response mechanisms, but the developers are most interested in the planned response portion of a system. From here on, we will concentrate on the transportable planned response system.

Our definitions of interactive and planned response systems by themselves will help you distinguish true requirements. Freed from the confusion caused by redundant, overlapping, and inconsistent ideas about the system, you can now find the true system requirements with less effort. Nevertheless, the framework introduced in this chapter isn't strong enough to fully support the search for a system's essence. We need to add two more major components; these are the essential and the implementation characteristics of systems.

Chapter 3

The Essence
of a System:
Logical Requirements

However helpful it may be, understanding what a planned response system is does not by itself enable you to define the true requirements of a system. To find these true requirements — called the system's essence — you need a framework to organize your understanding of them. We begin this framework by defining *essence* as all characteristics of a planned response system that would exist if the system were implemented with perfect technology. In this chapter, we discuss this definition in detail.

3.1 The concept of perfect technology

The great advantage of our definition is that it does not depend upon vague notions like "functions" or "policy." The logical version of a system is distinguished from the physical version by asking, What would be left of the system if it were implemented with perfect technology? The system that would exist no matter what particular technology is used to implement it is the system's essence.

To clarify the concept of perfect technology, we first must define the term *technology*. It is the means humans use to achieve a desired end. The technology used to implement systems has two basic components: *processors* that carry out activities and *containers* that convey data to processors and hold data for use by processors.

If its technology were perfect, a system would have a perfect processor and a perfect container. A perfect processor would be able to do anything and everything instantly; that is, it would have infinite capabilities and infinite workload capacity. It would cost nothing, consume no energy, take no space, generate no heat, never make a mistake, and never break down.

A perfect container would have many of the same virtues. It wouldn't cost anything, and it would be able to hold an infinite amount of data. Any processor would be able to access conveniently the data it carried.

All of this is intriguing to think about, but it probably sounds more than a bit absurd. System developers don't have such technology now, and they aren't likely to have it soon. So, why do we spend time pondering what perfect technology would be like? The reason is this: Thinking about the concept of perfect implementation technology is useful for identifying the essence of a system. By imagining how a particular system would look if you could implement it using such a perfect technology, you can distinguish the essential features of that system. This is a difficult but fascinating thought experiment that will serve as a good introduction to essential modeling.

3.2 The components of a system's essence

The essence of a system consists of the essential activities and the essential memory. The *essential activities* are all those tasks that the system would have to do even if you could implement it using perfect technology. If you think about the functions of a typical software system, you'll see that few of them really are essential. For example, most applications systems perform auditing checks to make sure they do not produce incorrect output. These checks would be unnecessary in a technologically perfect system, because such a system would never make a mistake.

The *essential memory* consists of all the data that the system would have to remember if all it did was carry out the essential activities. In other words, essential memory contains all the data that a technologically perfect system would have to remember. To continue with the example of auditing activities, many systems keep audit trail files and other extraneous data that an infallible and incorruptible system wouldn't need.

We distinguish between two kinds of essential activity: fundamental activities and custodial activities. Because fundamental activities are more important, they are discussed first. Custodial activities and essential memory are intimately connected, so they are not explained until after the discussion of essential memory.

3.2.1 *Fundamental activities*

An essential activity is also a *fundamental activity* if it helps to justify the existence of the system. In other words, a fundamental activity performs a task that is part of the system's stated purpose. Consider a system that pays hourly workers, like the one shown in Figure 3.1. Each week, employees tell the system who they are and how many hours they worked that week, and it produces their paycheck. Now suppose that you must purchase a software package that carries out this function. You interview several sales representatives from competing software firms, and they each list the features of their product.

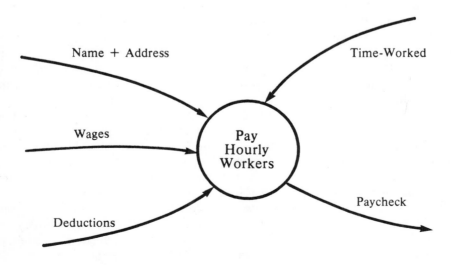

Figure 3.1. Fundamental activity of a payroll system.

Which features impress you? Are you impressed by a product that keeps track of employees, their wages, their withholding taxes, and their other deductions? Maybe, but you certainly cannot be satisfied — because the salesperson has not promised the very feature that you want most from the system: the generation of paychecks.

To put it another way, the reason that this system exists is to pay the hourly workers. It tracks employees, wages, and deductions in order to pay hourly workers, *not* because the organization needs this information for its own sake. Therefore, the fundamental activity of this system is the production of paychecks for hourly workers.

Each fundamental activity consists of two parts: a planned response and a definition of the activity's stimulus. The planned response is the set of actions carried out by the system in order to perform the activity. As in all planned response systems, the response is initiated by an external or temporal event. The activity begins when it recognizes a stimulus sent by the outside event whenever the event occurs. The definition of the stimulus allows the activity to recognize the arrival of a stimulus. In other words, the definition gives the circumstance under which the activity is performed. In the payroll system, the stimulus is the time card submitted by an employee and the response is the generation of a paycheck.

3.2.2 Essential memory

In order to produce a proper response, the fundamental activity must have at its disposal all sorts of information. For example, the fundamental activity Pay Hourly Workers needs the identity of the employee; that employee's wage rate, year-to-date pay, and withholding tax status; the amount of time worked during the pay period; and information about other payroll deductions. Some of this information comes directly from the environment, while the rest comes from essential memory. When an event occurs, the fundamental activity usually obtains some data from the system's environment in order to respond appropriately. In the example above, the time card gives the system the employee's identification number and the number of hours worked during the pay period. These pieces of information are components of the stimulus, coming from the initiating event. In fact, all data acquired by any essential activity at the time of that activity's initiating event are by definition part of the stimulus.

The fundamental activity also often requires information from another source: It needs pieces of data produced by itself or by another fundamental activity in the same system in response to some earlier event. The activity Pay Hourly Workers may need to know the total payments made to an employee since the beginning of the year so that the year-to-date pay can be shown on the paycheck. Many systems remember this kind of information.

The total of all data elements remembered by the system and required by its fundamental activities is the essential memory. In real-world systems, the stored data is kept in all manner of physical containers, from index cards to magnetic disk records. The physical format of the system's memory depends upon the processor that uses the information: People use paper forms; computers use magnetic tape records. However, a perfect processor wouldn't need to rely on any of these physical forms to read the system's memory. Rather, a perfect processor could use any data storage format whatever, including — in the extreme case — a set of disembodied data elements.

Systems must remember this data because of the imperfections of the system's environment. Sometimes the outside world just isn't reliable enough to supply accurate data. In an obvious example, no sane employer would adopt a plan to have employees remember their own wage and fill it in themselves on their time card each week. The certainty that one or two pranksters would give themselves an attractive compensation rate makes this an unreliable way to remember data from the past event "employee is given a raise." The results are similar when the outside world cannot supply accurate information because it is incapable of remembering it.

Another imperfection of the system's environment is that sometimes the outside world simply won't cooperate. Even when you could depend upon the people in the system's environment to remember past information reliably, they might grumble at having to repeatedly supply information just because the system won't remember it. Consider voluntary deductions, the amounts withheld from your paycheck because you want them to be used for some other purpose, like a charity or an investment program. Information concerning the destination and the amount of each deduction must be available to the fundamental activity of the payroll system. Because the disposition of the deduction is totally up to the employee, the company owning the payroll system might not care about the reliability of having the employees supply this information every time paychecks are produced. But the employees might well become annoyed if they were asked to declare their deductions over and over again.

Because of these imperfections in the environment, many systems must store data that weren't produced by the fundamental activities of the system, but that will be needed by some fundamental activity in the future. A payroll system remembers the employee's name and address, as well as the employee's wage rate, which was produced by the personnel system. These pieces of data are made available to the system by outside events ("personnel submits new hire list," "personnel raises wages"). The system captures the data when these events occur, and preserves them for the future use of the fundamental activities.

You may think that perfect technology would avoid environmental problems that force the system to store data. Unfortunately, the concept of perfect technology covers only the system itself, not its environment. If perfect technology extended to the world outside a system, the need for many, if not most, systems would vanish. After all, why worry about a payroll system when everyone works for free?

In summary, the system's essential memory consists of data produced by the system or captured from the outside world for use by the system's fundamental activities. The essential memory also includes statements of how the fundamental activities must be able to access stored data. For example, the payroll activity must be able to look up the wage rate for an employee who is identified, say, by social security number. Figure 3.2 shows how you would indicate the essential memory of the Pay Hourly Workers system on a data flow diagram. The named parallel line segments are called data stores; each one represents a subset of the system's essential memory. The name of each data store appears in the data dictionary along with a definition of the data elements in the store.

For the essential memory to fulfill its purpose, these two questions must be answered:

- How does the system acquire the data to be remembered?

- How does the system make sure the data is sufficiently up-to-date to serve the fundamental activities?

These tasks are performed by the other kind of essential activity: the custodial activity.

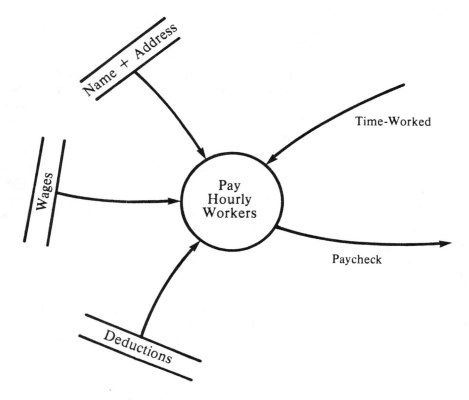

Figure 3.2. Essential memory of a payroll system.

3.2.3 Custodial activities

The custodial activities establish and maintain the system's essential memory by acquiring and storing the information needed by the fundamental activities. They also update the stored information so that it remains correct.

We have already mentioned several custodial activities: those that create and maintain the information in our payroll example. The essence of the hourly payroll system must include activities to keep track of changes to the set of employees and to their wages and deductions. These custodial activities are shown in Figure 3.3. Like fundamental activities, custodial activities consist of one or more planned responses and a definition of a stimulus. In the case of the custodial activities, however, the response consists of an update to the system's essential memory instead of an output to the outside world.

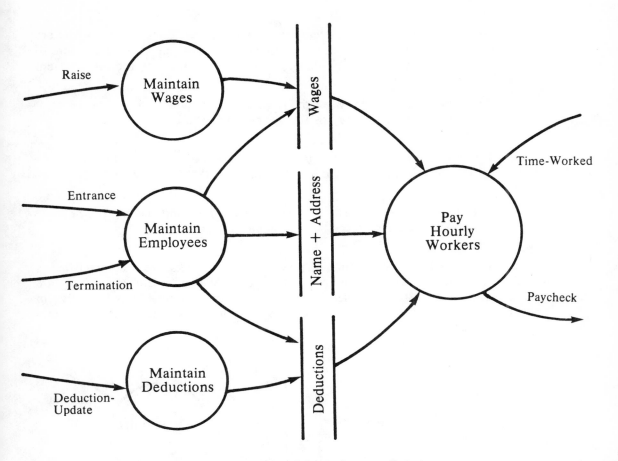

Figure 3.3. Custodial activities of a payroll system.

3.2.4 Compound essential activities

We have described two types of essential activities, but we don't want you to think that each essential activity is *either* fundamental *or* custodial. Many essential activities are both: These activities carry out fundamental planned responses, while also maintaining the system's essential memory.

The payroll activity in Figure 3.1 is shown as a pure fundamental activity. In real life, these are relatively rare, since most real fundamental activities must remember what they do so that their results can be used at a later time by themselves or by other essential activities. For example, the payroll activity needs to remember the amounts earned by, withheld from, and paid to each employee in order to produce an annual statement of earnings for income tax purposes. Like all compound essential activities, the payroll activity would both update the system's essential memory and send a response to the outside world.

3.3 Summary

The essential components of a system are shown in Figure 3.4. Although they are all equally essential, the components are not all equal: Parts of the essence do more to accomplish the system's purpose than others. In order of importance, the three system components are these:

- The system exists to carry out the fundamental activities. They are the only truly desired features of the system, and the only ones that are justified by themselves.

- The essential memory stores the data items between the time they become available and the time they are used by the fundamental activities.

- One or more custodial activities perform the care and feeding of the system's essential memory. They acquire the data items for memory and keep them up-to-date.

Viewed in this light, the essence of a system almost looks like a contract that is filled with tricky fine print. The only parts of the essence that benefit you directly are the fundamental activities. The essential memory and the custodial activities serve in supporting roles. Nevertheless, all three components must be present, even in the imaginary world of perfect technology.

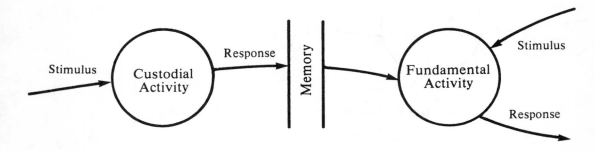

Figure 3.4. Components of the essence of a system.

Chapter 4

The Incarnation of a System: Physical Characteristics

The sum of people, wires, paper clips, carbon paper, pencils, typewriters, computer terminals, office furniture, file cabinets, offices, telephones, CPUs, and so forth that are used to implement the essential activities and memory of a system are what we call its *incarnation*. Whereas the essence of a system is an intangible set of ideas, the incarnation is something that you can actually see, hear, and feel. Some people call this collection of tangible elements the implementation of a system. The teller machine dispensing money, the terminals used in homes, the banks of disk drives enclosed behind glass walls, the salesperson taking an order, and the supervisor shifting worker assignments are all common components of incarnations.

The term *incarnation* is used to represent the embodiment of a concept; it reminds us that the tangible system embodies its conceptual form, the essence. This reminder is needed, because it is difficult to discern the essence of a system when looking at its incarnation.

Let's look at an example to understand this difficulty. Suppose you want to discover a particular system's essence, but all you know about the system is that Bob and Jack are processors who carry out the activities of a system. In Figure 4.1, are Bob and Jack implementing the essence of a payroll system? Are they preparing the annual statement for a publishing house? Maybe Bob and Jack are collating the results of experiments conducted on a linear accelerator. Bob and Jack could be implementing any of many different systems. The same holds true for job steps QR010 and QR020 in the second system of Figure 4.1. Are they job steps in a computerized order entry system? Or are they job steps in a program that schedules payments in an accounts payable system? If you want to know the system's essence that either Bob and Jack or QR010 and QR020 are implementing, you will have to do additional research. These examples demonstrate that characteristics of an incarnation of a system don't necessarily tell you anything about its essence.

There are three reasons why system developers have so much trouble discerning the essence when examining the incarnation:

- The system's essence must be implemented with tangible technological components that have no intrinsic connection to the essence.

- The incarnation also executes responses other than the transportable planned responses developers are interested in.

- The technology that humans use to incarnate the essence of a system is imperfect.

In this chapter, we describe the concept of incarnation by elaborating on these three reasons.

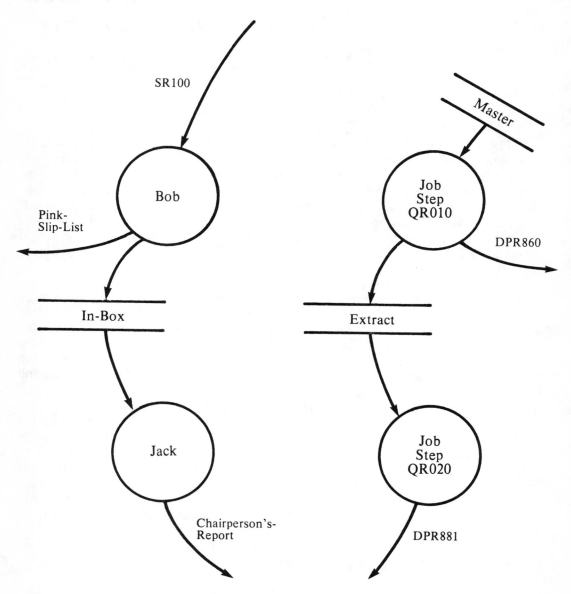

Figure 4.1. Incarnations of two systems.

4.1 The incarnation of essential features

As stated before, the essence of a system is an intangible set of ideas. In order to have a system that carries out essential activities, the essence must be incarnated with physical objects. Essential activities are executed by people and machines, and essential memory is stored on paper, in neurons, and with magnetically charged particles. Since you can't perceive the system's essence, you must make deductions from observing the system and from talking to others who know the system. In effect, you are playing the game of charades with the object to deduce the essence of the system from the physical clues provided.

4.2 The planned response system within the implementation

In Chapter 2, we said that an interactive system carries out two basic kinds of responses to an event: an ad hoc response, which is generated spontaneously; and a planned response, which is determined in advance. The first major problem in distinguishing the essence of a system is that the typical implementation executes both kinds of responses. The second major problem is that the processors of most implementations carry out responses for several interactive systems. From a scientific work station that controls several different experiments at one time to a large time-sharing computer system that edits source code while producing invoices, implementations that carry out the essential activities of many planned response systems are the rule rather than the exception.

Figure 4.2 shows an implementation that contains activities from four systems. If this abstract implementation were located in a bank branch, system 1 (the system under study) might be a savings account system. Systems 2 through 4, respectively, could be a checking account system, a commercial loan system, and a trust system. Notice that all interactive systems in this implementation carry out both ad hoc and planned responses. To find the essence of the system under study, you will have to remove those aspects of the implementation that make ad hoc responses and those aspects that carry out planned responses for other interactive systems.

Yogi Berra is reported to have said, "You can observe a lot just by watching." But merely observing the interactions of the system with its environment will not help you discover and remove ad hoc responses. Many human processors pride themselves on being able to respond quickly to the unexpected, without anyone realizing that the response wasn't planned. Airline pilots are especially respected for this ability, and many gifted musicians can also deviate from their score when the singer forgets a verse. In a fine hotel, the desk clerk usually won't even blink when you ask for a trapeze and a set of bolts for attaching it to the ceiling of your room. So, when you see an incarnation, you must realize that the essence of the planned response system is camouflaged by those aspects of the incarnation that produce ad hoc responses. Figure 4.3 shows how removing the ad hoc component simplifies the model of the incarnation in Figure 4.2.

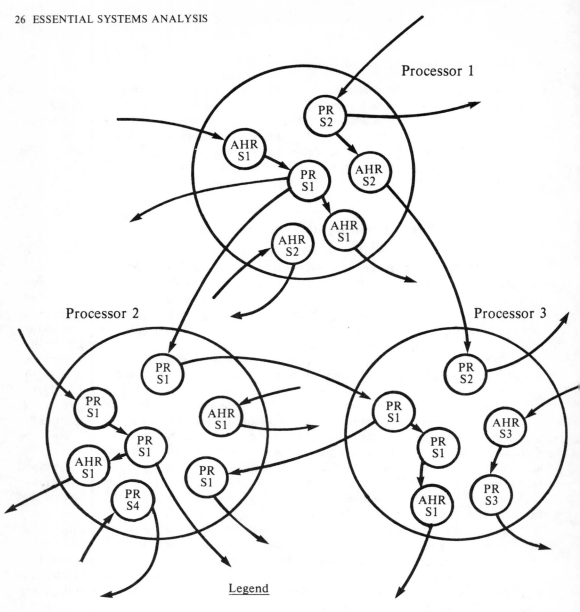

Legend

PR	—	Planned Response
AHR	—	Ad Hoc Response
S1	—	System 1 (the system under study)
S2-S4	—	Other Systems

Figure 4.2. A typical implementation.

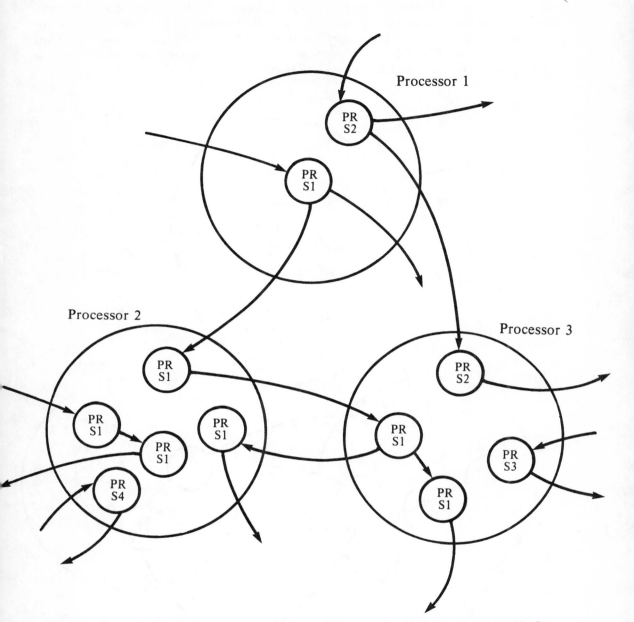

Figure 4.3. A typical implementation after the ad hoc responses have been removed.

Once you've isolated the planned response portion of the incarnation, it will still be difficult to find the essence of the particular system under study. First, you must remove the other planned response systems from your model of the incarnation. After doing this, the picture of the incarnation, as shown in Figure 4.4, looks much closer to the picture of the essence of the particular system.

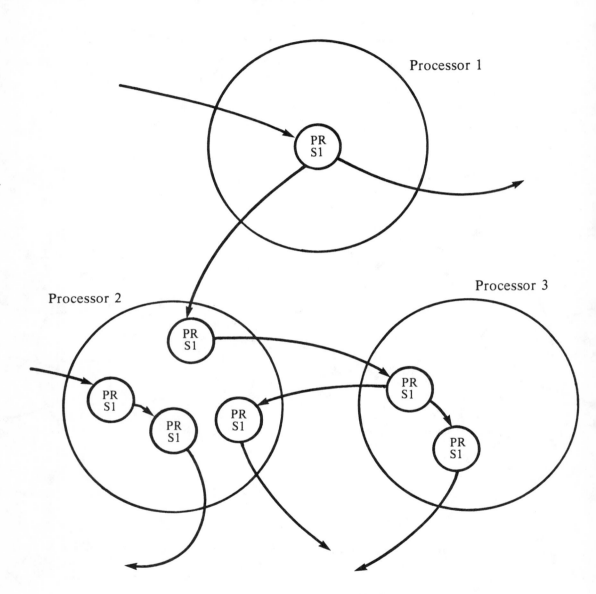

Figure 4.4. A typical implementation after the ad hoc responses
and planned responses by irrelevant systems have been removed.

Even though the model is now much simpler, the stripped-down version of the incarnation appears very different from its corresponding essence. The main reason is that the technology available for incarnating the essence of a planned response system is imperfect. The basic differences between the two that are due to imperfect technology remain intact throughout these manipulations.

4.3 The impact of imperfect technology

In order to understand the incarnation fully, you must understand the limitations of technology. Imperfect technology is the result of the search for ways to implement essence that will meet the ideal goals of a system. Someone will always try to develop the nonstop, goof-proof system with infinite capacity and instantaneous response, and indeed many advances in that direction have already been made. Yet — let's be honest — we humans will never reach that ideal, and we will always have to adjust our objectives from the ideal to the practical. We will have to work within the constraints of imperfect technology in order to transform the essence into a working system.

Technology has four limitations that greatly affect the shape of the systems being built. The first limitation is *cost*. All human technologies use costly, scarce resources. When developing a system, you must decide what mix of technology you can afford to buy. Presumably, you will try to strike a balance between the benefit you get from a system and the cost of the technology used to incarnate it.

Capacity is the second limitation system developers must cope with. No processor yet invented can do an infinite amount of work. Nor are there containers that can hold or convey an infinite amount of data. All working systems store only so much information and take a certain amount of time to carry out their activities.

The third limitation of system development technology is its *capabilities*. It would be beneficial to have a processor that both calculates complicated astrophysics equations without error and successfully sells kitchen gadgets. To date, however, most processors excel in only a subset of all the capabilities a system needs; some people compute well, others communicate well, and some can even tolerate watching for traffic violators in the Brooklyn Battery Tunnel. Computers similarly have different skills: Some are designed to crunch numbers, others to control large telecommunications networks.

The final limitation with technology is that it is *fallible*. Wouldn't it be wonderful if incarnation technology didn't make mistakes, break down, or commit crimes? Imagine correct phone and credit card bills, accurate prediction of whether Halley's comet will strike earth, and no need for security checks. Unfortunately, right now and probably forever, Murphy's Law has a tremendous impact on our incarnations. Security systems also are responsible for many of the characteristics of the incarnation.

These limitations in system implementation technology determine many of the characteristics of the systems you build. Six of these characteristics in particular make the search for a system's essence difficult. They are fragmentation, redundancy, extraneousness, convolution, conglomeration, and vastness. The following subsections detail these characteristics and show how they result from imperfect technology.

4.3.1 Fragmentation

When different parts of an essential activity are carried out by different processors, that activity is fragmented. For example, a factory as a whole may carry out one essential activity; perhaps it produces cars. Yet, that one activity is implemented by many workers and machines in assembly lines, and each worker or machine carries out only a small fragment of the entire activity. The particular fragments were chosen because of the imperfections of the technology in use. Similar assembly line arrangements can be found in most systems. In data processing systems, for instance, the "products" on the "conveyor belts" are pieces of data, instead of mechanical or electronic components.

Fragmentation hides the essence in the same way that breaking a delicate piece of crystal destroys its original shape. In fact, activities in a typical system are fragmented several times and the fragments are thoroughly mixed, making it especially difficult to reconstruct activities in their essential form.

4.3.2 Redundancy

Thanks to technological advances that enable the computerization of large databases, system developers are familiar with the stored data redundancy in most systems. The redundancy of activities and processors is even more apparent. Banks have many loan officers and tellers, for example, and spacecraft have a main computer plus typically at least two backup computers. In each case, many processors carry out the same essential activity. Another type of redundancy is dataflows or files carrying the same pieces of data.

Like a detective interviewing witnesses to a crime, you have the difficult task of examining all of the redundancies in the incarnation. You must determine whether redundant components are part of the system's essence or whether they perform tasks mandated by imperfect technology. If the redundant components are essential, you need to ask if they perform the same essential activities or different ones. Processors or containers may appear identical yet carry out different activities or store different data.

4.3.3 Extraneousness

Extraneous activities and data are those that do not belong to a system's essence; their purpose is to overcome the limitations of current technology. The incarnation of an essence contains much extraneous material, which falls into two basic categories:

- transportation-related activities and data
- administrative activities and data

The incarnation of the system requires all manner of transportation facilities to move both data and material to and from the processors that carry out essential activities. For example, data entry clerks take data generated by other departments within an organization and change its representation from one medium to another. As long as the other departments belong to the same system, data entry clerks are not part of the essence, because the activities that they perform would be unnecessary if the system had perfect technology. A single perfect processor executing the essential responses would have no other processors within the same system to communicate with, and therefore would not need transportation-related activities.

A processor often receives not only the data it needs to perform its work but also data that it doesn't need. These data may be included along with needed data on a form or in a physical file record sent to the processor from within the system. For an example, look at Figure 4.5. The Overdue-Report makes no mention of the specific PO-Items on an overdue purchase order. Yet, the processor carrying out the activity Report On Late Deliveries will obtain PO-Items when it obtains a PO-Record. This unused data is extraneous. It arises because it is too expensive to provide the additional data channels, extra file accesses, and data access restrictions that would prevent a processor from receiving unnecessary data. In the world of perfect technology, a processor would get the data it needs, nothing more and nothing less.

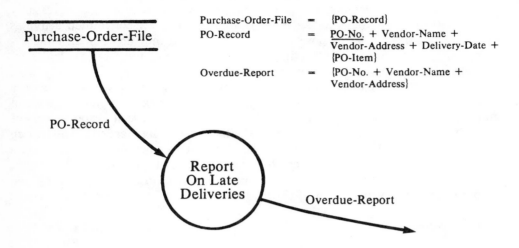

Figure 4.5. A file access containing extraneous data elements.

Many different kinds of data exist solely to aid the transportation of data and material within the incarnation. For example, the name and address of the data entry department in a bank would be written on the interoffice mail envelope used to transport a packet to the right department. Again, perfect technology would eliminate the need for these pieces of data. With only one perfect processor, there wouldn't be any other senders and receivers who could lose a transmission.

The incarnation requires an administrative facility to make sure that the processors correctly perform the essential activities assigned to them. When a senior manager of an advertising agency approves a junior member's purchasing decisions, for instance, the senior member is performing an administrative activity. Another example is the edits that check the data and material that a processor receives from another processor within the same system. These activities are extraneous to the essence. Since a perfect processor would never make a mistake, why should you worry about approving its work?

Some pieces of data within the incarnation are part of this administrative function, such as the signature of the ad agency's manager and the count, type, and source of an error. When perfect technology makes checking the work of a processor unnecessary, these pieces of data will also disappear from the incarnation. Extraneous activities and data are usually repeated throughout the system, adding to its redundancy and thereby compounding the difficulty of isolating the essence.

4.3.4 Convolution

The instructions in a road rallye are intentionally convoluted. The drivers must decipher cryptic clues that lead them in a roundabout way to other clues that eventually enable them to find the checkpoint. Otherwise, they may get completely lost.

Like road rallye instructions, the procedures in a system are often unnecessarily convoluted. Unlike the extraneous activities and data, however, convoluted activities and data serve a valid essential purpose. The problem is that they do so in an unnecessarily complicated way. Such activities may call for calculations, file accesses, and sorts that are made necessary only because of imperfect technology.

For example, in order to find the total amount on order for a given stock item, a clerk may have to look through all of the purchase orders and keep a running total of the amount on order. This convoluted search is necessary because it is too expensive for imperfect stored data technology to provide the exact data needed. The search would be much easier if the system provided the clerk with a stock file that contained a piece of data called the amount on order. Similarly convoluted procedures for sorting, extracting, and inserting data are present in most incarnations, thus obscuring the essential procedure for obtaining and using this information.

4.3.5 Conglomeration

When a business acquires enough firms that sell unrelated products, a conglomerate is formed, as in the case of Textron, DuPont, and Exxon. Conglomeration also occurs in the incarnation of systems when fragments of unrelated essential activities are assigned to the same processor.

Conglomeration is the other side of the fragmentation of essence described above. Fragmentation of essential activities and memory provides the pieces that make up a conglomerate activity or data store. The result is an organization of activity within the incarnation that thoroughly confuses efforts to see the essence. These unrelated activities may even be from totally different systems. For example, in a publishing house, the same person who invoices a customer's book purchases may also check that supplies delivered match those billed for. Pieces of essential memory can also form conglomerates. More often than not, unrelated pieces of essential memory are allocated to the same container. So, in the incarnation of a company that sells minicomputers, you may find data elements about the salespeople mixed in with data elements describing the hardware mixed in with data elements documenting the maintenance calls.

4.3.6 Vastness

One thing you can say about New York City without causing an argument is that it is vast. The city is so huge that it is doubtful anyone can know it as they might a smaller city or town.

Systems share this characteristic of vastness, often spanning many departments and geographic locations and consisting of hundreds of thousands of lines of code. Not many people can understand the incarnation of a typical system in its entirety. The essence of such a system may be relatively small, and yet the size of the incarnation makes it difficult to find.

4.4 Summary

Finding the essence of a system within its incarnation is difficult. The ability of the response mechanism to execute more than just the planned responses of one system blocks the path to the system's essence. The analyst must first sort out and remove those portions of the system that carry out ad hoc responses and responses from other systems. Imperfect technology makes the search for essence even more difficult. It adds many nonessential features to the system and makes the system large and convoluted. Essential activities are fragmented, and fragments of unrelated essential activities are assigned to the same processor, a phenomenon called conglomeration. Essential memory can also be fragmented and formed into conglomerates. Finally, the same essential components can be repeated throughout the incarnation.

Chapter 5

The Essence
of System Development

In the preceding chapters, we explored planned response systems and their essential and incarnation features. In order to use these concepts to define essential system requirements accurately, however, you need some strategies and procedures. Since defining system requirements is the first step in building a planned response system, those strategies and procedures should be an integral part of an overall method for developing such a system.

Compared to the currently available collections of system development strategies and procedures, widely known as methodologies, our approach is unusual. We specify two types of strategies and procedures for the development of systems in general and the definition of essential requirements in particular: technical strategies and managerial strategies. Technical strategies map out the technical activities that must be performed on any system development project, regardless of physical constraints such as system size, system developer skills, job descriptions, project budget, or schedule. Defining the essence of a system is an example of a technical activity.

The physical constraints ignored by the technical strategies are the domain of the managerial strategies. They tell how to carry out the technical strategies within the physical or managerial constraints of a system development project.

Most methodologies are not much help to system developers because they try to deal with the technical activities and all possible managerial constraints in the same package, thus supplying a grab bag of techniques specified at a superficial level. The effect of this mixed approach is that neither the technical nor the managerial side of system development receives proper treatment. A pinch of structured analysis, a tablespoon of normalization, and just a touch of structured design do not make a complete, well-integrated technical approach to system development.

Categorizing the strategies into separate types, by contrast, offers certain benefits. First, each strategy is individually more complete, because separating the technical strategy from the managerial uncovers questions on each side that are hidden by the mixed approach. When these questions are answered, each strategy is made more complete. For example, a managerial strategy leads developers to establish boundaries between the systems and programming tasks and the data administration tasks. Because we did not set up these boundaries, the technical question of how to model essential memory became central and we were able to concentrate on it. On the other hand, by not considering a particular technical approach when deciding managerial strategies, we were able to highlight the managerial question of how to expedite the process of systems analysis regardless of the technical approach used.

The second benefit of separating the two strategies is that each strategy becomes less complicated. By dividing system development into two strategies, we eliminated interference by each strategy on the other, thereby streamlining both of them. Finally, this approach removes the technological bias from the technical strategies. Since our technical strategies ignore managerial constraints, the same technical strategies can be used on many projects, no matter whether you are building systems using pencil-and-paper technology or an applications generator. Thus, as the system development technology changes, only your choice of managerial strategy need change. In effect, these are some of the same benefits that a well-designed system provides: completeness, understandability, and portability.

System development efforts can themselves be considered interactive systems whose purpose is to develop other systems. In this chapter, we take a new look at system development, using the notions of essence and incarnation. We want to discover the essence of system development by asking, What system development activities must be performed even with perfect system development technology? The technical strategies explain how to perform essential system development activities, activities that would be needed even if there were perfect system development technology. The managerial strategies tell how to use the imperfect technology available for building systems — the supply of workers and equipment that give rise to many of the managerial constraints — to carry out essential system development activities.

By the end of this chapter, we will have arrived at a definition of system development that is equally valid and applicable regardless of the circumstances of a particular project. Such a universal view of system development is ideal for our purpose: to explain techniques for defining the essence of a system that can be performed by all projects, regardless of their managerial constraints. We begin this quest with a discussion of the typical managerial constraints on system development activities.

5.1 Imperfect system development technology

Managerial constraints on system development projects arise from the use of imperfect system development technology. In fact, since any project can be thought of as an interactive system whose objective is to build a system, the same problems of imperfect technology discussed in the previous chapter apply to system development projects. The only difference between the two kinds of systems is that system development projects produce implemented systems whereas other interactive systems don't necessarily.

Just as with system implementation technology, system development technology consists of processors and containers. In discussing the effects of imperfect system development technology on the development process, we concentrate on the limitations of the processors, humans in this case, that are usually employed to develop software systems.

5.1.1 Imperfect system development processors

We humans constitute the largest part of system development technology, since we do most of the actual work of developing systems. While compilers, text editors, data dictionary packages, and other automated aids offer valuable support and carry out an increasing percentage of system development activities, we still bear most of the burden of creating software systems.

Humans obviously are imperfect. We have limited skills and we make mistakes. We can't do tasks instantly and we can't work endlessly. Moreover, we demand some sort of payment in return for our time-consuming, error-prone labor. In fact, we have exactly the same limitations as the technology discussed in Chapter 4: cost, limited capacity, limited capability, and fallibility.

These imperfections affect how we develop systems in general and software systems in particular. Specifically, they impose managerial constraints that make system development efforts more complex and difficult. Among the most important of these constraints are the need to employ multiple system development processors, the need to plan and control system development efforts, and the need to carry out quality assurance activities. The following subsections describe how each of these constraints results from a human imperfection and how different circumstances require different managerial strategies.

5.1.2 Multiple system development processors

The simplest way to allocate system development activities is to give all the work to one person, who acts as user, analyst, software designer, and programmer. Two human limitations — finite skills and finite workload capacity — force most software development projects to employ more than one worker.

Because users rarely possess the skills to develop their own software systems, even the smallest project typically employs at least two processors: a user to provide a description of the essence of the system and a software developer to select and construct the incarnation of the essence. If implementing the new system requires more skills than a single technician has, specialized workers must be added to the project. Systems analysts, system designers, database designers, system programmers, regular programmers, and EDP auditors often work side by side on projects that are not exceptionally large or complex.

Even if someone had every system development skill necessary, he or she would not be able to develop a large system alone and certainly not in a reasonable amount of time. Limited human workload capacity also dictates a team approach to system development.

The number of people on a team gives rise to at least two managerial constraints. The more important is that each participant in the system development process must be assigned a specific set of activities to carry out. The second constraint is that project participants must be provided with the means to communicate information to one another.

5.1.3 Planning and control activities

Because system developers cost money, take time to do tasks, and possess limited skills, a system development effort can be a major investment of both time and money. In order to use their time most effectively, developers individually must plan their own activities — making an estimate and establishing and refining a schedule. After finishing a task, they each review the original estimate to see how well they predicted the effort required. In this way, they continually revise and improve their ability to manage their own efforts.

When a project requires so many developers that an organization can no longer rely on individual self-management to deal with human imperfections effectively, it employs managers, who coordinate the efforts of the developers. The purpose of management is to extract high-quality products from imperfect processors. Management plans development activities, estimates the time and money needed, and monitors the project to keep it on track.

What strategies managers use to overcome human limitations depends upon several factors, among them the size of the effort, the nature of the project, and the skills of the developers. A single-developer project usually needs only very informal planning and monitoring activities. Planning large projects is a much more substantial task because these projects need a chain of command and formal planning and monitoring standards. A project with a legally binding development contract must be more carefully managed than an internal effort. However, no matter what factors apply to a project, imperfect system development technology requires some sort of estimating, planning, and monitoring.

5.1.4 Quality control activities

The last human imperfection to discuss is fallibility. As just about any machine would tell you if it could, we humans are a notoriously undependable bunch. Since software development employs humans so extensively, the detection and correction of human errors is the purpose of many of the activities in the typical development life cycle: phase reviews, design walkthroughs, and code inspections, to name a few.

The ways in which system developers choose to compensate for human fallibility depend upon the circumstances of a particular development effort. System developers working alone may rely upon themselves to check the specification, the design, and the code, or they may seek the assistance of an outside party. However, system developers have learned not to depend upon an individual's attempts to check his or her own work. They have found that such desk checking, while useful, is just as fallible as the original activity under scrutiny. In a larger project, colleagues will review each other's work. They must decide what products to review, how many people to include in such sessions, and how to resolve problems that are found. In other cases, developers may delegate some of the responsibility for detecting errors to a quality assurance group or an EDP auditor. They may institute rigorous protocols for walkthroughs or inspections, in which a group of people take responsibility for the correctness of a product.

Quality control in a very large development project is more complicated. Management may have to establish more than one level of quality assurance activity. Of course, the members of a first-level working group will review their products to their own satisfaction. But upper levels of project management will undoubtedly want to check for consistency of style among the groups and for the uniformity of interfaces between groups. They may ask that first-line managers review their products *again* among themselves or with their second-line manager.

The three major managerial constraints we've discussed — the need for multiple processors, planning and control, and quality assurance — lead to additional activities for system developers, activities not part of the essence of system development. Many more activities result from other constraints, like having to deal with internal politics or outside vendors. Clearly, then, a system development strategy that tries to deal with all of these constraints, or any major subset of them, cannot possibly do an adequate job. We want to remove these extra activities in order to focus on the purely technical ones. To do this, we isolate the activities that a system developer with perfect technology would perform.

5.2 Essential system development activities

Perfect system development technology would consist of a perfect processor to carry out the activities and a perfect container for all information to be stored during the project. This processor and container would have no human fallibilities and would therefore perform none of the activities that compensate for those fallibilities. Consequently, perfect system development technology would remove the need for job categories, project management, estimating, planning, resource allocation, and quality assurance. Yet, a perfect system development processor would still carry out the essential activities of system development, which we earlier called the technical activities.

The system development process is composed of three essential activities:

1. *defining the essence of the system* — Development managers must determine the purpose of the system, the events to which it must respond, the fundamental activities, the information both from current and past events (essential memory) that the system must store in order to carry out its responses, and any custodial activities required to establish and maintain the essential memory.

2. *selecting an incarnation of the essence* — System developers select a set of processors and containers to implement the system's essential activities and memory.

3. *constructing the physical system* — System developers establish the actual configuration of the processors and containers chosen during the previous activity.

All system development efforts *must* accomplish these three tasks in one way or another. Whether the system produced is automated or manual, whether the project is carried out by one person or one hundred, whether the effort takes one day or two years, these activities are fundamental. They are the activities that a perfect system developer would have to perform to produce an incarnation. They are the essence of system development.

Each of the essential activities can be described as a set of much more detailed activities. Those descriptions, however, must wait until later chapters. At this point, we have stripped away the managerial constraints that stand in the way of making a complete and understandable strategy for system development. In later chapters, we elaborate on this strategy, concentrating on the definition of essence.

5.3 Summary

Technical and managerial strategies for system development are more effective if they are distinct strategies. In this chapter, we describe a technical strategy that maps out only those activities needed by all projects; these activities are the essence of system development. To find the essential activities, we eliminate all activities that are made necessary by system development technology. Because the most important component of that technology is human developers, that means removing all activities performed to compensate for human fallibilities. We are left with three essential system development activities: defining the essence of the system, selecting an incarnation for the system, and constructing the system. It is on the first of these activities that our strategy concentrates.

Chapter 6

System Modeling

In Chapter 5, we supplied the three activities that would be performed by a perfect system developer and that are therefore found in all development efforts. The problem is that we have gone a bit farther than necessary. Through our imaginary perfect system developer, we have transcended not only the specific problems of a hundred-person project and the rather different problems of a solo effort, but also the problems that plague *all* human endeavors. Our model of the system development process is so devoid of technological limitations that it applies not only to all human system development efforts, but also to all system development projects carried out anywhere by any form of life.

In this chapter, we come back to earth. Because all system development efforts at this time rely primarily on human labor, we can incorporate certain effects of using human processors into our strategies without making them any less universal. Therefore, we reintroduce into the process of system development an accommodation to two human drawbacks that hamper all system development efforts: limited capacity to understand concepts and poor memory. Because people are easily confused by too much information, they can only concentrate on a few things at a time. Furthermore, they don't always do a perfect job of remembering information that they don't have to think about right away. The simple fact is that people can't keep track of a large amount of information unless they store it somewhere, and unless the stored information is represented in an easy-to-understand way.

Unlike the perfect processor, human system developers cannot create and remember the essential activities and memory relying solely on their memory, nor can they memorize an entire software design during the construction of programs. Instead, system developers write things down; they create an external, tangible replica of their mental image of a system's essence or incarnation. These replicas are called *system models*.

In this chapter, we discuss the concept of modeling systems. We do not yet introduce the tools and techniques for building system models. For now, we concentrate only on how models help system developers surmount their intellectual limitations and on what characteristics contribute to a model's success.

6.1 The benefits of modeling

Because of human intellectual limitations, system developers usually encounter at least three significant difficulties when trying to build systems. First, systems are usually difficult to understand, since they are often large, complex, and confusing. Just figuring out the goal of a system can sometimes be a major accomplishment. It is also difficult to remember what you know about a system. If the system is anything but trivial, defining its essence requires a great amount of information — far more than any person can comfortably and reliably remember.

Finally, it is difficult to know if your conception of a system is correct. How can you verify that you have correctly identified the essential activities of a system? One obvious way is to review your ideas with someone who can detect mistakes. However, because it is difficult to communicate your ideas to others, you can waste hours discussing some aspect of a system with a co-worker before realizing that each of you had a different idea of what was said.

An effectively constructed model helps you to overcome these difficulties. Even if you are performing a one-person development effort, you gain many advantages from building models. In addition to augmenting your own memory, models help you understand systems and visualize them. They also help you verify your plans for the system with other people. By providing one consistent view of the system to all project members, models allow everyone to discuss the system without misunderstandings.

Once you decide to exploit the benefits of using effective models, you are faced with a problem: Which of the many available modeling approaches should you use to help you define the essence and incarnation of the system you plan to build? The following section recommends a practical approach.

6.2 The choice of a modeling approach

Many different models can be made of a given object. Since models are imperfect replicas, they do not show everything about an object. The selectivity of models makes it possible to design a model that highlights certain features of an object and ignores others. In order to choose the best model for your purpose, you need to decide which features should be highlighted in order to fulfill that purpose. You then choose the kind of model that best depicts those features.

In the case of systems analysis, you should build models of the essence and the incarnation of the system to be developed. Figure 6.1 shows the three essential activities needed to produce the models and build the system. The modeling technique you choose should highlight the system's essential features and physical components. Specifically, you need a modeling approach that shows

- interactions between the system and its environment

- activities that the system executes in response to specific events

- interactions among the essential activities of the system

- essential memory that the system needs to support these activities

- allocation of the essential features to the components of the incarnation technology

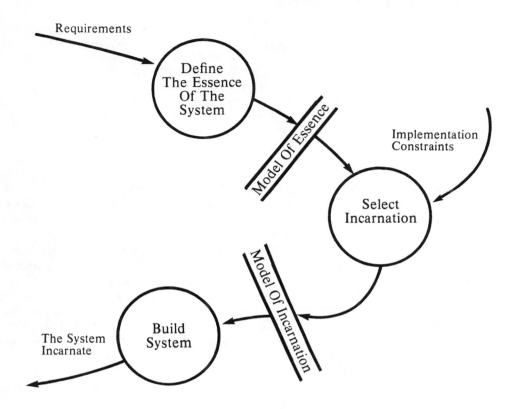

Figure 6.1. Essential system development activities and models.

Since the tools of structured analysis and information modeling do an excellent job of presenting all of these features directly and clearly, they are the modeling approaches of choice for systems analysis. The data flow diagram (DFD) declares activities, interactions between activities, interactions between the system and its environment, and essential memory. Data dictionary definitions define the detailed composition of interactions and essential memory, and minispecifications specify the details of essential activities, often in the form of structured English. All three tools work well together to show the allocation of essential features to the components of incarnation technology.

The tools of information modeling also support modeling objectives. An entity-relationship diagram effectively portrays a system's essential memory. The object, relationship, and data element definition conventions of information modeling supplement structured analysis data dictionary conventions.

No matter how appropriate these modeling tools are for our purposes, however, indiscriminate use of them significantly lowers the quality of the definition of either essence or incarnation. Since most of this book is concerned with the definition of essence, the next section supplies guidelines that will help you produce the best models of essence.

6.3 The principles of essential modeling

In this section, we introduce four principles that guide the building of models of a system's essence. Understanding these principles is important because much of the strategy for modeling essence given in this book is devoted to carrying them out. The principles are called the principle of the budget for complexity, the principle of technological neutrality, the principle of perfect internal technology, and the principle of the minimal essential model. The first principle, because it applies to all models, can be used to develop both essential and incarnation models, whereas the other three pertain specifically to essence.

6.3.1 The budget for complexity

When building a model of a system's essence, you must take into account the limited ability we humans have to absorb information. If the essential model is too complex, the reader won't be able to understand and verify the essential specification. As a result, you or someone else can omit or misstate essential requirements.

These human information processing limits are in effect a budget for a model's complexity. That is, essential models can be just so complex before they overwhelm the reader, overspending the budget for complexity. However, you won't get any bonus points for not spending the entire budget. Taking complexity out of one component of the essential model means putting that information elsewhere in the model, usually in components newly created for the purpose. It doesn't take long before these additional components make the model overly complex. So when building models of a system's essence, you want to spend the entire budget for complexity without ever overstepping the reader's limits.

The concept of a budget for complexity raises an interesting question: What is the maximum amount of complexity for a given model component? We don't have any scientifically established limits. Instead, we use the informal guidelines that have long been a part of structured analysis: A minispec should take up no more than one page and a DFD should contain seven (plus or minus two) activities. When there are no firm answers to this question, you can try to balance the following complexity factors, shifting information from one factor to another so that no portion of the model is too complicated to understand:

- *the number of components of the model* — How many separate components is the reader looking at? For example, a given DFD might have six activities and four data stores.

- *the complexity of each component* — How much is done in a single process shown on the DFD? Is the process a single, conceptually whole function, a set of related activities, or a group of relatively foreign activities?

- *the depth of leveling of each component* — How many levels are there between a parent bubble and the minispecs that describe that process? between a high-level dataflow definition and its constituent data elements? between the outermost "if" and the innermost action in a structured English group of nested "ifs"?

- *the complexity of the interfaces among components* — How tightly interconnected are the pieces of the model? For example, how many data elements pass between two processes or between a process and a data store?

- *the quality of the names on the model* — The ability to understand a model depends greatly upon how the pieces are labeled. The names on the model should give as much information in as few words as possible. Consider a process to sell books: A name such as "Process Order," while honest, is not the most effective one. Both "Process" and "Order" overstate what the activity is. "Process" could mean "sell" or it could mean "make." "Order" could mean "books" or "cars." "Sell Books" is a more meaningful name.

- *the clarity of the model's layout* — If the model is still difficult to understand after you have made sure that none of the above factors have made your model too complex, then it is probably poorly formatted. Without changing its content, reorganize the model until it looks better. Redraft your DFDs so that you cross as few lines as possible and provide a reasonable reading flow. Indent or outline activities on the minispecs so that they can be understood unambiguously.

By striking an acceptable balance between all these factors, you will effectively use your reader's budget for complexity and the resulting model will express the system requirements as concisely as possible. A concise model helps you establish the correct requirements by making it easier to spot omissions or misstatements of requirements.

6.3.2 Technological neutrality

When you build a model of a system's essence, you want it to contain only the essence of the system, nothing more. You do not want the essential model to bias the subsequent implementation effort toward one technology or another. You want the model to be technologically neutral.

Physical fossils and premature implementation modeling are two forms of technological bias likely to slip into an essential model. Physical fossils are remnants of the technology used to implement the current system. For example, if the division between manual and automated activities in the existing system is evident from the model, the model is not technologically neutral. In another case, the so-called essential model may include activities, such as sorting input transactions, that are performed only because the existing system uses sequential file structures. In both of these cases, the model of the essence of a new system is inappropriately influenced by the imperfect implementation technology now in use. These fossils obscure essential features of the system, jeopardizing the completeness and correctness of the essential model. They also impose unnecessary constraints on the selection of the new incarnation. Finally, if physical fossils end up in the new system, they decrease system reliability and maintainability by increasing the number of components liable to fail and by making it more difficult to find the source of a problem.

The other form of technological bias is premature implementation modeling. During consultations, we often detect a physical detail in an otherwise logical model. When we point it out to the author, we all too frequently receive the response, "Well, we already know that we're going to be using CICS under MVS with COBOL and VSAM, so I thought I might as well go ahead and include all the appropriate details in the logical model.'' This is a mistake. The purpose of logical modeling is to separate the declaration of what activities must be carried out from the determination of how to carry them out with the available technology. You separate these two activities for the same reasons that you partition programs into modules: They are very different and relatively independent from one another, and it is easier to do both of them correctly if you do them separately. Furthermore, if you do include technological information in the logical model, you may find that the details of the new technology have changed by the time you are ready to use it.

We have been careful to limit the kind of technology we want you to omit from the model of essence. The statement of a system's essence should contain no hint of the technology used to implement the system itself. Unfortunately, many people misconstrue this to mean that they should eliminate all technological influence, whether it is part of the implementation of the system or part of the technology used by the outside world that interacts with the system. To make our position quite clear on this matter, we state a third principle: the principle of perfect *internal* technology.

6.3.3 Perfect internal technology

We defined the essence of the system as the activities and memory that would have to be in the system even if the system were implemented using perfect technology. The issue now is, Do you assume that perfect technology exists everywhere or do you confine it to the system you are studying? When you build logical models, do you in fact imagine a technologically perfect universe?

No, you do not. When you try to discover the essence of a system, you assume that the technology only within the system itself is perfect. You make no such assumption about the correctness, speed, skills, capacity, or cost of the technology of systems outside the context of your study. This is the principle of perfect internal implementation technology.

To see the reasoning behind this principle, you need to consider the purpose of essential modeling and the consequences of assuming that external implementation technology also is perfect. The requirements imposed upon the system by the limitations of the technology outside the system are independent of the technology used to implement the system itself. A mortgage holder can fail to make a payment or can pay an incorrect amount whether the mortgage system uses batch or online computer technology. No matter what technology is used to implement that mortgage system, the system still must contend with these environmental imperfections. Because every possible incarnation would still have to deal with the same external technological constraints, external constraints don't bias the essential model toward any particular form of incarnation technology.

More important, assuming perfect technology in both the system and the environment destroys the purpose of many if not most interactive systems. If the environment for an air traffic control system were perfect, for example, every plane would have enough runway space to land without any chance of collision. No one would have to take out mortgages because everyone would have enough money to buy his or her dream house. Removing the imperfections of external technology takes the idea of perfect technology so far that it no longer helps define the essence of a system that is needed in the real world.

When you apply the concept of perfect technology, you eliminate any feature of technology that might be used to implement the essence of a system. However, you must retain any feature of the system mandated by the imperfect technology of the system's environment. The principle of internal perfect technology places useful limits on the concept of perfect technology. It keeps you from throwing the baby out with the bath water by discarding the reason for the system along with the implementation technology.

6.3.4 The minimal essential model

Our insistence on distinguishing between internal and external implementation technology is an example of how carefully you must apply the concept of perfect technology. However, you can distort the definition of essence if you don't follow yet another guideline, the principle of the minimal essential model. The purpose of this new principle is to prevent overspecifying the essential features.

Even though you apply the principle of perfect internal technology correctly, you may find more than one way to define an essential requirement, but only one of these possible definitions meets the standards of the minimal essential model: The least complex one is the definition to choose. Consider this example: A blood bank is presented with a request for a certain number of units of blood of a certain type. Since the blood may not be immediately available from donors at the time the request arrives, the blood bank, even with perfect internal technology, needs to maintain an essential data store of donated blood. In addition, the blood bank prefers to give out its oldest blood first. The essential activity Issue Blood probably has a requirement for accessing essential memory that is something like this: "Given a blood type, find the oldest usable units of blood of that type."

Other people might contend that the only access capability requirement is that the essential activity be able to get at all the units of blood from essential memory, without regard to type and age. Then, after having obtained every unit of blood available, the essential activity would select the oldest of the correct type. These people reason that since they are imagining a system running with perfect internal technology, the system would take no time at all to obtain all units of blood, look through all of them, select those that are of the right type, sort them by age, and pick out the requested number of units, starting with the oldest. Since this set of activities "gets the job done" with no performance degradation or additional cost (because of perfect technology), it must be logical.

We agree. Nonetheless, although both versions of the activity are essential, there is a difference. The second version says that the activity will obtain all blood units and then perform extract activities to get what it wants. The required access capability for this version has been made simpler, but at the cost of making the rest of the activity more complicated. The first version of the activity merely specifies what the activity needs from essential memory. This version assumes that somehow perfect technology will obtain the required data, and that the technology will take care of whatever sorts and extracts are required. It will give us precisely the units of blood asked for — nothing more, nothing less. In short, the first version is less complex than the second.

Why should you choose the simpler version? After all, with perfect internal technology, it shouldn't matter how complex or convoluted an activity is. Perfect technology can easily handle the most complex way of specifying an essential activity — as long as that way accomplishes what it has to, no matter how the essence is incarnated.

The problem is that an essential model with this extra complexity can't pass this spartan objective. Although you imagine the system to have perfect internal technology, you will ultimately have to implement the essence with imperfect technology. Now as much as you don't want to cater to any specific technology, all selections of incarnation technology will have to deal with any extra complexity that perfect internal technology allows you to put into the essential model. As useful as the principle of perfect technology is, you aren't willing to increase the complexity of selecting the incarnation because of it. So, your essential model has to pass another test besides the one for perfect technology. Given any number of ways to express an essential activity, you want to choose the way that specifies the minimum number of activities and data pieces and that minimizes the complexity of each activity. In that way, you make sure that you start the process of selecting any incarnation with a precise statement of what you must implement. This is the principle of the minimal essential model.

6.3.5 The benefits of the modeling principles

These last three essential modeling principles form a set of checks and balances that help you make sure that you develop the complete model of a system's essence. The perfect internal technology principle prevents you from overusing the principle of technological neutrality. It keeps technological requirements that are the result of an external system in the essential model by removing such requirements from the domain of the principle of technological neutrality. The minimal essential model principle helps you avoid overdoing your application of the principle of perfect internal technology. It stops you from writing specifications that are overly complicated, simply because perfect internal technology can easily handle complicated requirements.

The budget of complexity principle is independent of the other principles. That is, you must worry about achieving an understandable model whether or not you assume perfect technology or choose the minimum requirement. For this reason, the budget of complexity principle applies equally well to either essential or incarnation models.

6.4 Summary

System developers need to build models in order to remember, understand, and confer about systems. We advocate building models of the essence and the incarnation of the new system. The modeling tools of structured analysis and information modeling do an excellent job of presenting the most important features of these models.

We introduce four principles to follow when building models of essence. Following the principle of the budget for complexity ensures that no component of a model is too complex for the reader to understand easily. The principle of technological neutrality states that the model of essence should contain no information about the technology used to implement the system. According to this principle, you imagine that the technology inside the system is perfect, but another principle — that of perfect internal technology — says that you can make no such assumption about the world outside the system. The fourth and last principle is that of the minimal essential model; it states that when there is more than one way to specify an essential requirement, you choose the simplest way. These principles are crucial for deciding the content and representation of an essential model.

Part Two

Essential Modeling Tools and Strategies

You will not be able to comprehend the essence of a system unless you partition it. To help you do this, we propose the use of partitioning themes: precisely defined ways to partition essential features in such a way that the principles of essential modeling are observed. In Chapter 7, we introduce two partitioning themes: event partitioning for essential activities and object partitioning for essential memory. Chapters 8 and 9 each define one of the themes in detail and show how to represent the resulting systems with the tools of structured analysis.

All of the essential activities in our system development strategy are complex, and there is no one correct way to carry out the associated tasks. In Chapter 10, we introduce the two major technical strategies for defining the essence of a new system: creating the essence from scratch and deriving the essence of an existing system, then adding new essential features to it. The chapter concludes with a discussion of how to choose between these strategies for a particular project.

Chapter 7

Partitioning
Themes for
Essential Models

To plan the essence of a system, system developers require more than the modeling tools of structured analysis and information modeling and more than the set of modeling principles discussed in Chapter 6. They need to know how to partition the system and its memory into pieces that satisfy those modeling principles. These pieces should depict the true requirements of the system — meaning, the pieces should contain no technological bias and should not be too complex. In our work with users of structured analysis, we found not simply a single theme of partitioning that satisfies both goals but *two* themes that work together. The first theme, called event partitioning, is used to partition the portion of the system that produces responses; it uses events to partition the system into essential activities. The second theme, called object partitioning, is applied to partition essential memory by assigning stored data elements to the objects they describe. In this chapter, we explain the themes' underlying concepts to serve as an introduction to the details given in Chapters 8 and 9.

7.1 Partitioning the system into essential activities

In Chapter 2, we saw that a planned response system interacts with the world around it by recognizing a particular stimulus and executing a set of prepared instructions in response. This set of instructions is called the essential activity; it responds to a single event, which can be either external or temporal. These concepts are basic to our use of the event partitioning theme. In the following subsections, we describe in turn the proper identification of an event, the system that emerges from event partitioning, and the benefits offered by this scheme.

7.1.1 Recognizing events

To use event partitioning correctly, you must be able to recognize what an event is. This requires you to consider both the system and its environment, because events occur in the system's environment and both temporal and external events are beyond the system's control. Many analysts confuse events with processes inside the system because they do not make a formal distinction between a system and its environment. This causes problems.

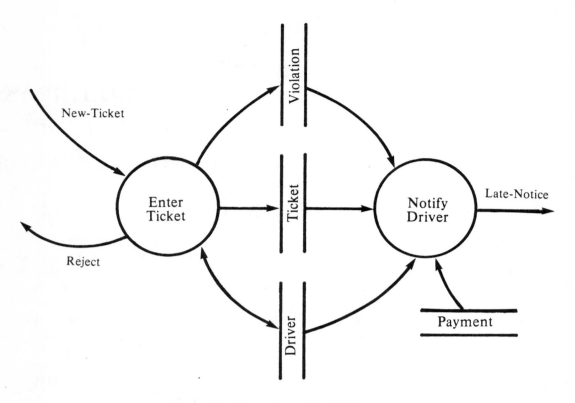

Figure 7.1. Part of a traffic violations system.

The first problem is that analysts sometimes think an activity within the system is an event. For example, in the traffic violations system partially shown in Figure 7.1, Enter Ticket is not an event but is an activity within the system, even though it is the first part of a response. Even worse would be for analysts to decide that "issue late notice" is an event, since it is a system response and as such is the consequence of an event. Worst of all would be to say that "stored data update" is an event, since it is a set of activities that are parts of responses to many events. This kind of error is made when analysts look for events within the essence of the system instead of within its environment.

A second problem results when analysts look at events from the viewpoint of entities outside the system. "Driver receives late notice" is an event, but it is an event from the driver's viewpoint. From the viewpoint of the traffic violations system, it is a response going to an outside entity. The analysts' perspective must not shift from that of the system to that of an entity outside the system.

Analysts often assume incorrectly that any point in time that triggers an activity is a temporal event. Such a conclusion leads to temporal events like "time to produce end-of-run reports" or "time to ship new subscription transactions to the computer center." These events are not temporal events because they are dependent upon a particular type of system implementation technology. With perfect technology, many reports could be produced before the run is over. New subscription transactions would be processed instantaneously, so there would be no need to hold them for processing later by the batch computer system. True temporal events trigger activities regardless of the technology used to implement the system. In a payroll system, for instance, the users wish to produce paychecks periodically regardless of whether technology would enable them to pay every two microseconds or every day.

Problems with recognizing events are easily corrected. In fact, the examples of events in this section and throughout the book should be about all you'll need to learn to do so properly. In any case, you should watch out for non-events.

7.1.2 The event-partitioned system

By now, you are used to seeing data flow diagrams that show a system partitioned into responses according to events. On these DFDs, each process symbol, or bubble, represents an essential activity. We call a DFD that exhibits this characteristic an event-partitioned data flow diagram. For another example of such a diagram, look at Figure 7.2, which shows the essence of another part of the traffic violations system.

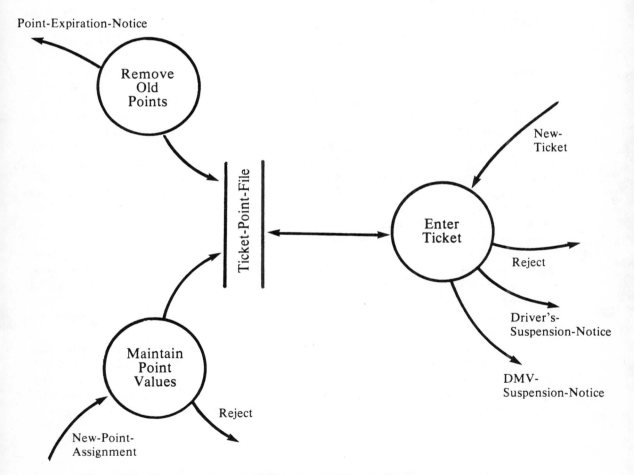

Figure 7.2. Event-partitioned DFD for part of the traffic violations system.

In the diagram, each bubble represents the system's *entire* planned response to a single event. You can see this by comparing the diagram to the following list of events:

- Traffic officer submits new ticket.

- Department of Motor Vehicles establishes new point values for violations.

- It is time to remove old points.

To qualify as a true essential activity, each activity in the event-partitioned system must pass two tests based on a more formal definition of an essential activity. To pass the first test, the activity must contain the actions that would be carried out in response to one and only one event if the system were implemented using perfect technology. The activity in Figure 7.3 flunks this test, because the activity shown responds to two events: "traffic officer submits traffic ticket" and "court issues judgment." It therefore has activities that belong to more than one essential activity and is an example of incorrect event partitioning.

The second test is designed to make sure that an essential activity is complete. When all the activities that make up the essential activity have been performed, the system *must become idle* until the event in question occurs again or until a different event occurs. If the system can immediately begin carrying out another activity, even though no other event has occurred, then the so-called essential activity doesn't contain all of the proper activities. However, if the system, even with perfect technology, simply has *nothing left to do,* then the essential activity contains all the activities it should.

The two diagrams in Figure 7.4 demonstrate how this second test ensures the completeness of an essential activity. In Figure 7.4a, you see two essential activities: Issue Material and Reorder Stock. Your first clue that these activities are improperly partitioned comes from examining the response Purchase-Order. Since a supplier will accept a purchase order as soon as it is generated, a purchase order could be generated directly upon receipt of the external stimulus Material-Request, which lowers inventory below an amount designated as the inventory reorder point. Why then are Issue Material and Reorder Stock linked by a data store? Answering this question reveals more problems with the partitioning in Figure 7.4a.

Since it receives no external stimulus, Reorder Stock appears to respond to a temporal event. However, in fact, Reorder Stock is triggered when the inventory level drops below a fixed amount. The activity itself detects this drop through continuous sampling of the inventory data store. Reorder Stock is therefore triggered neither by an external nor a temporal event. This is a second clue that these activities have been improperly partitioned.

Figure 7.4b shows the result of applying the completeness test. Along with Issue Material, Reorder Stock is part of an essential activity called Satisfy Material Request. There is no reason for a perfect processor to stop processing when Satisfy Material Request receives a request for material that forces inventory below the reorder point. Instead, processing could continue with the immediate generation of a purchase order.

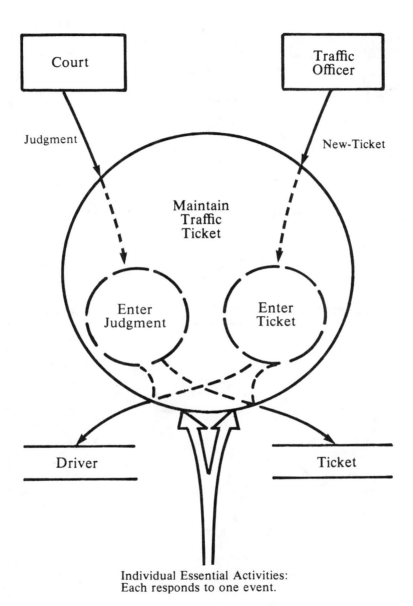

Individual Essential Activities:
Each responds to one event.

Figure 7.3. Poorly partitioned essential activity: two responses packaged together.

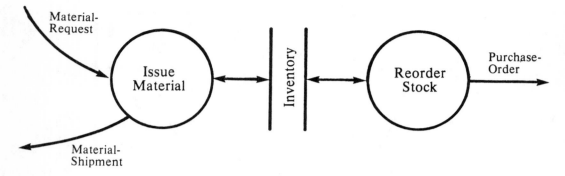

Figure 7.4a. Poorly partitioned essential activity: one response packaged as two responses.

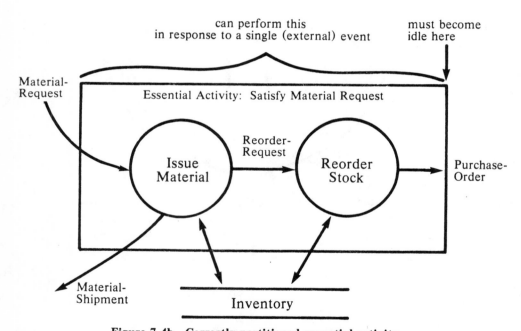

Figure 7.4b. Correctly partitioned essential activity.

Event partitioning produces essential activities that are either pure custodial activities, pure fundamental activities, or a mix of the two. The pure custodial activity is perhaps the most common kind of essential activity. After all, think of the number of events that trigger only a change to a system's essential memory. By contrast, pure fundamental activities are rare. Fundamental activities often produce responses that the system needs to remember, and therefore most events that trigger fundamental activities trigger custodial activities as well.

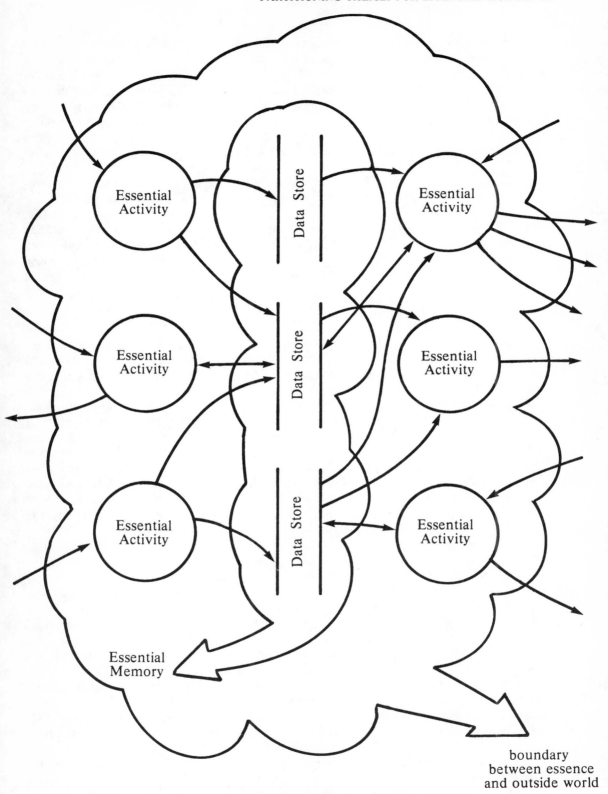

Figure 7.5. Characteristic shape of an event-partitioned DFD.

Event partitioning has a characteristic effect on the way essential activities interface with one another. All essential activities are connected through essential memory. An activity sends information to essential memory, and another activity obtains that information when needed from memory. There can be a delay of any amount of time between when an activity sends information to a data store and when another activity receives it. This means of communication accounts for the characteristic shape of an event-partitioned DFD, as shown in Figure 7.5. Each essential activity communicates with the outside world directly, and each sends dataflows to the data stores that connect them.

That essential activities interface exclusively through essential memory is no accident. It is a result of our partitioning theme. Because each activity contains the complete response to an event, all the direct, immediate communication among the subactivities lies *within* each activity. Otherwise, if one of these activities could pass data directly to another, it would mean that the event response was not complete, and would therefore imply a partitioning error. So once you succeed in getting an entire response to an event into each activity, everything the activities have to say to one another must be remembered between events.

7.1.3 The advantages of event partitioning

Event partitioning helps you tremendously in your quest for a system's true requirements because it produces reasonably uniform results no matter who partitions the system, it results in concise modeling, and it is faithful to the system's essence. Let's look at each of these advantages in turn.

An oft-heard comment is that structured analysis should be so rigorous and clear in its guidelines that ten different analysts analyzing the same system would produce identical DFDs. Although we don't think this degree of uniformity is either desirable or possible, analysts should agree at least roughly on the essential activities. Event partitioning fosters this rough degree of uniformity and hence minimizes unproductive arguments, since different analysts derive very similar activities from using the approach. This result contrasts sharply with the activity models of systems partitioned without event partitioning. These models vary widely from analyst to analyst, increasing the difficulty of finding true requirements.

Event partitioning is a major contributor to the analysts' ability to model concisely. Before event partitioning, analysts could choose to partition at the lowest level, dividing the system into the smallest possible activities and thus creating so many activities that no one person could understand them all. If instead analysts chose to partition at a higher level, creating fewer, larger activities, everyone fought over the boundaries between the activities because they lacked a formal, objective method to create them. Event partitioning provides a solution: It gives results that are precise enough for everyone to agree upon, and because it partitions at a middle level, the results do not overwhelm reviewers with too many details.

Concise modeling requires each activity to be as independent as possible from other activities, so that they can be understood and verified individually. Analysts achieve this objective by creating the fewest number of interfaces between two activities. Event partitioning typically yields a DFD that satisfies the goal of minimum interfaces.

Event partitioning also helps analysts to produce a faithful model of the true requirements, because it clearly shows that the system's responses to different events can be performed in any sequence. In a payroll system, for example, some actions obviously must be performed before others: The employees must be hired before they can be paid. However, over the whole set of employees, some will be hired before others, some will be fired before others are hired, and some will be given paychecks before others are hired. Event partitioning reflects that systems typically don't interact with entities in the environment in any set sequence.

7.2 Partitioning essential memory

The essential memory for any but the most trivial systems must also be partitioned into understandable, verifiable pieces according to a theme. The result is a model of essential memory that is easy to understand and free of technological influence. Unfortunately, the event partitioning approach cannot be used to create this model, for the reasons given in Subsection 7.3.3. Analysts must therefore use a different theme, one called *object partitioning.*

The concept of object partitioning, or entity-relationship analysis, is hardly new to structured analysis, existing as part of various system development techniques since the 1960s. For example, DeMarco incorporated object partitioning of data stores in his books *Structured Analysis and System Specification* and *Controlling Software Projects: Management, Measurement & Estimation* [11, 12]. As old as this approach is, however, it is still often poorly understood. So in this section, we explain object partitioning and why we use it.

7.2.1 Objects

The theme used to group the many data elements in essential memory comes from how we humans normally understand and communicate information about things. Creatures with limited brain power that we are, we don't have the ability to identify uniquely each thing we see. So, we group things that share certain characteristics into a set, and we give the set a name. In this way, one noun, like *computer,* can stand for the many distinct things that make up the set. The process of grouping things with common characteristics and naming the set is the basis for our theme of partitioning essential memory. We call any such named set of things an *object,* and anything within the set an object *instance* or object *occurrence.* The object Engine is a set of many specific instances of engines; the object Student is a set of many specific occurrences of students.

Real objects are tangible things, like jet engines, cars, and books. There are also named sets of things with no physical existence called artificial objects. Artificial objects are ideas shared by a group of people. For example, the artificial object "line of credit," if anything, is a faith, shared by a banker, a buyer, and a seller that the buyer will pay for a purchase. Many artificial objects are so familiar to us that they almost seem to be physically real, but they are in actuality shared ideas.

The data elements kept in essential memory are representations of the characteristics of both real and artificial objects. To paraphrase DeMarco, essential memory is a simulation of things whose true existence lies outside system boundaries [11]. Essential memory is the film within an imaginary camera that we train on these objects, recording their features with data instead of silver bromide crystals or electronic pulses. At some later point, the data in essential memory will be projected through an essential access in a fundamental activity. In Figure 7.6, the process of storing and retrieving objects is depicted as the taking and showing of a photograph.

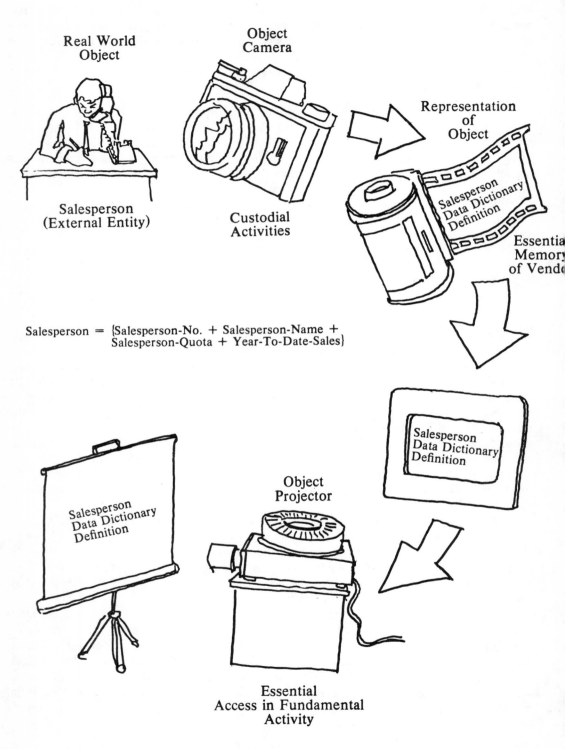

Real World
Object

Salesperson
(External Entity)

Object
Camera

Custodial
Activities

Representation
of
Object

Salesperson
Data Dictionary
Definition

Essentia
Memory
of Vendo

Salesperson = {Salesperson-No. + Salesperson-Name +
Salesperson-Quota + Year-To-Date-Sales}

Salesperson
Data Dictionary
Definition

Salesperson
Data Dictionary
Definition

Object
Projector

Essential
Access in Fundamental
Activity

Figure 7.6. An object in essential memory likened to film in a camera.

Since the essential memory simulates both real and artificial objects in the world outside the system, we can use these objects to partition essential memory. There are many ways to do object partitioning. In this section, we outline a theoretical approach; in Chapter 20, we offer practical suggestions for using object partitioning.

7.2.2 Object partitioning

To begin object partitioning, analysts must first identify the data elements in the system's essential memory, and then proceed as follows:

1. The analysts identify the objects outside the system that are most closely described by the data elements in essential memory.

2. They group the data elements that led them to choose a particular object.

3. They give each group of data elements the name of its corresponding object.

Table 7.1 provides an example of object partitioning. Before beginning object partitioning, you establish a list of the required essential memory data elements. In Step 1, you identify the objects Automobile and Renter as being most closely described by the essential memory elements. Step 2 results in two groupings of data elements, one for each object. In Step 3, you name the groups for the objects you chose.

Table 7.1
Car Rental Example of Object Partitioning

Preliminary Step	Step 1
Essential Memory Data Elements:	*Objects:*
Mileage Make-Of-Car Renter-Name Renter-Address Automobile-ID Driver's-License-Number Credit-Card-Number Model-Year	Automobile Renter
Step 2	**Step 3**
Groups:	*Automobile:*
Automobile-ID, Mileage, Model-Year, Make-Of-Car	Automobile-ID, Mileage, Model-Year, Make-Of-Car
	Renter:
Renter-Name, Driver's-License- Number, Credit-Card-Number, Renter-Address	Renter-Name, Driver's-License- Number, Credit-Card-Number, Renter-Address

After completing this procedure, the analysts organize the data elements within each object group. The analysts put together all the data elements that refer to the same occurrence of an object.* Because most activities need only some of these groups of data elements at any one time, the system needs a way to identify individual object occurrences. The analysts typically assign a data element to identify each occurrence. These identifiers are used to find data elements within essential memory that describe a particular occurrence.

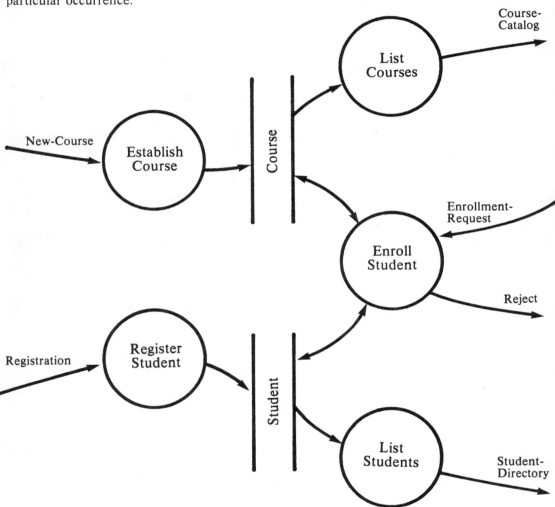

Figure 7.7. Object partitioning of essential memory data stores.

In summary, we can take all of the data elements that describe an object and organize them into simulations of the object occurrences outside the system. Each object occurrence grouping is composed of the identifier data element for the occurrence and all of the data elements that describe that particular occurrence. Figure 7.7 shows an event-partitioned DFD with object-partitioned data stores. Student and Course are the

---

*You might call these occurrences "logical records." However, the physical connotations of the word *record* make us cringe, so we prefer object occurrences.

objects in this case. This DFD illustrates the benefits of object partitioning: It has data stores that are easy to understand and that are free from technological influence. Notice that these data stores have names that are singular, not plural. This doesn't mean that they each contain only one object occurrence; always assume that a data store contains multiple occurrences of the same object.

Next, we discuss an issue that baffles many system developers: Why do we choose to use object partitioning as our theme for partitioning essential memory? To show why, we compare object partitioning to other possible ways of organizing stored data on an essential DFD.

7.3 Problems with other partitioning themes

All other themes for partitioning essential stored data elements fall somewhere at or between two extremes: allocating all data elements to the same data store or allocating each data element to its own data store. In the following subsections, we consider the problems caused by each of these schemes and by one alternative that falls between the extremes.

7.3.1 The single data store partitioning solution

If you establish a single data store for all the essential memory elements, you'll end up with something like the diagram in Figure 7.8. Our criticism of this diagram has nothing to do with whether the single data store approach is technologically neutral; we have no objection on that issue. But we're not happy with the quality of the model for two reasons. First, this solution distributes complexity poorly. The diagram will be a total mess if the system is anything but trivial. The great quantity and diversity of accesses to stored data will result in a rat's nest of arrows into and out of the Essential Memory data store.

Our second criticism is that this partitioning theme, which really doesn't partition at all, hides useful information about the kind of data that must be stored by the system, and it also prevents the reader from telling which essential activities use which subsets of the stored data elements. Most system memories contain many different objects, and a particular essential activity rarely requires information about *all* of them. We'd like to be able to see which subset of stored information each activity requires. In addition, when there's only one data store, its name will tell the reader of the model very little about the essential memory.

Every data flow diagram, at every level, should show as much useful detail as it can without straining the reader. So if you can afford to show more than one data store on any DFD, and if the additional detail gives useful insights about the essence of the system, you ought to do so.

7.3.2 The data store for each element partitioning solution

At the opposite end of the spectrum of partitioning strategies is the scheme of allocating a private data store for *each* data element, as in Figure 7.9. This arrangement solves the problems of the single data store scheme, but it introduces a problem of its own. We all know that the average database contains 160 data elements [U.S. Bureau of Facts]. There is no way to fit 160 data store symbols on an essential DFD without creating an unreasonably complicated diagram. For this reason alone, this partitioning method is unacceptable.

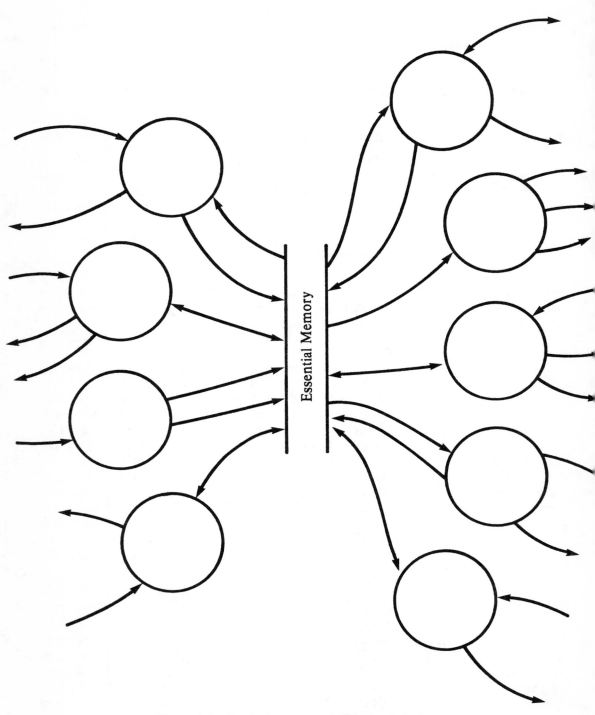

Figure 7.8. Single data store partitioning solution.

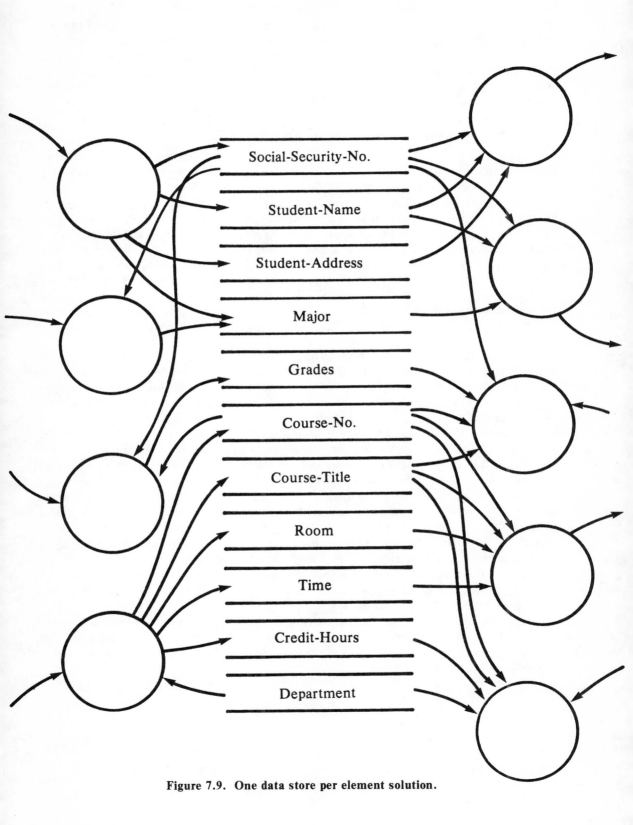

Figure 7.9. One data store per element solution.

As in the previous case, we do not object to this solution on the ground that it is technologically prejudiced, but rather because the model produced is not concise. Neither of these two extreme forms of distributing essential data elements among data stores provides an appropriately detailed view of the system's essential memory while still producing an easy-to-understand diagram. We now examine a theme that falls somewhere between these two, to see if it works better.

7.3.3 The private component file partitioning solution

Suppose you actually did try to draw a data flow diagram with one data store for each essential stored data element, and you immediately recognized the problem of having far too many data stores. What would you do next to reduce the complexity of the model? You'd probably try out a few different ways of combining the single element stores, but without introducing physical prejudice into the model.

One way might be to group all the data elements accessed by a single essential activity into one data store, something like the DFD in Figure 7.10. This partitioning theme improves the diagram by reducing the number of data stores, and at the same time, there doesn't seem to be anything too physical about the way the data elements have been grouped. By contrast, the DFD in Figure 7.11 is much more physical, because a partitioning between batch and online reflects technologically limited processors. If the system had a cost-free, instantaneous processor, all essential memory could be kept online.

Despite its appeal, there is something insidiously wrong with this approach. To discover the problem, think about the contents of each of these data stores. How will you determine where to store a data element that is accessed by more than one essential activity? For example, the data element Hourly-Pay-Rate in a payroll system is used by more than one essential activity. Will you find this element in the data store associated with the custodial activity that updates it, like Record Newly Hired Worker, or will you find it in the data store for a fundamental activity, like Pay Hourly Workers, that makes use of it? Neither answer is as disturbing as a third possibility: You will find the data element in *both* data stores.

Modeling the same data element in more than one data store introduces a subtle kind of false requirement into the specification. It is a violation of the minimal modeling principle. Perfect technology could implement an essential memory that repeated data elements. However, you want as little data element redundancy as possible, because one copy of a data element may be updated differently from another copy. For this reason, the model of a system's essential memory must not contain duplicate data elements, regardless of how you choose to group data elements into data stores for a concise, technologically neutral model.

The private component file partitioning approach in effect corresponds to event partitioning of essential memory. As we have shown, the partitioning theme that proves effective for essential activities has some serious flaws when applied to essential memory.

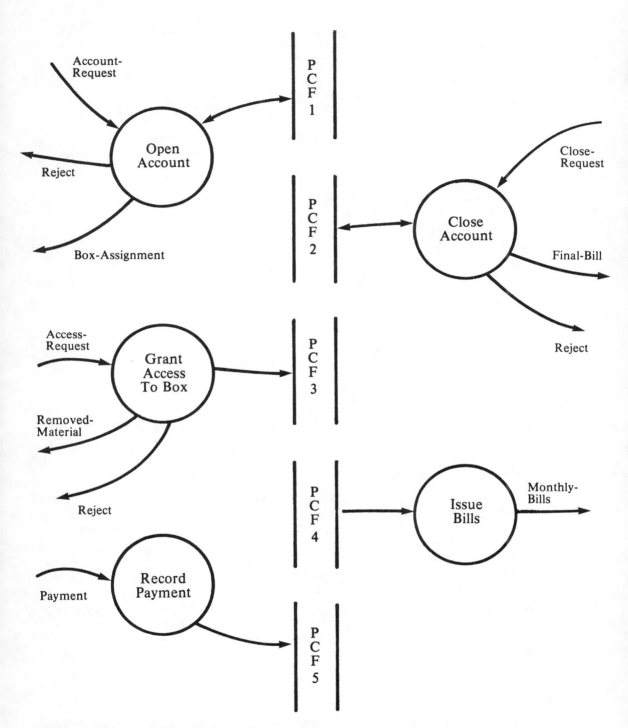

Figure 7.10. Private component file solution for the essence of a safe deposit system.

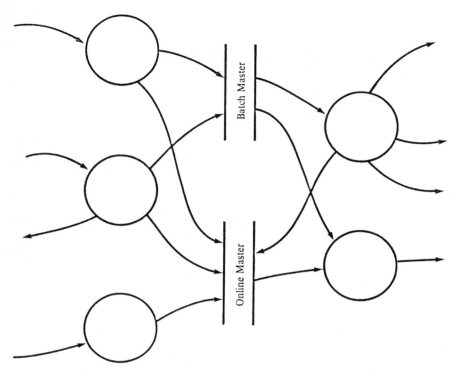

Figure 7.11. Batch data vs. online data store solution.

7.3.4 The benefits of object partitioning

By exploring different options for partitioning essential memory, we see three benefits of object partitioning. First, object partitioning produces a model of essential memory that is concise. Grouping data into objects helps you to see more of the contents of essential memory and the memory needs of essential activities, but does not overwhelm you with information. Second, an object-partitioned memory is technologically neutral. Object partitioning is independent of the constraints imposed by internal imperfect technology. Finally, object partitioning minimizes the repetition of stored data elements. With the proper strategies for refining a model, you should be able to achieve your goal easily of a nonredundant model of essential memory. In sum, all of these benefits lead to the elimination of false requirements from the model of essential memory, as well as to a complete and correct specification of the true requirements for essential memory.

7.4 Leveling

Both event partitioning and object partitioning have a limitation. Most systems that you are likely to study respond to dozens, if not hundreds, of events and contain perhaps dozens of objects. Using our partitioning themes leads to a DFD with literally dozens, if not hundreds, of essential activities and dozens of objects. While some advocates of structured analysis, notably Gane and Sarson, permit as many as fifty processes on a DFD, we believe, along with DeMarco and many others, that an effective data flow diagram should not contain many more than seven bubbles. Exceeding this limit by a few — having, say, a dozen processes on one DFD — probably will not greatly harm the final product. But diagrams of thirty or fifty or eighty processes cannot be used effectively by anybody, because, at the very least, they are confusing.

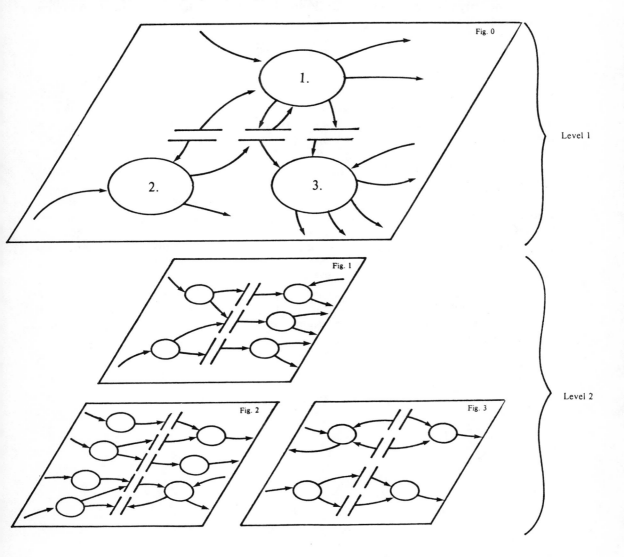

Figure 7.12. Leveled set of essential DFDs.

A good way to limit the complexity of every DFD to seven (plus or minus two) bubbles is by using the concept of leveling. If your event-partitioned DFD contains several dozen essential activities and objects, you create higher-level diagrams containing bubbles that are groups of essential activities and objects. So, in a typical essential model, the highest-level diagram does not depict the individual essential activities, but shows instead aggregates of essential activities. In Chapter 24, we provide some guide-lines for grouping essential activities into upper-level bubbles.

Figure 7.12 shows a set of essential activities that have been formed into three groups to make a higher level. To find such a system's planned response to an individual event, you must look at a lower-level diagram. If the system is medium size, you would probably find the response on the second-level diagrams. In the model of a large system, you might have to look as far down as the third or fourth level before finding an essential activity.

7.5 Summary

The themes of event and object partitioning are valuable in that they help you to structure your understanding of the essential activities and memory into concise, tech-nologically neutral, and minimal essential models. Event partitioning means dividing the system into essential activities. Each activity groups the actions that would be car-ried out in response to one and only one event if the system used perfect technology. After an essential activity has been performed, the system must become idle until an event occurs; if it doesn't, the essential activity isn't complete. All essential activities in an event-partitioned system communicate through essential memory.

Essential memory is partitioned into objects. Each object in essential memory corresponds to a real or artificial object in the real world and consists of data elements that describe the object's characteristics. All of the data elements that refer to the same occurrence of an object are grouped as a unit, and one data element is assigned to iden-tify the occurrence. Comparing object partitioning to other ways of partitioning memory reveals that it is superior.

Even with both these themes for partitioning essence, however, you face a prob-lem: No one tool of either structured analysis or information modeling can model all of the essential features, and so many modeling tools must be used. These modeling tools and their use with the partitioning themes are the topics of the next two chapters.

Chapter 8

Modeling
Essential Activities

The essential activities that you discover using event partitioning are modeled with the three main tools of structured analysis: the data flow diagram, the data dictionary, and the minispecification. Each tool portrays one or more parts of an essential activity: a definition of the stimulus that signals an occurrence of the initiating event, the appropriate response to that event, and the results of that response. Although they are part of an activity's response to an event, we treat accesses to essential memory as a separate, fourth aspect of an essential activity. This chapter describes how to model each of these aspects.

8.1 Modeling the stimulus

A stimulus that initiates an essential activity arises from an event, which can be either external or temporal. How you model a stimulus depends upon which type of event produces it.

8.1.1 A stimulus from an external event

An external event, like "employee reports time worked," usually produces an incoming dataflow that crosses the boundary of the system. This dataflow appears on the context diagram. The context diagram of Figure 8.1 and the data flow diagram of the essential activities in Figure 8.2 together declare the existence of the stimuli from external events. The context diagram shows the entity responsible for an external event (the employee in "employee reports time worked") with a box.

But none of the DFDs in the essential model tells what data elements make up the stimulus dataflow. This information is provided in the form of a data dictionary definition, perhaps like this one for the Time-Worked dataflow:

Time-Worked = Social-Security-No. + Regular-Hours + (Overtime-Hours)

This definition identifies the data elements and relationships between data elements that make up the initiating dataflow. If some of these data elements are really groups of data elements, the data element group must also appear in the data dictionary, divided into subcomponents until all the data elements have been identified. The individual data elements must be declared in the dictionary as well, and for those that can take on more than one meaning, both meanings must be given.

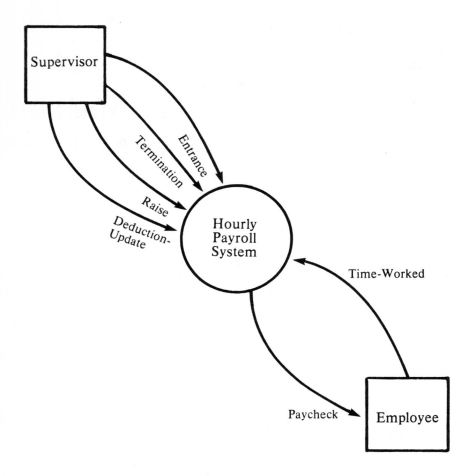

Figure 8.1. Stimuli on the context diagram of an hourly payroll system.

All of this is a straightforward use of the structured analysis modeling tools, particularly the data dictionary. The only special problem with modeling external stimuli results from applying the principle of the minimal essential model, as described in Chapter 6. According to this principle, the finished logical model should be a statement of only those requirements that are absolutely necessary even if the system were implemented with perfect technology.

This principle affects how you define an incoming dataflow: You must be sure to exclude any data element that is not absolutely necessary to the system. For example, suppose that the business form used to report the hours worked by an employee contains a data element that is not actually used for this purpose, but is necessary to some unrelated function of the form. Should it be included in the definition of the stimulus dataflow Time-Worked? Absolutely not. To do so would create a false requirement, transmitting to the implementors of the new system the need for a data element that is in fact extraneous.

Aside from this complication, defining stimulus dataflows that arise from external events is fairly simple, as long as you have a good grasp of the basic modeling tools, especially the data dictionary. Modeling temporal events, however, is more challenging.

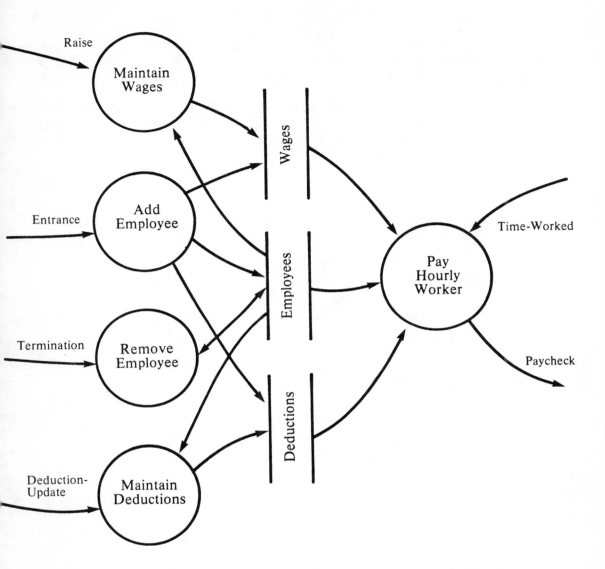

Figure 8.2. Essential activities that respond to stimuli in an hourly payroll system.

8.1.2 A stimulus from a temporal event

A temporal event — the arrival of a particular time — does not result in a dataflow crossing the boundary of the system, mainly because the passage of time does not happen in any one specific place. You could say that it happens everywhere, or that it is independent of location. Nevertheless, we often see data flow diagrams that try to represent temporal events as dataflows, such as the one in Figure 8.3.

The arrow labeled Payday is not really a dataflow as we use the term. It is not used or processed or acted upon by the system; instead, it *acts on the system itself* by controlling the execution of the essential activity. We like to draw an analogy between a DFD and a washing machine. Dataflows are most like the clothes, the water, and the detergent: They are acted upon and often altered by the process. Control flows are more like the dials and buttons that control the washing process. Data flow diagrams are usually restricted to showing dataflows, not control flows. Since you won't be showing control flows on the DFD, how will you define the stimuli that characterize temporal events?

In fact, the data flow diagram will not even indicate the presence of a temporal event, as you can see in Figure 8.4. In this DFD, the process seems to just leap into action spontaneously, since its bubble hangs directly off the data store, without any apparent contact with the outside world.

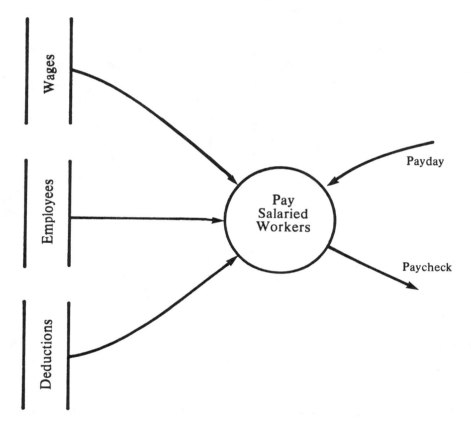

Figure 8.3. Improper representation of a temporal event.

You will not see any mention of the temporal event until you read the minispecification for this essential activity, a part of which might look like the one below:

On the 15th and last day of each month,
    do the following:

To define a temporal event that initiates an essential activity, you state it in the form of structured English at the beginning of the minispecification for the activity. You can specify the exact time that the event occurs, such as "6:00 p.m.," "Christmas," or "the first day of the month." Or you can define the event indirectly using a variable name such as "issue date" or "billing period" and define the name in the data dictionary. In sum, you rely primarily upon the data dictionary to define stimuli arising from external events and on the minispecification to define temporal stimuli.

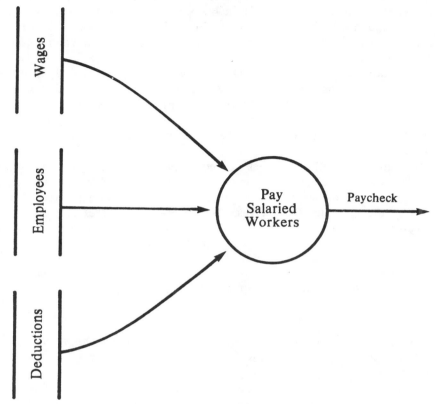

Figure 8.4. Proper representation of an essential activity that responds to a temporal event.

8.2 Modeling the planned response

The model of a system's planned response to a defined stimulus specifies the criteria by which the system *recognizes* the occurrence of an initiating event and the actions it performs in response to the event. The most detailed description of a planned response is called a minispecification. To write a minispecification, you rely primarily upon the basic specification tools of structured analysis: structured English, decision trees, decision tables, or even disciplined narrative text. In practice, people generally write most of their minispecifications in structured English.

As before in modeling the stimulus, you need to apply the goals of essential modeling to create a model of the system's planned response that is both an effective and accurate representation of the system's true requirements. You encounter two potential problems in achieving these goals: The minispecifications may be too complex and may show activities occurring in unnecessary order. Fortunately, you can use the modeling tools of structured analysis to avoid them.

8.2.1 Controlling the complexity of minispecifications

Since the minispecs are part of the essential model, you want to make them as easy to understand as possible, just as you did the DFDs. You want each minispec to be mentally "bite-size," small enough to be digested comfortably, but substantial enough to communicate a useful amount of information. Toward this end, you employ the structured analysis guideline that each minispec should be no more than one page long and not much less than one-half page long. But what if you write a minispec for a particular essential activity that ends up being three or four or twenty pages of structured English? What should you do?

The solution is to draw a lower-level data flow diagram that divides the essential activity into subfunctions so they each can be specified in a minispec between one-half page and a whole page in length. If your system contains essential activities that are long and complicated, you may need to draw one or two levels of data flow diagrams below the essential activity level. Such essential activities may each require a dozen or more minispecifications.

Fortunately, essential activities are not usually very complicated. We have seen many system models, small, large, and gigantic, and the difference between small systems and large systems seems to be the *quantity* of essential activities, rather than the *complexity* of the activities themselves. So, an essential model of a very large system usually consists of many levels of data flow diagrams *above* the essential activity level, because the system responds to many dozens of events, but many of those essential activities are defined in only a few minispecs each. In the typical system, an essential activity can be specified by one lower-level DFD and a half-dozen or so minispecs, together with the associated definitions from the data dictionary.

8.2.2 Avoiding false requirements

Minispecifications are the parts of the essential model where false requirements are most likely to appear. The primary culprit is the English language, and most other languages for that matter. Regardless of whether you write structured English or formal English, playscripts or pseudocode, or just plain narrative text, you are using an inherently *sequential* form of communication. That is, the actions stated in a structured English minispecification usually appear to take place in order, and only one action is performed at any given time.

This characteristic presents no problem as long as the actions that make up a planned response *must* be performed one after another, regardless of the implementation technology. But if some subfunctions of an essential activity can be performed simultaneously, a sequential modeling tool forces you to choose an arbitrary sequence among the actions, thereby introducing a false requirement into the specification.

Consider the minispecification below:

> If Quantity-Ordered ≤ Quantity-Available,
> > Subtract Quantity-Ordered from Quantity-Available.
> > Add Quantity-Ordered to Quantity-Shipped.

In this example, there is no essential requirement that the quantity ordered be added to the quantity shipped after it is subtracted from the quantity available. The order of these two statements is unimportant. With a nonsequential specification tool, such as a data flow diagram, we could show these two operations separately. But as it stands, this structured English minispecification states a requirement that the addition occur after the subtraction. Clearly, we do not *intend* to convey such a constraint to the implementors of the system, but our modeling tool has, figuratively speaking, put the words in our mouth.

Admittedly, the unnecessary sequence in this minispecification would not cause much damage. Indeed, probably no experienced reader of the specification would infer such a requirement, even though it is implicitly stated. Nonetheless, we care about this issue because not all such artificially introduced sequences are so innocuous. Forcing the system to perform actions in sequence for no essential reason can add expensive complexity to the system and degrade its performance.

Fortunately, you can take steps to solve the problem. There are two main methods of specifying planned responses without implying a sequence of action: redefining the structured English constructs and choosing a nonsequential tool for writing minispecs. Because these solutions are easy to use and do not delay the development effort, you don't have to put up with false requirements in the essential model no matter how small or seemingly harmless. Why should you settle for a specification that tells this sort of little white lie if you can do better?

8.2.3 Redefining structured English

Structured English uses the same three basic procedural constructs as structured programming: sequence, n-way decision, and conditional repetition. We are concerned here with only the sequence construct. To avoid the problem of implying a nonessential order among the actions in a minispec, you could redefine the sequence construct, in effect taking the sequence out. You could declare that in all essential minispecs, the order in which statements appear *does not* imply an essential order among them.

We hope you're at least mildly dissatisfied with this proposal, for in some cases, there *is* an essential order among the actions that make up an essential activity and you've got to have some way of stating it. By redefining the sequence construct, you are simply trading one problem for another.

We recommend, instead, the adoption of a new structured English construct: the *parallel* construct. This construct would mean "do the following things in no particular order," as in the example below.

> If Quantity-Ordered $\leq$ Quantity-Available,
>> do the following in no particular order:
>>> Subtract Quantity-Ordered from Quantity-Available.
>>> Add Quantity-Ordered to Quantity-Shipped.

Through this convention, albeit awkward, you can avoid even slightly implying an artificial sequence in the essential system requirements model. But sometimes, this problem can be solved more effectively by abandoning structured English for another modeling tool.

8.2.4 Using a nonsequential modeling tool

When you find the parallel construct awkward, you should consider using another tool for writing the minispec in question, such as a data flow diagram, a decision table, or a decision tree. Don't forget entirely about data flow diagrams just because you have partitioned the essential activity into pieces small enough to fit on a one-page minispec. If the activity contains many parallel actions, consider drawing a lower-level DFD, even if the resulting minispecs will be very small. Sometimes the benefits of avoiding an artificial sequence outweigh the cost of building and using a DFD with a number of trivial, but nonsequential, minispecs.

The essential activity you are modeling may evaluate many conditions in order to decide what action to take. If that is the case, you may prefer to use a decision table or a decision tree. These tools specify the policy without imposing an artificial sequence on the order in which the conditions are evaluated.

You have the choice of several tools for modeling a planned response — DFDs, decision tables, decision trees, modified structured English, or perhaps some other modeling tool. Which tool you choose is much less important than insuring that your final definition of a planned response communicates effectively and does not imply a nonessential order among its actions.

8.3 Modeling planned response results

The context diagram and the event-partitioned DFD declare the results of carrying out the planned response, just as they show the external stimulus dataflow. The context diagram shows results that cross the boundary of the system. In Figure 8.1, you see Paycheck, the result of an essential activity, coming from the system and going to the employee. As with external stimuli, the context diagram represents the entities outside the system toward whom a response is directed. Again, the box is used to represent these outside entities, as with the employee who receives a paycheck in Figure 8.1.

Essential activities can have two kinds of results: One kind is produced by the fundamental activities and the other by the custodial activities. The result of a fundamental activity is a message to the outside world. The DFD shows it as a dataflow leaving the system, such as Salary-Report in Figure 8.5. The result of a custodial activity is an update to the system's essential memory, as in Figure 8.6. Many essential activities carry out both fundamental and custodial roles and, as a result, produce both external responses and essential memory updates. Figure 8.7 is an example of this type of essential activity. As with the stimuli from external events, the context diagram and the event-partitioned DFD only declare the response results. To define the detailed features of results, you can use the data dictionary or the minispecification, depending upon the nature of the result. The detailed modeling of external results and essential memory updates is discussed separately in the following subsections.

Figure 8.5. The external response, Salary-Report.

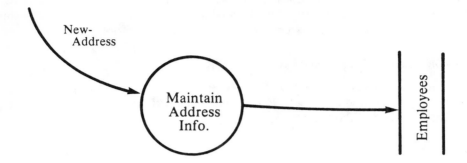

Figure 8.6. An internal response: an update to essential memory.

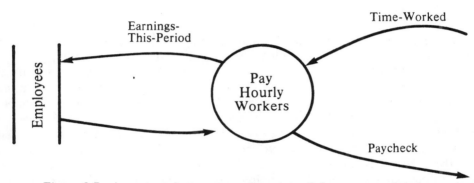

Figure 8.7. An external response and an essential memory update.

8.3.1 Modeling external results

You define the system's message to the outside world in the same way that you define stimuli arising from external events — as dataflow entries in the data dictionary. For the external response Salary-Report, shown in Figure 8.5, you might produce a definition like this:

Salary-Report =   {Company-Name + Date-Of-Report +
                   {EEO-Minority-Code + No.-Of-Employees + Average-Salary} +
                   Total-Employees + Overall-Average-Salary}

As in the case of the external stimulus dataflow, the use of the data dictionary is straightforward: Each flow has a name and a data dictionary definition showing its component data elements and the relationships between data elements, and, ultimately, each data element in the result is defined. Your experience with basic structured analysis should get you by, with one possible exception.

As explained in Chapter 6, the environment of the system has an important effect on its essence. You can assume that perfect technology only exists within the context of study as shown on the context diagram. Dataflows leaving the system pass from the domain of perfect technology into the technologically imperfect world. So, your definitions of the external responses must take into account constraints imposed by imperfect external technology. In practical terms, this means that the essential definition of dataflows for the external result must include all data elements and relationships between elements required by the external entity receiving the response.

8.3.2 Modeling essential memory updates

To define the updates to essential memory performed by a custodial activity, you use the minispec as the modeling tool. For the custodial activity depicted in Figure 8.6, you might find a minispec like the one below:

> Find the entry in Employees data store
> that has the same Social-Security-No. as New-Address.
> If found,
> replace Employee-Address in Employees data store
> with Employee-Address in New-Address.
> Otherwise,
> reject New-Address.

The minispec tells the content of the update by specifying the data elements and relationships that are being added, modified, or deleted from essential memory.

You might also choose to label the dataflow from the custodial activity to the essential data store and to define the flow in the data dictionary. Keep in mind that dataflows between a process and a data store do not have to be named and defined, provided you can determine the content of the dataflow by reading the minispec and the definition of the data store. Naming every dataflow to essential memory often clutters diagrams with meaningless and contrived names for perfectly reasonable accesses to stored data. If you decide to name a dataflow between a process and a data store, make sure that doing so adds useful information and does not make the diagram too complex.

So far, we have presented most of our conventions for modeling essential activities, and are fast approaching the treacherous topics of essential memory modeling and of integrating essential activity modeling with essential memory modeling. The difficulty analysts have in modeling essential memory is explained in a number of ways: Perhaps because most analysts have a programming background, they favor a procedure-oriented approach over a data-oriented one; or perhaps the reason is that system developers are usually processors and so have trouble dealing with the data point of view. We don't know which one is the real reason and we don't really care. But to find the true requirements of a system, you must feel comfortable with modeling both activities and memory, and you must be able to unify the two within the essential model. To prepare for our discussion of essential memory modeling, we finish our presentation of how to model the essential activities by giving a few conventions for modeling the interaction between essential activities and essential memory.

8.4 Modeling essential memory accesses

Every aspect of a system's planned response to an event — recognizing a stimulus and producing the response — may require information about past events or may store information for future use. In other words, essential activities may need to interact with the system's essential memory. These interactions, which we call accesses, are part of the model of each essential activity. In this section, we discuss the tools and techniques for modeling these accesses for a particular essential activity. First, however, we detail exactly what we mean by the term *essential memory access.*

8.4.1 Defining an essential memory access

Essential memory accesses are instructions within an essential activity to obtain, add, modify, or delete data elements and relationships within essential memory. These instructions obtain data from essential memory to support a fundamental activity; they add, modify, or delete data from essential memory as part of a custodial activity.

Essential memory accesses typically modify, add, or delete an individual object occurrence, which it selects from one or more objects. To accomplish this selection, essential accesses use a combination of identifiers and relationships between object occurrences. The following examples reveal the features of essential accesses:

> Find the Address and Phone-Number of the Pilot associated with the Pilot's Pilot-Number.

> Add Proficiency-Rating to essential memory, associating it with the Pilot's Pilot-Number.

> Modify the Top-Speed of the Plane identified by Plane-Number.

> Remove Pilot-Number and the Name, Address, and Phone-Number associated with that Pilot-Number.

Each of these accesses selects data from a single object; Plane is the object in the third example and Pilot in the others. Each example shows the essential memory elements required by the access and shows the identifiers used to select the particular object occurrence that the desired elements are part of. In these examples, the identifier is either Pilot-Number or Plane-Number; it is supplied by the stimulus that initiates the activity.

Essential memory accesses can be much more complicated than these examples. Here are a few such more complicated accesses:

> Find all the Planes that a particular Pilot is qualified to fly.

> Update the Last-Date that a particular Pilot flew a particular Plane.

> Add a new association between a Mechanic, a Plane, and all of the Plane's Maintenance-Changes.

The first access obtains an entire object occurrence, rather than giving a detailed statement of the data elements obtained from an object occurrence. For this representation of the access to be correct, the rest of the essential minispec containing this access must make it clear which elements of all those that characterize Planes are required for this essential activity. The second example shows the updating of a date that can only be identified meaningfully with two identifiers, Pilot and Plane. If Pilot and Plane are both objects, the Last-Date is a piece of information that characterizes the relationship between a particular pilot and a plane. The third example shows how to represent the creation of a relationship between a particular occurrence of Plane, Mechanic, and Maintenance-Changes. Like the one preceding it, this access is a custodial activity. It creates a relationship that some other essential access needs in order to identify a particular object occurrence.

No matter how complex an essential memory access is, you must find the simplest way to express it. If you don't, you violate the minimal essential modeling principle. For example, perfect internal technology would allow the system to obtain all pilots, then find the one it's interested in, then find all airports, then find the one where the pilot is stationed. However, stating the requirement in that way introduces false requirements. The same access can be stated more concisely: Find the airport where a particular pilot is stationed. That is the best choice for a statement of the access requirement.

8.4.2 Modeling essential memory accesses in minispecifications

The primary tool for modeling the access capabilities required by an essential activity is the minispecification, with some help from the data dictionary. Examine the minispec below:

> For each Student,
>> issue Schedule-Header.
>> For each Course associated with Student,
>>> issue Schedule-Detail.

It clearly states that this essential activity needs to extract certain essential stored data elements that describe a particular student and the courses the student is taking or took in the past. To discover the data elements that describe the objects Student and Course, you look them up in the data dictionary. There you will also find that the student in question is identified by his or her social security number.

In this minispec, the relationship between students and courses is very general. In a different minispec, the access requirement could be "Find the courses that the student is now enrolled in" or "Find all the courses that the student has ever taken" or "has passed" or a number of other possibilities. Naturally, you want to describe the relationship as precisely as possible so that you express the system's true requirements.

8.4.3 Modeling essential memory accesses on the DFD

Earlier in this chapter, we mentioned that the DFD is used to represent updates to essential memory. It should come as no surprise that this tool also shows the retrieval of data from essential memory. After all, the DFD is a more abstract view of the essential activities specified by the minispecs. However, the conventions for representing essential memory accesses on DFDs do not depict many of their aspects.

First of all, an essential DFD does not show each access from an activity to a data store. An essential activity can ask to see any number of different pieces of a data store at different times. However, we usually summarize all of these accesses into a single, one-directional dataflow from the process to the data store. Similarly, an essential activity could make several modifications to the data elements in a particular object data store, and all of those updates would be represented by a single dataflow from the activity to the data store. If an essential activity both produces and consumes essential memory, a double-headed dataflow would be drawn between the activity and the data store.

Many accesses require related data elements from two or more data stores, as in our course and student example. However, the DFD intentionally does not show these access relationships.

Finally, the DFD omits the specific data elements accessed by the activity. Some analysts write all of the data elements accessed by an activity from a data store next to the dataflow between the activity and the data store. We don't recommend this. In fact, as we mentioned before, we usually don't even give the flows between a process and a data store a name. So, there is virtually no way to find out the specific data content of an access from the DFD alone.

We omit such information about individual accesses, interobject relationships, and specific data elements from the DFD because it otherwise would become too complex. Since all of these aspects of essential memory accesses are defined in the minispec, we don't lose information about the essence.

What, then, does the DFD show about essential accesses? The DFD shows only that an access exists, and it does that by having an arrow between a given activity and the object data store that the activity accesses. This convention works easily for an access involving both identifier and nonidentifier data elements, such as "Find the Pilot-Address associated with a given Pilot-Number." However, it isn't so obvious how the DFD handles an access like "Modify the number of Items associated with all Parcels, within a particular Consignment, on a particular Ship." This access uses the objects Ship and Consignment as a means to find the desired parcels. As such, it is typical of a whole class of accesses to objects that merely use the relationship of the object accessed to another object to locate a desired data element.

Figure 8.8 shows how this convention would work for the Ship, Consignment, and Parcel example. We show an arrow between the process and the data store whether we access the object to find another object or to find a particular object occurrence within it. Again, if an essential activity uses an object for any reason, even if only to access another object, an arrow should be drawn between the activity and the object.

As to the direction of the arrow, the arrow should be drawn from the activity to the object if the access adds, modifies, or deletes essential memory associated with that object. Again, if an access is creating a relationship between object occurrences, arrows should be drawn to each of the objects linked by the relationship. The essential activity Add New Parcel in Figure 8.8 provides an example of this kind of representation. It creates a new parcel object occurrence, and it establishes a relationship between that new occurrence and the consignment it is a part of. That's why the activity updates both Parcel and Consignment.

The conventions we've just introduced are perfectly workable for representing essential accesses. However, there is another option to consider.

8.4.4 Modeling accesses with the data dictionary

Since you can think of access capabilities as an aspect of essential memory, it's not unreasonable to look for them in the data dictionary. And if you're not fond of defining them in the minispecifications, you might choose to put them in the dictionary. To make this work, you must follow two rules as you develop the essential data flow diagrams. First, on the lowest-level DFDs, you must name every dataflow between a process and a data store. An access must have a name so that it can be defined in the data dictionary. The second rule is that you have to draw a separate arrow for each stored data access, even if two or more accesses originate from the same process and access the same data store. In other words, you will no longer be able to represent more than one access with each dataflow. You must identify each memory access in the data dictionary.

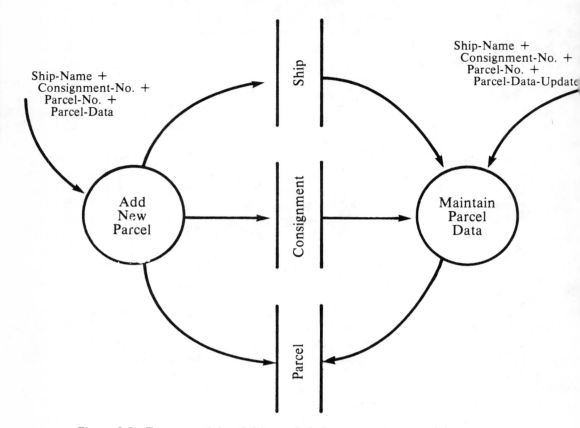

Figure 8.8. Two essential activities and their accesses to essential memory.

Once you've named every dataflow into and out of the essential data stores, you can use the data dictionary to define the required access capabilities for each of them. You will have to make extensive use of the comments feature of the structured analysis data dictionary, since the standard dictionary constructs (selection, iteration, optional, and so on) are not ideal for describing access paths. You write a standard dictionary definition of the data being read out of or written into the data store and supplement it with a plain English description of the required access capability. An example of this approach appears below, and Figure 8.9 shows the accesses defined here.

Course	=	Course-No. + Course-Title + Credit-Hours + Room + Time
Student	=	{Social-Security-No. + Student-Name + Student-Address}
Student-Schedule	=	Schedule-Header + {Schedule-Detail}
Schedule-Header	=	Student-Name + Student-Address
Schedule-Detail	=	*Obtained for each course associated with student* Course-Title + Room + Time

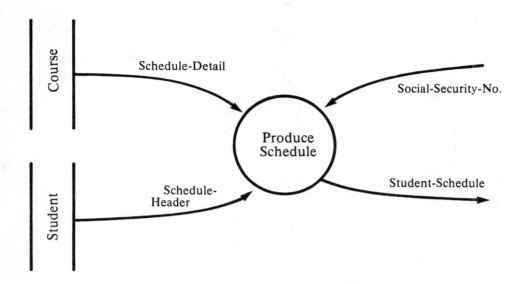

Figure 8.9. DFD declaring essential accesses that are defined in the data dictionary.

Although this is a perfectly valid approach, we favor using the minispecification simply because it decreases the complexity of the DFD and of the dictionary. First, it's not easy to make industrial-strength data flow diagrams that convey a large and complex system without becoming hopelessly tangled. You don't want to add more complexity by drawing and naming a dataflow every time a process accesses a data store. Not only are such dataflows often difficult to name well, but they also clutter the DFD significantly. Second, most data dictionaries are so large they're scary. We cannot recommend a course of action that would add hundreds of definitions to the dictionary. On the other hand, the overall number of minispecs will not rise significantly if you use them to specify access capability requirements, especially since most minispecs already say something about stored data accesses, as you saw in the Produce Schedule example in Subsection 8.4.2. In the end, which of these two approaches you adopt is entirely up to you, since either of them can be made to work.

There is one situation in which you should consider using the dictionary instead of the minispec. Figure 8.10 shows an activity whose only purpose is to retrieve information and send it on. In this case, you can determine the required access capability unambiguously from the definitions of the ingoing and outgoing dataflows and the data stores. The requirement is so simple that it really doesn't have to be stated in the minispecification. You might decide to do so anyway, just to make the minispec more readable, but this does create a slight degree of redundancy in your specification, which is not without cost to its maintainability. So, whenever you find an activity that just "passes through" stored data — no calculations, derivations, or decisions are made on it — the data dictionary is the preferred modeling tool.

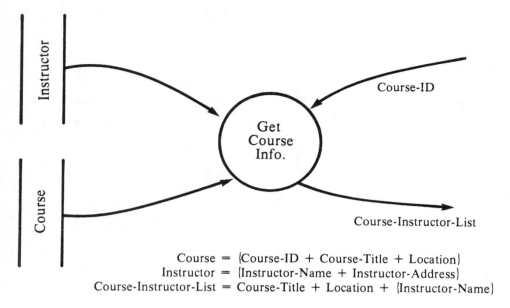

Course = {Course-ID + Course-Title + Location}
Instructor = {Instructor-Name + Instructor-Address}
Course-Instructor-List = Course-Title + Location + {Instructor-Name}

Figure 8.10. A simple retrieval access.

8.5 Summary

In this chapter, we describe how a set of old and new documentation conventions can be used to model the complete response to a particular event. The complete response consists of a stimulus, the planned response, and the results of a planned response. The dataflow representing the stimulus from an external event is shown on the DFD and modeled in the data dictionary. Stimuli from temporal events do not appear on the DFD; they are given in the minispecification.

The most detailed description of a planned response is given by the minispecification. If a planned response is too complicated to fit within a one-page minispec, the response is partitioned into several lower-level activities, each of which has its own minispec. You must be careful not to imply an unnecessary sequence of activity in a minispec — you can use the parallel construct in the minispec or a different modeling tool, such as a decision table, decision tree, or DFD.

The results of a planned response are shown as dataflows on the DFD. If a result is a message to the outside world, it is defined in the data dictionary. If it is an update to essential memory, it is defined in the minispec. Accesses to essential memory are also indicated by arrows on the DFD and are detailed in the minispec.

Chapter 9

Modeling
Essential Memory

As you know, essential memory is composed of all of the information about past events and system responses to those events needed to carry out the system's essential activities. Because of its crucial role as the interface between responses to different events, essential memory serves as the glue that holds the individual essential activities together. While any one essential activity responds to only one event, essential memory takes part in most if not all the responses of the essential system.

To build a model of essential memory, you use modeling tools borrowed from structured analysis and from information modeling. Unlike the tools for modeling essential activities, you probably will not use all these tools for any one project. In this chapter, we introduce you to these modeling tools and show how they are used together with object partitioning to build a proper model of essential memory. First, we explain their use in modeling three components of essential memory: data elements, objects, and relationships between objects. Then we compare the strengths and weaknesses of the tools so that you may select the ones that are best for your own situation.

9.1 Modeling data elements

You model essential memory data elements in the data dictionary, defining each data element under its own name. This much is easy; more difficult is deciding what information the definition should contain.

Since you are modeling essential memory, you don't have to worry about the physical format of the data, including the number of bytes of storage or the storage structure within that space. Instead, following some of Matt Flavin's information modeling conventions, you define a data element by giving both a semantic definition and a data content definition [15]. DeMarco provided a good model for semantic definitions [11]. He said that an element definition should consist of the class that the data element belongs to and the features that distinguish it from all other members of the class. To us, that means the definition specifies the object that a data element describes and tells what characteristic of that object the data element conveys information about.

For example, the definition of Course-No. on the next page says that the course number is a member of a class of unique identifiers. Course-No. is the unique identifier of unique occurrences of the Course object. The semantic definition also gives the origin of the Course-No. — namely, it is generated by the system.

85

Course-No. = *semantic definition:
> a unique identifier for each unique course
> Course-No. is generated by the system

data content definition:
> Course-No. can be any integer greater than or equal to 1*

Notice that the definition of Course-No. uses symbols from both DeMarco and Flavin's conventions for defining data elements. The "is composed of" symbol (=) and the comment symbol (*) are from DeMarco's notation [11]. The individual fact symbol (>) and the division of the definition's internal structure into a semantic and data content section are from Flavin's notation [15].

The data content definition states the set of logical values that a data element can take on. These values can be made up of letters, numbers, or other symbols. Do not confuse them with the codes and abbreviations that many system implementations use to represent values, which are designed to save storage space used by programs, input forms, output reports, and physical data. Right now, you are only interested in the real meaning that these physical values represent. In addition, you need to define the unit of measure of the values, such as yards, meters, ergs, or leagues per fortnight.

9.2 Modeling objects

Although objects can be modeled with several modeling tools, we concentrate in this section on two of them: the data flow diagram and the data dictionary. The DFD merely declares the existence of the object data stores and shows which essential activities access which data stores but does not show their relationships or component data elements. For example, the DFD in Figure 9.1 shows that in order to produce a student's schedule, the essential activity Produce Schedule needs essential memory about the student and about the courses he or she takes. Therefore, the DFD shows accesses from the data stores Student and Course to the essential activity.

Modeling objects would be incomplete with only the DFD, for you really need to define objects in the data dictionary. The basic objectives of object definitions are the same as those of data element definitions: Each object should be defined under the name of the object. The definitions themselves always use the symbolic conventions described by DeMarco for group data dictionary definitions. The definitions can be enhanced by using some of Flavin's conventions for object definitions. Two objects, Course and Student, would be defined as follows using this set of conventions:

Course = * > a course is a program of instruction for
university students
> a course is designed to meet instructional objectives*
{Course-No. + Course-Title + Credit-Hours + Room + Time}

Student = * > a person registered with the university
> students must take a minimum amount of instruction
as measured by course credit hours*
{Social-Security-No. + Student-Name + Student-Address}

Notice that object definitions are structured similarly to data element definitions, except that the second part of the definition portrays both the data content and the data structure of the object. Data content refers to the data elements attributed to the object, and data structure refers to the organization of those elements. The most important characteristic of object structure is that objects are collections of object occurrences. The braces ({ }) around the data elements attributed to the object mean that the object is composed of repetitions of the data elements within the braces.

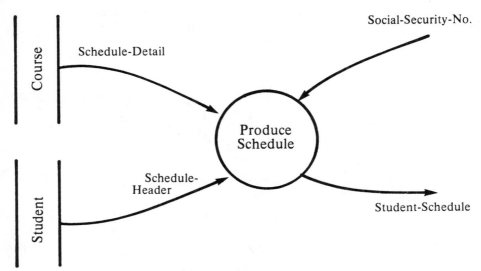

**Figure 9.1. DFD for Produce Schedule,
declaring that objects Course and Student must be accessed.**

9.2.1 Modeling large objects

To keep the data dictionary concise, you need a way to handle objects that group scores, if not hundreds, of data elements. You *could* write the definition for a three-hundred-data-element object called Employee by simply listing all three hundred elements, one after another, like so:

$$\text{Employee} = \{E1 + E2 + E3 + \ldots + E271 + \ldots\}$$

Although this scheme is technologically neutral, it produces a definition too complex to be easily understood.

We recommend partitioning any large object into a reasonably small number of groups. These groups then become first-level components of the definition of the large object. They are themselves defined under their own names elsewhere in the data dictionary. This scheme is called top-down definition of data groups, and is shown in the example below:

$$\text{Employee} = \{ \underline{\text{Employee-ID}} + \text{Personal-Description} + \text{Family} + \text{Salary} + \ldots \}$$

This definition defines Employee with data groups like Personal-Description and Family, which contain many data elements and possibly other groups and must be defined as well.

Writing definitions of essential data stores in a top-down fashion allows you to control the complexity of each portion of the definition so that the reader is never overwhelmed with information. Nonetheless, in choosing your upper-level groups of data elements, you have plenty of opportunities to violate another modeling principle: that of technological neutrality.

Because you don't want to reduce the complexity of an object definition at the expense of introducing a technological bias, you have to exercise care when deciding how to group data elements. Figures 9.2a through 9.2c show examples of data stores whose partitioning introduces bias toward a particular technology. These groupings of stored data elements are not acceptable in an essential model since they all accommodate the limitations of a certain implementation technology.

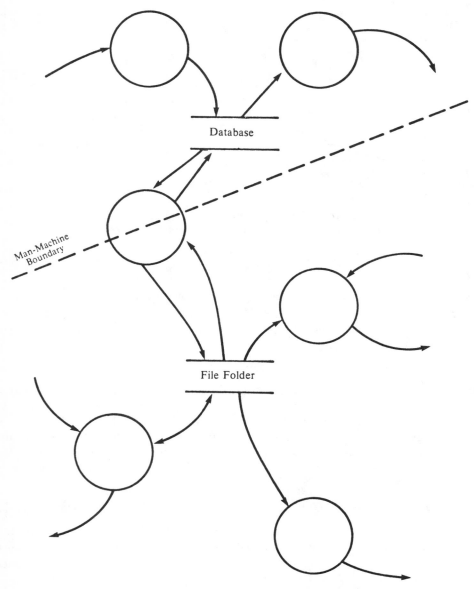

Figure 9.2a. Data store groupings based on whether data is computerized or maintained manually.

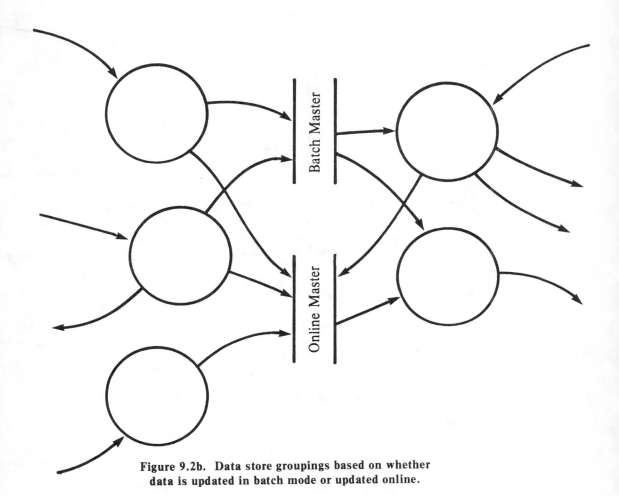

**Figure 9.2b. Data store groupings based on whether
data is updated in batch mode or updated online.**

Less obvious, however, are the alternatives that will both control complexity and
avoid technological prejudice. We suggest that you choose one or two ways of grouping
the data elements and then evaluate whether they satisfy our two criteria. Consider the
breakdown of the Employee data store in Figure 9.3. Here the elements have been assigned to groups according to what letter of the alphabet they begin with. There is
nothing particularly implementation dependent about this grouping, but it does not
result in a concise definition. Although this definition controls complexity by limiting
the definition of the Employee data store to only three first-level components, the
names of these components are not meaningful.

You and everyone else who reads the specification will probably get a better
understanding from a partitioning like the one in Figure 9.4. We call this kind of partitioning subtopic partitioning. Subtopic partitioning is performed exactly as you would
perform object partitioning except that you do not create a separate stand-alone object
as you do in object partitioning. Instead, you partition within an object. This partition-

ing is not too complex, provides meaningful names for the subcomponents, and still does not prejudice the model toward a particular technology.

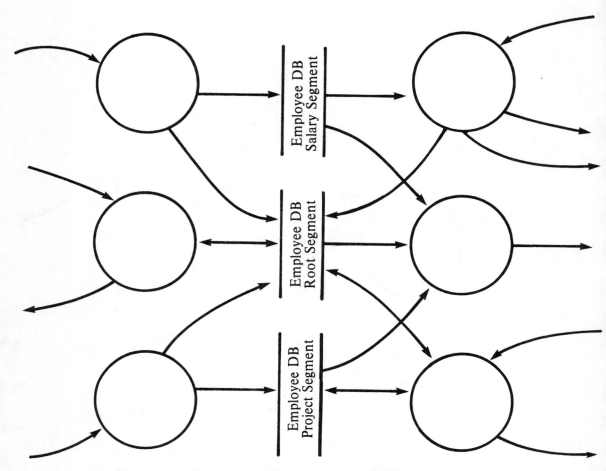

Figure 9.2c. Data store groupings based on IMS database segments.

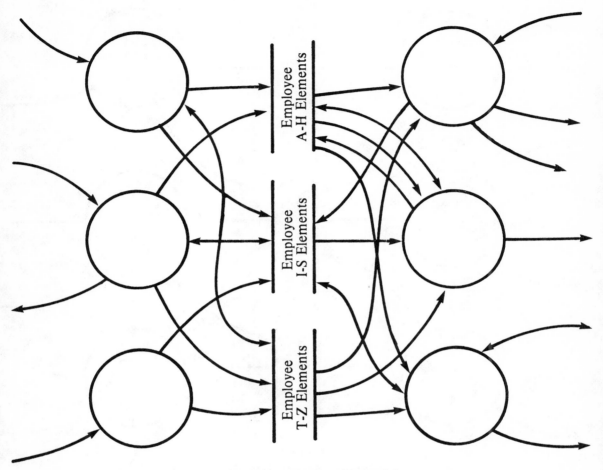

Figure 9.3. Arbitrary partitioning of essential memory based on letters of the alphabet.

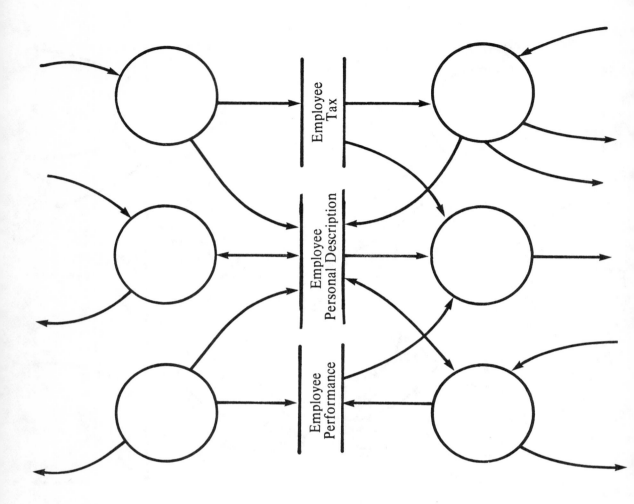

Figure 9.4. Data groups derived from a top-down partitioning of the Employee object.

9.2.2 Intra-object relationships

An object groups not only data elements, but also relationships between the data elements inside those objects. These relationships typically occur among the data elements of a particular object occurrence, and they are called *intra-object* relationships. The most important of these relationships is represented by the symbol "+," which indicates that the data elements on each side of the symbol always occur together within an object occurrence or a subtopic. Another relationship within an object occurrence is the "uniquely identifies" relationship. For example, a particular Course-No. uniquely identifies all data elements within a particular object occurrence of Course. We represent this intra-object relationship by underlining the unique identifier. The relationship "this data element is optional within each object occurrence" is expressed by placing the optional data element or group within parentheses. For example,

$$\text{Employee} = \{\underline{\text{Employee-ID}} + \ldots + (\text{Payroll-Direct-Deposit-No.})\}$$

Again, we use DeMarco's data dictionary notation to represent intra-object relationships, which was based on work in data structure definition by Michael Jackson [19].

Like the data elements, these relationships exist because the essential activities demand them. An intra-object relationship is justified if at least one essential activity needs that relationship to carry out its work. For example, if an essential access needs to uniquely identify object occurrences, the "uniquely identifies" relationship must be present within the object occurrence.

9.3 Modeling interobject relationships

Relationships between data elements assigned to different objects are the third important concern of essential memory modeling. Again, as with the intra-object relationships, these *interobject* relationships exist only because they are needed by the essential accesses within the essential activities. There are many ways to model these interobject relationships, but we suspect they may be very different from the ways a lot of you use now. Were you uneasy when you read the definitions of the Student and Course data stores in the previous section? Did you perhaps wonder how it is possible to access the Course file from the Student file when there is no apparent link between them? Did you expect to see a list of course numbers in the Student file to facilitate the access, or maybe a copy of each student's social security number in the Course file?

If your answer to any of these questions is yes, you are not alone. Many people are convinced that there should be some visible connection in the data dictionary between one essential data store and another if an access capability requires a relationship between the two. Unfortunately, this approach to modeling interobject relationships has a problem: It introduces major technological prejudice.

The problem becomes clear as you think about the answer to the question, What do you add to the definition of the Student data store to represent the requirement to access the Course data store? The first choice is the key to the Course file, Course-No., as shown below.

Course   =   {Course-No. + Course-Title + Credit-Hours + Room + Time}

Student =   {Social-Security-No. + Student-Name +
             Student-Address + Course-No.}

Adding the course number shows the relationship, but it also communicates something more: the *requirement* that you store the course number in more than one place and that you use the course number as the link between the two data stores.

Such redundant identifiers are false requirements. Although this approach is widely used to implement interobject relationships, it is not the only one. Of the numerous possible ways to organize a file, many do not involve the redundant storage of identifiers. For example, you could use absolute hexadecimal pointers, relative record numbers, or database locator files, or you could duplicate the required data rather than the key. You definitely do not wish to limit the new system to only one way of implementing interobject relationships, but that is exactly what you do when you show the connection between two data stores in the data dictionary.

A second problem with putting the identifier in both files is that it violates the minimal modeling principle. You can remove this redundancy, and therefore streamline the model of essential memory, if you take another approach to modeling interobject relationships. In fact, there are three better approaches for modeling such relationships: the data structure diagram, the entity-relationship diagram, and the minispec.

All three tools are used with the data dictionary and DFDs, and you use the same DFD and data dictionary conventions regardless of which tool you choose. Neither the data dictionary definitions nor the DFDs show links between objects. At the most, you could add a comment in the data dictionary definition that mentions the other objects that relate directly to the one you are defining. However, this modeling frill is one of the first to be removed if you have to streamline your model. The reason you don't show links between objects in the data dictionary or DFDs is that the three tools discussed below can do a much better job documenting these relationships.

9.3.1 Data structure diagrams

The data structure diagram (DSD) is a graphic modeling tool for depicting interobject relationships. Figure 9.5 shows a data structure diagram for the essential activity model in Figure 9.1. The corresponding object definitions are given in Section 9.2. Each of the object data stores declared in the DFD and defined in the data dictionary appears here as a square.

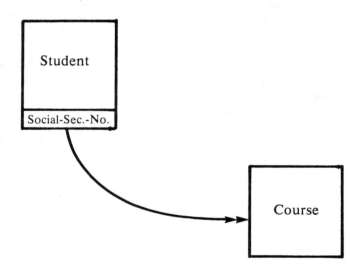

Figure 9.5. Data structure diagram of the essential accesses required by Produce Schedule.

The small rectangle within the Student box denotes that the identifier of the Student object is the starting point for at least one essential access. Actually, the absence of such a box says more than the presence of one. If any object is accessed exclusively through the identifiers of other objects (Course in this example), its box in the DSD will not have a small rectangle.

The most useful feature of the data structure diagram is the arrow between the object boxes, representing an interobject relationship; the direction of the arrow indicates the direction of the relationship. For any arrow on the DSD, the end without an arrowhead comes from an object whose identifiers are known by the essential activity; the activity can therefore access the object data store directly. The end with an arrowhead points at an object containing unknown identifiers; information from this data store is available to the essential activity only by accessing the data store on the other end of the arrow. The arrow from Student to Course in Figure 9.5 means that at least one essential access is looking for the unknown courses that are associated with a known student.*

The arrows can have either one or two arrowheads. One arrowhead indicates that for each entry of the object at the beginning of the arrow, there is one entry in the object at the arrowhead end. Two arrowheads indicate that the object at the arrowhead end contains many entries for each entry at the other end. A one-to-many relationship on a DSD means that at least one essential access expects to find many occurrences of one object when it is accessed through another object occurrence. In our example, the one-to-many relationship between Student and Course reflects at least one minispec that says, "Given a student with social security number X, find all courses that the student is currently taking."

To summarize our points on DSDs, we elaborate on our example. If an access in our essential activities was labeled, "Given a Course-No., find all Students enrolled in that Course," our data structure diagram in Figure 9.5 would be incomplete. To accomplish the essential access, we need to model an additional relationship in essential memory using the DSD: We draw another arrow on the diagram, as shown in Figure 9.6. The arrow goes from Course to Student because the essential access knows a course number and doesn't know the students in the course. The arrow has two heads because the essential access expects to find many students enrolled in a given course.

9.3.2 Entity-relationship diagrams

The entity-relationship diagram (ERD) is a graphic tool for modeling objects, also known as entities, and for depicting the relationships between them. Figure 9.7 shows an entity-relationship diagram for our Student and Course example. As in the data structure diagram, boxes represent the object data stores. The diamond in the ERD names the relationship between the objects. The numbers and letters on each line going into and coming out from the diamond give the ratio of object occurrences between objects in this relationship. The letter m or n is used by convention to indicate *many* in the one-to-many and many-to-many object occurrence ratios.

As with the DSD, you can use the entity-relationship diagram to represent interobject relationships in essential memory. If you do, you can say more than in a DSD about each relationship because the diamond gives you space to name the relationship. On the other hand, a disadvantage is that you don't see the direction of the relationship, a feature that the DSD presents plainly. Both the DSD and the ERD show the ratio of object occurrences between objects in the relationship.

---

*Some analysts don't accept the notion that essential accesses determine the existence and characteristics of interobject relationships. They contend that there is an inherent many-to-many relationship between students and courses and furthermore that the essential access is determined by these inherent characteristics, not the other way around. We, on the other hand, agree with Flavin's point that objects and relationships are made, not born. They are made through decisions by whoever creates or modifies the essence of a planned response system and who therefore determines the system responses and the essential memory needed.

Put another way, if on the streets of Boston you see thousands of people wearing university T-shirts, can you distinguish by their inherent features who are students, who are graduates, and who are high school dropouts? If you do find out who the students are and put them in a group, could you tell by looking at them which courses they are taking? The point is that these students and their course enrollments are assigned attributes, assigned so that a university can teach a select set of people, collect tuition from them, schedule lectures for them, and so forth.

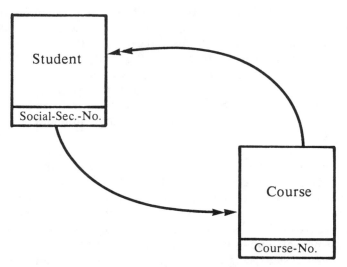

Figure 9.6. Data structure diagram showing multiple essential memory accesses.

Figure 9.7. Entity-relationship diagram for the essential activity Produce Schedule.

9.3.3 Minispecifications

The third tool for modeling interobject relationships is the minispec. In creating the minispec as part of the essential activity model, you have identified all essential accesses. Since you know from the custodial activities any interobject relationship used by another essential access, you have specified the total of interobject relationships by specifying all essential accesses. Therefore, the minispec models the interobject relationships that are required for a fundamental activity to select specific object occurrences and the interobject relationships that must be created by a custodial activity for other essential accesses. Thus, a DSD or an ERD duplicates what the minispec already models.

9.4 Choosing the right approach

Having read about several options for modeling interobject relationships, which one should you use? To help you decide, we discuss how the various tools compare on specifically five issues concerning interobject relationship modeling.

9.4.1 Interobject relationship modeling vs. essential access modeling

Throughout this chapter, we make a distinction between an essential access and essential memory. An essential access specifies a need for part of essential memory, some combination of data elements, intra-object relationships, and interobject relationships. Essential memory is the total of data elements and relationships between data elements that satisfies these requirements.

Each of the three tools described above for modeling interobject relationships typically models either essential accesses or essential memory. The data structure diagrams and entity-relationship diagrams model the relationships that link different objects. The minispec models essential accesses, among other activities.

This distinction affects your choice of modeling tool. Let's say the essence to be modeled has trivial essential activities and complicated interobject relationships. Minispecs might not be worth spending too much time on. Instead, you create a more effective model if you work from the viewpoint of essential memory. This means using a DSD or an ERD. Therefore, to choose the proper approach, assess the complexity of the essential activities, excluding the essential accesses, and if they are trivial, a DSD or an ERD is the better choice for modeling interobject relationships.

9.4.2 Global models vs. local models

Another major difference between essential accesses and essential memory is that an essential access is part of the minispec to one essential activity, while essential memory is accessible to all essential activities. Therefore, we say that essential accesses are *local* because they occur within the context of a specific essential activity. Essential memory is *global,* since it is available to any essential activity.

The essential accesses specified in minispecs give only a local view of interobject relationships. To obtain a global view of interobject relationships from minispecs, you would have to collect all of the minispecs for a given essence. On the other hand, a single data structure diagram or entity-relationship diagram can provide either a local view by showing the interobject relationships required by one essential activity or a global view by showing the interobject relationships required by all essential activities.

Sometimes, you want to get an overview of the global interobject relationships without wading through all the minispecs in order to do so. You would then draw either a data structure diagram or an entity-relationship diagram for all of essential memory. Here is a short list of the situations for which you would choose this modeling option:

- You are just starting the definition of essence and you want to record what you know about interobject relationships without making the considerable effort required to write all of the essential minispecs.

- You are reviewing an individual essential activity or a global DFD and you'd like to verify quickly that the proper custodial activities have been specified for the accesses. Checking a global essential memory model will accomplish this, because it should reflect all custodial activities.

- You are beginning your relationship with the implementors, particularly database designers, and you want to give them a general idea of the nature of your system's essential memory and to determine the probable scope of your need for their services.

- You are defining the essence of a portion of a larger system and you want a model of the interobject relationships in the larger system to make sure that your definition of essential memory will be consistent with a definition of the entire system's essential memory.

9.4.3 Specialized models vs. general models

The minispec specifies not only interobject relationships through access requirements, but also all of the other actions that must be carried out in response to an event. Since the data structure diagram and entity-relationship diagram specialize in showing interobject relationships, by choosing either diagram, you can concentrate on the essential access requirements without having the unnecessary information about the rest of the detailed activities that may interfere. Many choose to use DSDs and ERDs for that reason alone.

9.4.4 Requirement models vs. semantic models

Sometimes, especially when you begin to define essence, or even possibly before when you perform project preliminaries, it is desirable to have a model of interobject relationships that aren't necessarily part of the requirements for the new system. These relationships are the ways in which people express in words the relationships between objects, for example, Students take Courses, Students flunk Courses, and Students pass Courses. We call these semantic interobject relationships.

If you wish to model this kind of relationship, the entity-relationship diagram is the tool of choice, since it provides space for a semantic description of the relationship. Neither the data structure diagram nor the minispec will work, because these tools are designed to specify requirements.

9.4.5 Graphic models vs. narrative models

The graphic documentation of structured analysis offers certain advantages over narrative text documentation, certainly in clearly specifying complex multidimensional relationships. As an inherently linear medium, narrative text can fail to handle these relationships. For example, writing out and reading the directions to a certain destination can be very difficult without a map of the area. Many people are so thoroughly convinced of the advantages of graphic documentation that whenever they have to choose between graphic modeling tools (like the DSD and ERD) and narrative modeling tools (like the minispec), they know they should choose the graphic approach just because it is graphic. There's nothing wrong with choosing a modeling tool for this reason, but the newer graphic approach has a built-in limitation. The limitation is that most people are uncomfortable with using graphic tools and so prefer to specify interobject relationships with the minispec, rather than with the DSD or ERD. This basis for choosing a relationship modeling tool is also perfectly acceptable, since the minispecs alone do indeed form a complete, albeit indirect model of the essential interobject relationships.

When choosing a modeling approach for interobject relationships, you will find no clear-cut answer. Whatever tool you do choose, however, you must account for all of the interobject relationships required by the essential accesses and, for each relationship, you must show the objects linked by the relationship, the ratio of occurrences, and the direction of the relationship.

9.5 The need for a better convention

If you decide to use either the data structure diagram or the entity-relationship diagram to model interobject relationships, you should realize that neither tool has all of the conventions needed to do a complete job of specifying the essence of the system. The DSD is a bastardized tool, devised originally in the late 1960s as a Bachman diagram to model physical database structures, a very different problem indeed. The tool was brought to structured analysis by Ross and was quickly adopted, with a variety of inconsistent changes, by DeMarco, Gane and Sarson, and Weinberg [11, 17, 52].

The main problem with the DSD is that it cannot show relationships between more than two objects. For example, assume you have an essential access that reads, "Find the price of the part sold by a particular vendor to a particular customer." You'd be stumped if you tried to express this very typical kind of relationship with the DSD conventions from basic structured analysis. Our entry for a graphic convention to solve this problem is shown in Figure 9.8.

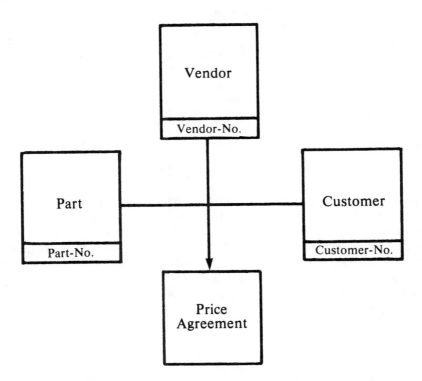

Figure 9.8. Data structure diagram showing an n-way essential access.

Entity-relationship diagrams come to us more recently from the work of Peter Chen and Flavin [4, 15]. Both of these sources use ERDs to model information or data in general, without restricting their focus to essential memory, and they both divide domains of information into object types (our objects plus more characteristics that aren't part of essential memory) and valid semantic relationships between object types. Since they are concerned with the semantic relationships between objects, ERDs don't provide a convention to document an important feature of an interobject relationship: the direction of the relationship. This problem could be remedied by adding arrowheads to the lines connecting diamonds with squares in the ERD, as in Figure 9.9.

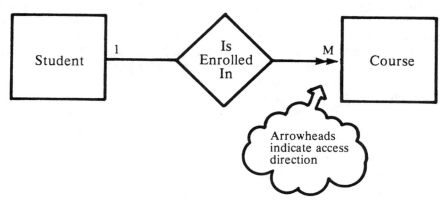

**Figure 9.9. Entity-relationship diagram modified to show
the direction of the relationship.**

Both of these conventions are quick fixes for the problem of developing a comprehensive graphic modeling approach for objects and interobject relationships in essential memory. In time, a good solution to this modeling dilemma will be found, probably an amalgam of a DSD and an ERD. Until then, analysts will have to do the best they can with what they've got.

9.6 Summary

This chapter presents the basics of essential memory modeling. The data dictionary can be used to model data elements, objects, and intra-object relationships. If the dictionary definition of an object is too long, the object should be broken into subtopics. Objects also appear on data flow diagrams. Interobject relationships can be modeled by data structure diagrams, entity-relationship diagrams, or minispecs. Which of these three tools you choose depends upon such factors as how complicated the interobject relationships are and whether you prefer a specialized tool over a general one or a graphic tool over a narrative one. Taken together, these documentation conventions allow you to rigorously define essential memory requirements declared on the DFD and in the minispecs.

Chapter 10

Strategies for Modeling the Essence of a System

In Chapter 5, we don't really say how to define the essence of a system; and in Figure 6.1, we are even pretty vague about the inputs to the process. Now, we describe two basic strategies for using the tools of structured analysis to build a model of a new system's essence.

The first strategy is used when there is no existing system or, if one exists, when its essence may not be worth studying. In either case, you have no choice but to create the essence of the new system directly. System users must supply the information about the essential responses for you to incorporate them into the essential model, as shown in Figure 10.1. The second strategy derives from classical structured analysis: First, define the essential responses of the existing system and then add new responses to them. Although you might only use one strategy at a time, you need to know how to use both strategies. After all, one strategy will not always be correct for all future projects. In fact, you may not always use only one strategy for a project. So, we present both strategies and a framework for deciding which one to use for a given system development project.

Here again are the two strategies:

1. System developers derive the essential model of a new system directly from information about what the users would like the new system to do.

2. System developers derive the essential model of a new system from a model of the essence of one or more existing systems.

We start with the first strategy by proposing a coherent set of steps for building a model of essence for a system that does not have an existing incarnation, or whose existing incarnation you have decided not to study. We divide this process into two phases: *creating* the essence — that is, deciding its contents — then *modeling* the essence, using the tools introduced in the last three chapters.

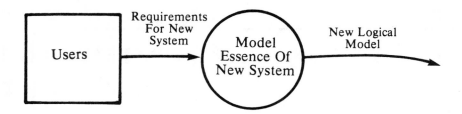

Figure 10.1. Creation of the essence of a new system.

10.1 Creating the new essence

You can define the essence of a new system in a number of ways, depending upon what aspects of the system you consider first, and how they lead you to other decisions and assumptions about the system. There are primarily four decisions, discussed in the following subsections, that will lead to the creation of a model of a system's essence.

10.1.1 First decision: identifying the system's purpose

To start the discovery process, the creator of the new system asks a fundamental question about the system: Why should it exist? In a general sense, the purpose of a system is to interact with its environment in a planned fashion to achieve some objective. The purpose arises from conditions in the environment, which exist before the new system and independently of it; these conditions drive system creators to create a particular essence. For example, in an environment where people own homes containing valuable possessions and where thieves sometimes break into those homes to steal the objects, someone long ago invented a burglar alarm system. That system's purpose is to create noise in order to scare burglars and to notify neighbors that a burglary is in progress.

An individual planned response may not fulfill the system's entire reason for being; it may be only one of many contributions toward its purpose. In a mortgage system, for instance, the payment of a municipal tax bill on a mortgaged property and the overdue payment notice to a homeowner are both responses that contribute to the same purpose: to earn money on loans for home purchases for which the home itself is collateral.

You may find it hard to relate to a definition of *purpose* as abstract as the one at the beginning of this subsection. For any real system, the purpose will be specific and multifaceted. For example, the purpose of a materials management system of course is to manage materials, but this purpose breaks down into the procurement of materials from outside sources, the distribution of materials to users within a company, the accounting for materials usage, and so forth. An airline reservation system also has a set of purposes: to provide customers with quick and accurate information about flight availability, to minimize the lack of availability of seats, and to minimize lost revenue due to no-shows. From these examples, you can see some of the characteristics of a system's purpose: It can be technologically independent, it can exist at many levels, and it can be a combination of smaller purposes.

When beginning the search for a system's essence, you can usually settle for a general statement of its purpose, such as one of these:

- Produce payroll for hourly personnel.

- Schedule and track preventive maintenance work.

- Rent roller skates.

Once armed with such a statement of the purpose of the system, you can proceed to define the essential components of that system.

10.1.2 Second decision: identifying fundamental activities

A system's fundamental activities are a subset of all its planned responses to events in the world around it. Since they carry on dialogues with external entities that help to fulfill the system's purpose, the fundamental activities could be defined as those that fulfill the outside world's *expectations* of the system. Figure 10.2 contains a fundamental activity, Issue Skates, that carries on a dialogue with the external entity Customer in response to the event "customer requests skates." The rented skates presented to the customer fulfill the customer's expectation of the system.

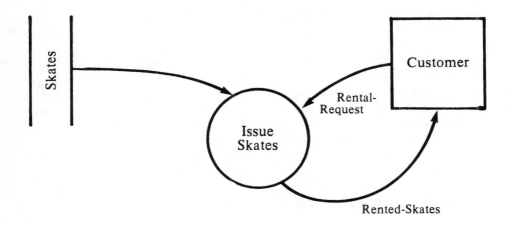

Figure 10.2. Dialogue of the fundamental activity Issue Skates with its environment.

The search for a system's fundamental activities begins with a search for the events that elicit the planned responses. The events are either external or temporal. Whenever you find an event, you create the response to it, the second component of a fundamental activity. Although there are a variety of responses that you could choose, you want to choose those that best fulfill the system's purpose.

For example, suppose you were creating the first roller skate rental operation, whose purpose was defined in the previous step. To create the fundamental activities, you first examine the environment of the future system. Imagine a park where people who own roller skates can enjoy skating on well-manicured paths. Many other people would enjoy skating, but don't have skates and would be willing to rent them. As an intrepid entrepreneur who realizes this, you start the Good Skate Company.

To identify the fundamental activities for the system that will be your company, first you decide what events your system will respond to, then you write the definition of the stimuli from those events and you create the responses to them. Each event that the Good Skate Company will react to is listed below in italics, followed by a brief description of the stimulus and the newly created response:

1. *Customer requests skates.* When the customer approaches the shop and asks to rent a pair of skates, you carry out a standard procedure.

2. *Customer returns skates.* When the customer brings back the skates, you determine how much the customer should pay according to how long the skates were rented.

3. *Customer pays for rental.* The customer pays the sum you charge for the rental of the skates.

4. *Time to close shop.* When the clock says 6:00 p.m., you deposit the day's receipts in the night deposit box and check for damaged and missing skates.

Each of the first three fundamental activities is initiated by an external event originated by the customer, one of the external entities with which a roller skate rental system must deal. (The other external entities might be a bank, a skate dealer, a skate repair shop, the police, and so forth.) The fourth activity is triggered by a temporal event: 6:00 p.m. on the clock.

The planned response component of a fundamental activity needs information to respond to the outside world. For example, the Good Skate Company has to tell the customer how long she rented her skates and how much she has to pay for that rental. Although a fundamental activity can derive information from other pieces of information, it cannot create information from nothing. All information produced by a fundamental activity must ultimately come either from the system's environment at the time an event occurs or from essential memory. The need to identify these information sources leads to the next step in creating essence.

10.1.3 Third decision: *identifying required information*

You begin the process of identifying the information a system needs in order to carry out its activities by examining a set of data elements produced by a fundamental response. First, you consider whether this information is derived from other pieces of information. In the Good Skate Company system, the rental charge presented to the customer is the product of the rental rate and the amount of time the skates were rented. The rental rate on the other hand is not derived at all: Good Skate has a single fixed rate for all rentals. The rental time does have to be derived. It is the amount of time elapsed from the start of the rental to the time of return rounded up to the nearest quarter hour. For any derived data required by a fundamental activity, you identify all of the components that are required to derive that data.

Assuming that you know all of the data needed to derive the fundamental response, you now establish the source of the information. There are two basic sources: The system gets the information either from its environment when an event occurs or from an essential activity when a response occurs. Skate size comes from the environment, while rental start time and end time come from the system.

Next, you determine what pieces of information, if any, should be kept in the system's essential memory. Essential memory exists because the system's environment can't be trusted to remember the information. For example, the system needs to know the time the skates were rented and the current time. The system can get the current time immediately from a clock, but how will it remember when the rental period began? It can't ask the customer because the customer may not know the correct time or may lie in order to save money. So in this case, there is no reasonable alternative to having the system store the time that each rental begins at the time the customer takes the skates. To fail to do so would jeopardize the system's ability to satisfy its objective.

To complete the specification of essential memory, you consider whether essential memory will remember one or many examples of a particular piece of information. Will the Good Skate Company system rent one pair of skates and therefore have one rental start time to remember, or will it rent many skates with many rental start times? In the latter case, you must establish some way for the system to identify which rental start time goes with which customer. The typical way is to select additional data to remember for identification purposes, such as a customer's driver's license number to identify the rental start time. Another option would be for the system to assign numbers that the renter must present when the skates are returned.

An interesting by-product often results from the specification of essential memory requirements. You may find that information required from essential memory is only available from events other than the ones you have identified so far. In this case, you simply identify the events, waiting until you discover the corresponding custodial activities to work further with them.

Having decided what information will come from essential memory, you must assume that other pieces of information will come from the system's environment. For a given fundamental activity triggered by an external event, this information is available from the stimulus that signals the event. For example, in order for the activity to issue the right size skate, the customer must present her skate size to the system when she asks to rent skates. For a fundamental activity triggered by a temporal event, the direct information requirements are very simple. You always have time in some form as the single piece of information presented to the essential activity when the event occurs.

10.1.4 Fourth decision: identifying custodial activities

You finish defining the essence of the new system by identifying the custodial activities; they will be responsible for storing and updating the essential memory required by the fundamental activities. For each memory requirement, you have to find precisely when a piece of information should first be stored, how current it must be to be useful to the fundamental activity, and when it should be removed from essential memory.

To define the time when a piece of information should first be placed in essential memory, you look again at the events and the fundamental activities that you have already identified, including those that you identified when establishing the required information. You check every event and every response to see if the information provided will be needed by any fundamental activity, whether or not that activity responds to this event. If the information is needed later, then you consider the event to be the point of origin for those pieces of essential memory, even when they are produced by the response. As you check the fundamental activity that responds to the event "customer requests skates," for example, you know that rental start time is available to the system as it executes this response. You also know that rental start time is required by the fundamental activity Charge Customer (the response to "customer returns skates"). Therefore, the event "customer requests skates" is the point of origin for rental start time in essential memory.

Once you find the point of origin for stored information, you add the appropriate custodial activity to enter this piece of information into essential memory. In our example, you must consider where to place the custodial activity that will store the rental start time. Finding no other activity capable of remembering rental start time and no other event capable of providing it, you would conclude that the custodial activity is correctly placed as part of the response to "customer requests skates." Any piece of essential memory that doesn't have its source in an event that triggers a fundamental activity is obtained from the additional events that you noticed when you defined the essential memory requirements for the fundamental activities. In either case, you have to create one or more custodial activities to record these pieces of essential memory.

Next, you have to consider how to update pieces of essential memory to keep them consistent with the needs of the fundamental activities. You inspect the events again to see if any of them cause a change in data already stored in memory. If so, you add a custodial activity to modify that piece of memory.

Finally, you consider how information will be removed from essential memory; that is, you must find out when a piece of stored information becomes inappropriate for a fundamental activity to use. Again, you look through the events you have identified so far to see if any of them make a particular item of memory obsolete. During this search, you watch for unidentified temporal and external events that may also cause obsolescence. When obsolescence-causing events are identified, you establish the custodial activities that will remove items from essential memory.

You are unlikely to come up with all the appropriate activities the first time you run through the four-step process for creating the new essence. Others will occur to you later, and you will have to repeat the steps, although not necessarily in the same order. For example, suppose you are thinking about the third activity in the Good Skate system, the response to "customer pays for rental." You realize that a customer might not pay for the rental and might not return the skates at all. Because you need to encourage customers to return skates, you decide to keep a deposit from each customer to be returned only when the skates are returned and paid for. To accomplish this, you first add a data access activity to the fundamental activity that responds to the event "customer pays for rental." The purpose of the access is to retrieve the deposit from wherever it is stored so that you can return it to the customer. Next, you must establish memory for where the deposit is stored and establish the custodial activity that places the deposit in storage. This custodial activity will be part of the system's first fundamental activity, which is the response to the event "customer requests skates." Finally, you redefine the stimulus for that event so that it includes the deposit.

10.2 Modeling the new essence

Once you've derived the system's essence, you record it using the modeling tools of structured analysis. In Figure 10.3, you can see an event- and object-partitioned data flow diagram for the essence that was created in the previous section. The data flow diagram shows the stimuli that arise from external events, the responses to all events (the fundamental and custodial activities), the results of the responses, and groups of essential memory elements. Again, event partitioning leads to one activity, or circle, for each event that the system responds to. Although a given event may trigger both fundamental and custodial activities, all activities triggered by that event are represented by a single circle. For example, the completed version of Issue Skates (last seen in Figure 10.2) emerges as a mix of fundamental and custodial activities after all the steps for creating essence have been completed: It now puts the rental start time and the deposit into the Rentals data store, as well as giving the skates to the customers. The data stores on the essential data flow diagram are essential memory elements grouped into objects.

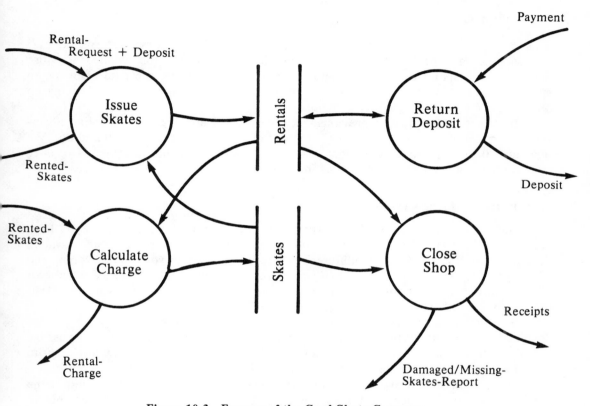

Figure 10.3. Essence of the Good Skate Company.

Here are some data dictionary definitions for Issue Skates, an essential activity in our example:

Rented-Skates = {Skate}
Rentals = {Customer-Name + No.-Of-Skates-Rented + Deposit + Rental-Start-Time + Payment}

The data dictionary defines the data elements that make up the stimulus, the response, and the essential memory. The data dictionary also shows relationships between elements within the stimuli and within the responses, such as showing that a set of elements may appear more than once in a data store or that an element is optional.

A minispec for the essential activities in Calculate Charge appears below:

Return Skates to shelves.
Locate Rental-Record based upon Customer-Name.
Determine Rental-Charge:

Duration   = elapsed time since Rental-Start-Time, rounded
                 up to the nearest quarter hour
Charge   = Duration × Hourly-Rate

Note Charge in Rental-Record.
Inform Customer of Charge.

This minispec describes the entire response to the event "customer returns skates." We use the minispec to show stimuli from temporal events; the detailed activities necessary to respond to an event; and the relationships among the data elements in stimuli, essential memory accesses, and responses, such as the rules for deriving an output data element from essential memory.

In this section and the previous one, we discuss creating and modeling a new system's essence based upon the user's notions of what the new system should do. We now turn our attention to the other major strategy for modeling a system's essence from a model of the essence of an existing system.

10.3 Deriving the new essence

Naturally, after months of hard work, you won't like to discover that all you have done is reinvent the wheel. So, it always makes sense at the beginning of a system development effort to assess the differences between the system you are developing and whatever systems already exist. When you find a similarity, you must decide whether it is better to develop that portion of the new system from scratch or to rework the existing system into what you want.

You must make the same choice again at several points during the system development process: When you write programs, you want to make the fullest possible use of existing code and commercial packages. When you design software, you want to refer to previous designs for similar systems. And when you specify the essence of a new system, you should exploit any similarity between the essence of the system being developed and the essences of existing systems. You do that by deriving the essence of an existing system first, creating from scratch only the new essential features that the existing system doesn't provide for.

In this section, we give our reasons for preferring this approach, and then explain why you need a model of the existing system's incarnation as well. Afterward, we offer some guidelines for deciding between the two major development strategies.

10.3.1 Why derive the essence?

There are three reasons that it makes sense to base the essential model of a new system on the essential model of an existing one when both systems closely resemble each other: First, if you copy an existing system, you are less likely to omit features that belong in the new system. Second, you are more likely to derive a function accurately from an existing system than you are to create it accurately from scratch. Third, it may be quicker and easier to work from an existing system.

All of these arguments are valid only if the new system *should* resemble the existing system. Of course, a new system can resemble an existing system in many different ways, but it is never an exact copy of the existing one. If it were, the development project would never have been started. The existing system satisfies some of your requirements, but it also has deficiencies that are great enough to justify the development of a new system to replace it. You want to isolate those aspects of the existing system that are appropriate to the new system, to model them exclusively, and to ignore as completely as possible the deficiencies and the irrelevant aspects of the existing system.

Most of the similarities between an existing system and a desired system are in the essence of the existing system, while most of the differences are in the ways the system employs technology to implement its essential activities and memory. In fact, we have seen the members of many development projects discover that the existing system actually carried out the organization's business policy accurately, but in a fashion that was unacceptable for purely technological reasons. Perhaps the existing system was too inefficient or too difficult to maintain. In these cases, a model of the existing system's essence could serve with very little modification as a definition of the new system's essence.

But how common is this phenomenon? And why does it occur? This phenomenon is very common, and we believe the reason has to do with the differing rates of change for incarnation technology and system essence. Since about 1945, computer technology especially has advanced at an astounding rate, while few truly new essences have appeared on the scene. If we look at how business applications systems changed from 1965 to 1980, we see they differed from their predecessors largely in the technology used. During this period, three generations of computers came and have almost gone, and data storage and retrieval technology also improved. Through the advances of online processing and telecommunications, data entry and routine inquiries are handled much more efficiently than they were by the batch processing technology of 1965.

How did the essences of these systems change between 1965 and 1980? One thing is certain: The essences did change, evolving continuously. However, the important point is not that these new systems had both different essences and different incarnation technology, but that incarnation technology changed much more rapidly and dramatically than did the essence of systems.

In most cases, the biggest difference between a new system and existing ones is that the new system will use a new, improved mix of technology, while the essence is probably the same. For this reason, studying the essence of an existing system is very often useful. Hence, our strategy for modeling the essence of a new system calls for first modeling the essence of an existing system, and then adding any new essential features that must be a part of the new system. Figure 10.4 illustrates this approach.

10.3.2 Why model the current incarnation?

The strategy described above implies that you derive a model of the essence of an existing system directly from information about current operations. No model is produced before the current essential model; the first product completely ignores the characteristics and limitations of the existing implementation technology. This is a lofty and admirable goal. After all, why bother to model anything that you will later eliminate during the logicalization process? That seems to be a complete waste of time. Consequently, many project managers attempt to speed the progress of their analysis efforts by skipping the current physical model and going straight to the current logical model. At least one company has gone so far as to eliminate any mention of current physical models from its structured analysis methodology. Unfortunately, there simply is no way to avoid a certain amount of current physical modeling.

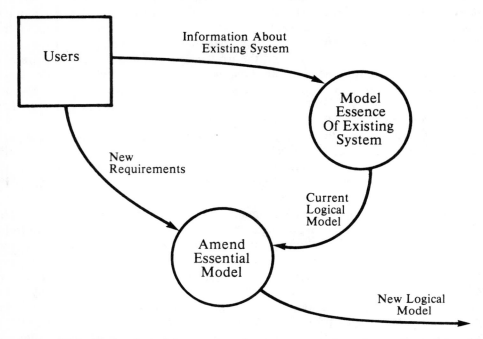

Figure 10.4. Derivation of the essence of a new system from the essence of the old.

There are two major reasons for always producing a model of the existing incarnation. First, human capacity to deal with complexity is severely limited. Whether or not you can readily distinguish essential features from incarnation features, you are not able to consider all the information about a system at one time. However, in order to isolate essential features, you need to understand all of the existing system in detail, since the essence of the system is fragmented and scattered throughout the incarnation of the existing system. This leads to a Catch-22 when deriving the existing essence directly: To model the essence of a system directly, you must work at a fairly high level of abstraction so that you can understand the entire system. But if you look at the system at this abstract a level, you miss most of the detailed, fragmented pieces of essence.

The second reason to develop a model of the existing system's incarnation is that you need to obtain verification from people who either cannot or will not be abstract in their thinking. Even if you think you understand the existing system's essence well, you still need to obtain independent corroboration of your understanding. However, the essential model of a system may look so completely different from the incarnation that you may have trouble obtaining this corroboration. Some people can only verify the current essence of a system by studying a model of its incarnation.

For these reasons, you must first develop a model of an existing incarnation and then rework that model to yield the current essential model. This more detailed set of activities is shown in Figure 10.5.

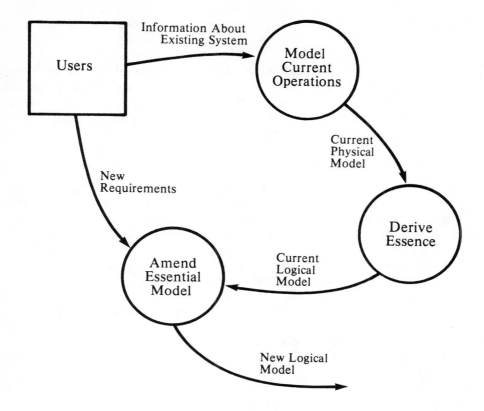

Figure 10.5. Derivation of the essence of a system using the current physical model.

Although building a current physical model before building an essential model is like having to eat your brussels sprouts before your ice cream, it has some additional benefits:

- You can use the current physical model to detect problems in the existing system. If it will take a long time to develop the new system, you may want to streamline and otherwise improve current operations in the meantime.

- You can use the current physical model as formal system documentation for users, analysts, designers, and managers who work with the system. For many organizations, the current physical model is the only understandable model of their operations that these people have ever seen. In one case, a user manager said that even if the development project failed, he would consider the analysis effort a success simply because it had produced a model that allowed him to visualize his operations for the first time.

- The current physical model can help you prepare for conversion to the new system. It serves as a kind of map of the existing terrain, helping you to choose where to modify the existing operations to incorporate the new system.

Useful as these applications are, they aren't by themselves sufficient justification for building a current physical model. You build this model primarily to learn about the essence of an existing system and to verify that knowledge. Only then can you build a model of the current system's essence that will help you to develop a model of the new system's essence and, ultimately, to build the new incarnation.

10.4 Choosing a modeling strategy

In the previous three sections, we discuss the two strategies for defining the system's essence, shown in Figures 10.1 and 10.5. In an ideal world, you might use both strategies, choosing the best portions of each resulting essential system model. In the real world, however, scarce resources of time and money force you to compromise — by choosing to perform only a subset of activities from these strategies, by decreasing the amount of detail that goes into a given activity, and by limiting the scope of the system to a small subset of environments and existing systems. The nature of the compromise is the responsibility of project management, and the set of activities management chooses constitutes a strategy that the project team will execute to develop the new system.

Of all the hybrid strategies that could be created, we devote the remainder of this chapter to a discussion of which circumstances require which of the two major strategies already discussed.

10.4.1 Factors in choosing a modeling strategy

So which strategy should you choose for your project? Should you follow a direct approach to modeling the essence of the new system? Or should you first model the essence of an existing system? To answer these questions, you must first ask two additional questions:

- Which strategy will lead to the most appropriate model of the new system's essence?

- Which strategy will make effective and efficient use of scarce project resources?

These two issues are strongly related. If a strategy is thorough enough to capture all the appropriate components of a system's existing essence, it may take so long to apply the strategy that the project's survival is jeopardized. On the other hand, some projects must produce systems whose appropriateness of essence is so critical that *only* such a thorough approach is justified. To untangle these issues, we discuss the factors that influence which strategy is efficient and which results in appropriate systems, and then describe the characteristics of the situations favoring each strategy.

10.4.1.1 *Factors that influence appropriateness*

Two basic factors determine which strategy produces the most appropriate new essence. First, you have to decide the tolerance of the environment in which your completed system must function. How much trouble will result if the system does not carry out the proper planned responses? Of course, nobody plans to produce bad systems, but reality dictates that some systems must make fewer mistakes than others. For example, an accounts receivable system has a higher tolerance for error than does an air traffic control system. Once you decide how accurate the new system must be, you next determine which strategy produces the most accurate system.

The second factor influencing appropriateness is which strategy will reduce the likelihood of errors in the essential modeling process itself. Is the essence of the new system conceptually complex? If so, modeling the essence of the existing system will prevent some error and give you at least some confirmation that your definition of the new essence is correct. On the other hand, you may not have to worry about making errors if the essence of the new system is relatively simple. Another consideration is that the essence of a similar existing system might be almost impossible to extract. In this case, you may make more mistakes extracting the essence than you would defining the essence from scratch.

10.4.1.2 *Factors that influence efficiency*

Your first priority may be to develop the right system, but you also must consider how long this is likely to take using each modeling strategy. It will do you no good at all to choose a strategy that yields a terrifically appropriate essence, but takes so long to complete that your project is canceled. Sometimes you must change your strategy in order to complete the system in a reasonable amount of time: You may have to create the essence of the new system from scratch if you don't have time to model the existing system. Occasionally, this means that you have to settle for a less appropriate essential model.

Whether you have the time to model the essence of an existing system depends upon a second major factor: the accessibility of the existing system's essence. If there is no existing system, the point is moot. Even if there is an existing system, almost certainly no essential model of it was ever created. (For the reasons discussed in Chapter 1, many structured analysis efforts have also failed to produce a true essential model.) It therefore falls to you to build the model of an existing system's essence — if you choose to do so.

You might not choose to model the existing system's essence if it is not accessible for other reasons. Perhaps nobody knows the system's essence, or those who do know won't take the time to tell you. The system itself may be indecipherable (all you've got are fifteen-year-old spaghetti tangles of Viatron assembler language code) and therefore inaccessible because it would take too much time to ferret out the essence.

You might decline to model the essence of an existing system even if the essence were accessible, based on evaluating how useful a completed model of the existing system's essence will be once you have it. Will the current essence have to be changed radically to produce the new essential model? What if the existing essence is exactly what the new essence needs, but additional essential features make up the overwhelming majority of the new essence? In either case, the time spent to model an accessible existing essence will probably be wasted.

10.4.1.3 *Situations that favor using each strategy*

In sum, when you choose to model the existing system, it is usually for the following reasons:

- There is a significant risk involved in building an essential model from scratch because it is especially important that the new system be free of error.

- Knowledge about the essence of existing systems is accessible.

- The existing essences are significantly similar to what you think will be the essence of the new system.

- There is sufficient time to study the essence of existing systems.

Although we favor studying an existing system whenever feasible, in some situations it is more appropriate to develop the essence of the new system from scratch. The conditions that favor this approach are the following:

- There is relatively little risk in specifying the essence of the new system from scratch.

- The appropriate existing systems are inaccessible.

- The essences of existing systems are only marginally related to the essence of the proposed new system.

- There is a time constraint.

Choosing a strategy is not a binary decision, and you should employ both strategies, each for different parts of a given project. Suppose that you are redeveloping an existing, largely manual inventory control system. A portion of its policy is carried out by a particularly abstruse commercial software package. The vendor never delivered the source code and subsequently went out of business. Fortunately, the package does perform well, and it has been in use for years with few problems. Given these circumstances, would you choose to model the existing system or would you start anew?

No one strategy is perfect for this situation. Certainly, the commercial software package is not an appetizing candidate for modeling. Given the absence of source code and contact with the developers, building a current physical or logical model of the package could prove unpleasant and, perhaps more important, very time-consuming. On the other hand, you have heard no such discouraging words about the system's manual portion, and the best tack to employ may be to build a current physical model for the manual portion. There is no reason not to employ both modeling strategies for such a project, studying the manual portion of the existing system and at the same time building a logical model of the software package functions from scratch.

It would be a big mistake to adopt such a rigid attitude that you end up using a modeling strategy inappropriate for a particular part of the system solely in the name of uniformity or consistency. The flexible approach does add to management's already significant burden, but the additional headaches will be rewarded by shortened projects and more accurate requirements models.

10.4.2 The reasons for our choice of strategy

We recommend that you study the essence of the existing system on most development projects for these four reasons: In most cases, there is an existing system; its essence is reasonably accessible; the essence of most new systems is pretty much the same as the essence of the existing system; and finally, there is often enough time to build a model of the existing system's essence. For the remainder of this book, therefore, we focus on the development of a model of a system's essence from knowledge of the incarnation of an existing system.

10.5 Summary

This chapter presents the two strategies for building the model of a new system's essence. The first strategy is to create the new essence from scratch, basing it directly upon information from users about what they would like the system to do. We offer a four-step procedure for creating such a model: identify the system's purpose, identify the fundamental activities, identify the information that must be stored in memory, and identify the custodial activities. Finally, you build the model using the tools of structured analysis.

The second strategy for building a new essential model is to derive that model from a model of the essence of an existing system. Because technology is changing much faster than business policy, the essence of a new system often closely resembles that of an existing system. In such a case, you will produce a more accurate model if you base it upon that of a current system's essence. To obtain a current essential model, you will also need to build some kind of current physical model.

Which strategy you choose depends upon how much time you have, how accurate the new system must be, how accessible the essence of the current system is, and how similar the new system is to the old. Because there is usually an accessible existing system whose essence resembles that of the new system, we recommend that you study the essence of the existing system on most development projects.

Part Three

The Anatomy
of Existing Systems

System incarnations are typically complex arrangements of components. Finding the essence of a system within such a complex structure is a tremendous problem, since the complexity forces you to do an extensive amount of analysis and synthesis to find the essential features. Fortunately, a basic pattern of activities occurs in most existing systems, and understanding that pattern can help you cope with this complexity. In Part Three, we describe this pattern, which you might think of as the anatomy of existing systems.

The anatomy of a physical system is based upon three fundamental ideas: First, a technologically imperfect processor is the crucial unit of the incarnation; second, the processor's technological imperfections lead to many new, nonessential activities and give the incarnation a complex organization; and third, incarnations can be viewed as an organization of nested processors. These three ideas are treated in the next three chapters.

Chapter 11

The Anatomy of
Single Processor Systems

The activities carried out by a given processor may fall into one or more of several categories: Some activities are a part of the system's essence; they would have to be performed regardless of the nature of the processor. Other activities serve only to help a processor perform to its fullest, such as doing its internal housekeeping and error-checking functions. Still other activities facilitate effective communication between two or more processors that work together to carry out an essential activity.

When deriving an existing system's essence, we know the importance of being able to recognize the portions of the system that actually are a part of the essence and those that result solely from the system's use of technology. If you cannot make this distinction, you are likely to produce a bad requirements model that may leave out some important parts of the existing system's essence, or include nonessential features that place unnecessary and expensive constraints on the new system.

The key to distinguishing the essence of an existing system from the remainder of its activities and memory is to know the anatomy of the system. The anatomy consists of the typical activities performed by a system and their location in the system. By learning the anatomical pattern that characterizes most planned response systems, you can focus on the most relevant parts of the system you are studying, thereby increasing the accuracy of your work while making less effort to accomplish it.

Since our view of planned response systems centers on the activities they perform, we focus our study of existing systems on the component that actually carries out both planned and ad hoc responses: the processor. In this chapter, we explore the anatomy of a single processor that carries out the activities of an entire system by itself.

11.1 Processors

We use the word *processor* in a very general way to include any device that carries out a response, or part of a response, to an event. Thus, a processor can be all manner of entities beyond the obvious example of a computer processor. It can be a human being, a purely mechanical device like a manual typewriter, an electromechanical device like a copying machine, a fully electronic device like a digital computer, or anything else that performs human-created responses. We also use the term to refer to groups of devices that carry out such responses. So, a network of minicomputers taken together is a processor, as is a department of human workers.

Processors are the major building blocks of real systems. This is only natural, since processors execute the planned or spontaneous activities of a system, thereby fulfilling its purpose.

Although processors have different forms, they usually have some components in common. A typical processor, shown in Figure 11.1, contains at least four types of components to carry out its interactions. They are a sensory facility, a central component, essential memory, and an external interface component.

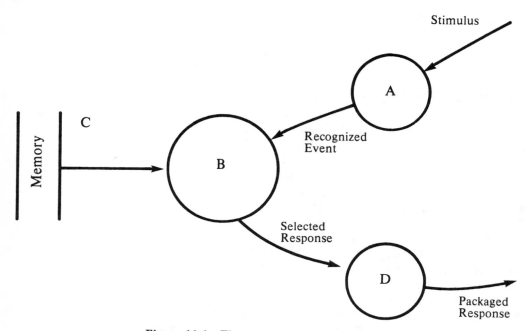

Figure 11.1. The anatomy of a processor.

The processor maintains a sensory facility (labeled A in Figure 11.1) through which it monitors its environment. Humans have five senses; burglar alarms have electric eyes and sound detectors; and computers have keyboards and disk drives, among other input devices. Through these, processors learn of external events that require a planned response. The sensory facility must discriminate among similar events, detecting erroneous input by referring to the processor's stored definition of the appropriate stimulus. For example, a processor employed by a car dealer must intercept requests for toaster ovens and other products that are not sold by the dealer. In addition to notifying the system that an event has occurred, the sensory facility may have to provide information about the event so that the processor can respond appropriately. For example, a processor registering hotel guests needs to know more than that a guest is attempting to check in; it must also know the guest's name in order to locate his or her reservation.

For each predefined event, the processor has at least one predefined response. The processor's central component (B in Figure 11.1) formulates a planned response to a recognized event. It selects among the alternative responses known to it, basing the selection upon the information it receives about the event. This component then builds the information that makes up the response, performing any necessary calculations in the process.

In order to select the right response, most processors must have essential memory consisting as always of information about the outside world that is not available at the time an event occurs, and information about the processor's own actions in the past — how it responded to previously detected events. This essential memory (C in the diagram) may reside within the processor itself (data that a clerk has memorized, for example) or it can be separate from the processor (such as a secretary's file cabinet).

The processor must communicate its responses to the outside world. To do so, it has a counterpart to the sensory facility for outgoing information. This external interface component (Figure 11.1's D) packages the selected response into the data containers used to communicate with whatever external entity is to receive the response. For example, if the processor transmits the response by telephone, the external interface component must carry out the procedure for making a telephone call. If the response is to be sent over a data communications network, this component formats the message and takes care of the network communications protocol.

Besides sharing these basic facilities, all processors have another characteristic: They are imperfect. Specifically, their facilities are imperfect in that processors are not universally skilled in all activities. Human processors possess different skills because they have different abilities and different educational backgrounds. Processors also are imperfect in their finite capacity to carry out activities. Every processor has limits to its power, its work capacity, and its stamina.

Additional evidence that processors are imperfect is that sooner or later processors make mistakes. Just about every processor eventually fails to carry out a planned response properly. In addition to general fallibility, human processors suffer from the limitation of corruptibility. Some of them make mistakes on purpose, and even lie, cheat, steal, and make mischief.

Processors are imperfectly slow. Certainly, some are faster than others, but no processor is so fast that it can carry out a planned response *instantly*.

However imperfect they are, processors nevertheless cost money, with the cost usually inversely proportional to the other limitations. Highly productive human beings who possess rare skills — such as a successful trial lawyer — cost more than unskilled laborers. Similarly, computers that are fast and reliable cost more than old clunkers that crash on the hour. Nonetheless, although higher cost means a better processor, cost is still a processor limitation.

Because of their imperfections, processors have the greatest influence on the appearance of a real system. They are the source of the characteristics of imperfect technology in the system. To understand how these imperfections give rise to new activities, we look at what happens when you allocate combinations of essential activities to different configurations of processors. We start by discussing a single processor that carries out a single essential activity.

11.2 Single processor/single essential activity systems

There are more single processor incarnations than you can imagine. Every one of us is a single processor incarnation, and so are microcomputers, minicomputers, maxicomputers, electric can openers, refrigerators, and elevators. Any single physical entity that implements human-designed planned responses qualifies as a single processor incarnation. The most basic of all incarnations is a single processor, either man or machine, that executes a single essential activity.

For example, Lou Pole prepares tax returns for his friends for free. We find him one evening, hunched over at a roll-top desk piled high with papers. He wears a green visor and a pair of thick-lensed glasses, which reflect the glare of the naked light bulb hanging from the ceiling. On top of Mr. Pole's desk are an ink well for his fountain pen, a stapler, boxes of paper clips and rubber bands, and a thick booklet titled *The Last Word on Preparing Income Tax Returns*. Next to his desk are some antique wooden file cabinets that hold the copies of the prior years' returns. On top of the cabinets are stacked some paper trays. He motions us to place all of our receipts, 1099s, and W-2s into the top tray marked IN, and he shoos us out so that he can return to work.

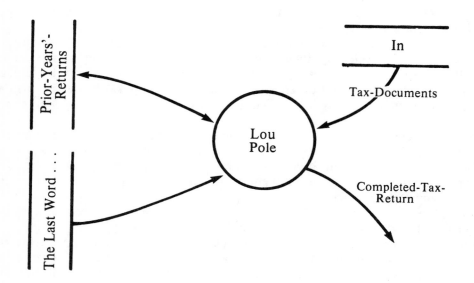

Figure 11.2. The Lou Pole system.

Figure 11.2 shows a data flow diagram of the incarnation we find. To consider the incarnation in its entirety, however, we would have to include Mr. Pole's office location, furniture, lighting, supplies, and a host of other characteristics of his workplace, because these physical things are certainly part of the system's incarnation. Yet, for this example, we ignore them. Instead, we concentrate on the activities Mr. Pole carries out, because it is in the activities that the patterns most widely observed among different system incarnations occur.

The lower-level DFD shown in Figure 11.3 gives a detailed view of Mr. Pole's activities. We classify these activities into three groups: essential activities, internal quality control activities, and internal transportation activities. Each group is discussed in the subsections below.

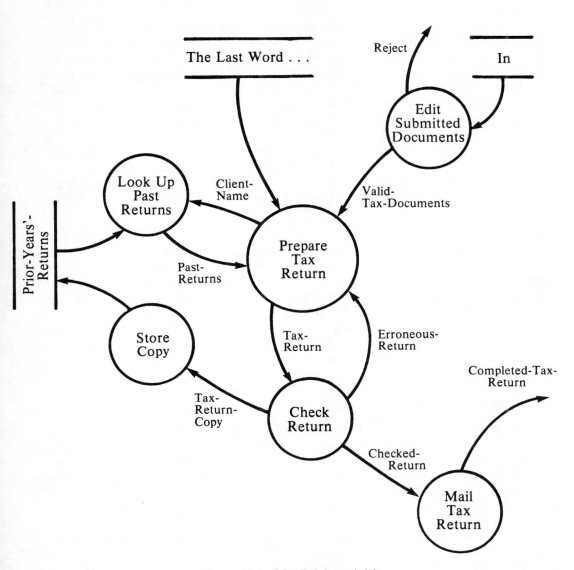

Figure 11.3. Mr. Pole's activities.

11.2.1 Essential activities

According to our definitions, the essential activities are those that the processor would have to carry out even if its internal technology were perfect. They are constrained by two types of external imperfections: the technological limitations of the entities outside the system and the limitations of the technology needed to communicate with those entities. The essential activities that are most insulated from technological concerns are called central essential activities; Prepare Tax Return is the sole central essential activity of the Lou Pole system. It must take into account the technological imperfections of the client and the Internal Revenue Service, but it is independent of any internal imperfections and of limitations in the mechanisms for communicating with the external entities.

External interface activities are the essential activities that cope with the imperfect technology of the external communications medium. There are two kinds of external interface activities: activities that receive information from external entities and activities that send responses to them. Edit Submitted Documents is an example of the first kind of activity. It receives documents from clients and checks that all the required forms have been submitted. Mail Tax Return is an example of the second type of external interface activity. It interfaces with the imperfect technology of the U.S. Postal Service as a means of getting the return to the IRS. Mail Tax Return selects the proper size envelope, labels the envelope, determines the proper postage, affixes the postage to the envelope, and drops the package into the mailbox. Both the central essential activities and the external interface activities are affected only by external technological imperfections, so they are a part of the system's essence.

11.2.2 Internal quality control activities

The first activity in our example that is not part of the system's essence is an internal quality control activity, Check Return. It reflects the imperfect internal technology that is Mr. Pole. Because he is fallible, Mr. Pole must check his work for errors. In performing the activity Check Return, Mr. Pole retraces all of his steps to generate the return. He double-checks the entries on the input documents he used, his choice of tables, the figures from the tables, and his calculations. If Mr. Pole finds an error, he must redo the activities to correct the return. Because this kind of activity approves or disapproves the results of a processor's activities, we call these activities *approval* processes.

11.2.3 Internal transportation activities

Mr. Pole carries out two internal transportation activities in which he does not carry out any part of an essential activity, but merely moves data to and from its internal storage location. In Figure 11.3, these two internal transportation activities are Look Up Past Returns and Store Copy. Mr. Pole performs the first activity when he needs to see a client's past returns, usually because the client has asked him to do income averaging. These files are kept by taxpayer name in manila folders stored inside the antique wooden file cabinets. With perfect technology, Mr. Pole would merely specify that he wants all returns for a given client. However, the imperfect technology at Mr. Pole's disposal requires him to execute some additional activities. First, he must find the right drawer, then search through it until he finds the right folder. He removes the folder and carries it back to his desk, opens it, and removes the past returns.

The other internal transportation activity, Store Copy, is nothing more than Mr. Pole's moving a copy of the checked return from his desk to the customer's manila folder in the file cabinet. Although the data store in this case is separate from the processor, these activities are still considered *internal* data transportation because Mr. Pole has sole control of the files.

Internal transportation activities are part of what we call the *intra-processor infrastructure*. The word *infrastructure* usually refers to all components of a society that transport people, goods, and information from one location to another. We use it in a similar way, but in this case we are focusing on a single processor. The intra-processor infrastructure moves information between its storage site and the site where it will be used. The other part of the internal transportation activities is the set of containers used by these activities. For example, the manila folder in which Pole stores prior years' returns is such a container.

The internal quality control activities are part of a larger class of activities called *administrative* activities, which check for errors and coordinate activity within the system. Again, our focus is restricted to a single processor that is carrying out a single essential activity. Both the infrastructure and the administrative activities compensate for the imperfect communications and fallibility that characterize imperfect technology.

Now we look at which processor components are responsible for which of the three types of activities in a single processor/single essential activity system. The processor's central component carries out the central essential activity, Prepare Tax Return, the internal quality control activity, Check Return, and both of the internal transportation activities, Look Up Past Returns and Store Copy. The peripheral components perform the external interface activities: Edit Submitted Documents is carried out by the sensory facility, and Mail Tax Return is carried out by the external interface component that exports the processor's response.

11.3 Optimizing processor performance

So far, our processor has organized its activities in a relatively straightforward way. However, unlike an infinitely fast perfect processor, processors that use imperfect technology take considerable time to do their work. So, they must find ways to reorganize their activities to speed up their performance.

If Mr. Pole reorganized his activities to improve his productivity, the result might be an arrangement like the one in Figure 11.4. Mr. Pole has used what is perhaps the most popular tactic for increasing the efficiency of a processor: batching. In batching, a processor collects a group of input before beginning an activity so that the activity can be executed repeatedly without pause. The processor works more efficiently, because it is doing one simple task over and over rather than many different tasks, one after the other. The repeated execution minimizes the time needed to gather all the information required to begin the batched activity, and the time needed to clean up after the activity is complete. In Figure 11.4, for example, Edit Submitted Documents is a set-up activity that can be performed on several returns before any of them goes on to the Prepare Tax Return process. The processor also batches the products of the activities Prepare Tax Return and Check Return.

Batching adds data stores to the system. In our example, Valid-Tax-Documents and Erroneous-Returns accumulate the products of an activity until all the documents in an earlier batch have been processed, or until there are enough documents to begin the next activity, or until it is time to begin the next activity regardless of the number of intermediate products.

Batching can also affect the organization of the data stores to be updated. In order to improve efficiency, the sequence of information in the data stores may be chosen to minimize the time required to search for a given record or file. Such schemes usually organize the data store being updated in the same way as the batch data store that supplies the updates. Sometimes, instead, a sort activity is added to reorder the batch data store to match the sequence of the data store being updated.

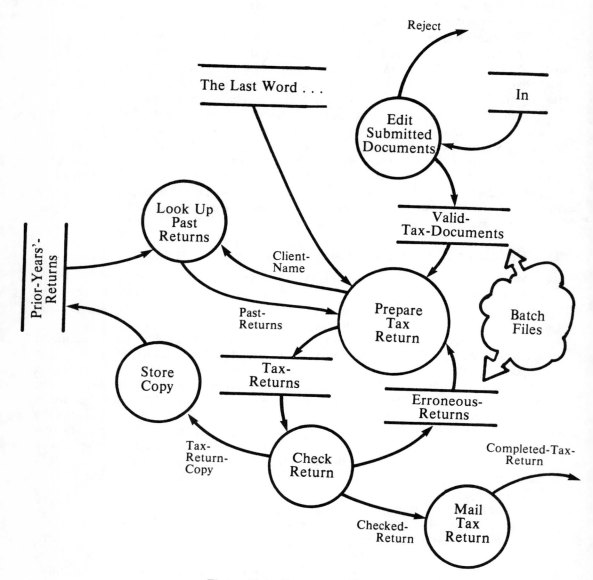

Figure 11.4. Batched activities.

Another optimization technique involves calculating data items in advance of when they are needed. Usually, this technique is employed either when the calculation is particularly time-consuming, or when the processor must carry out the essential activity with great speed. For instance, Mr. Pole might decide to keep a running total of each client's averageable income along with the prior years' returns. Instead of deriving that figure when a client elects to income average, Mr. Pole updates the figure each time he performs the process Store Copy. He can do nothing with the total averageable income figure *until* the client asks him to prepare a tax return using income averaging. The only advantage he gains is that if the client ever makes such a request, he will be able to respond more quickly. Again, such an advantage would be meaningless to a technologically perfect processor.

11.4 Single processor/multiple essential activity systems

Outside the world of technical books, most single processor systems perform more than one essential activity and thus have planned responses to more than one event. Any single proprietorship business is a good example of a single processor that has several essential activities. The command computer of an avionics control system is another, as is the average arcade video game.

Single processor systems that carry out multiple essential activities look very much like those that perform only one essential activity. They exhibit the same basic anatomical patterns, with only two differences worth mentioning: The pattern discussed above is *repeated* for each essential activity, and additional intra-processor administrative activities usually keep track of which essential activities are being performed at any one time and where the processor is in each of them. For a human, the multiple activity administrator may work with a things-to-do list; in a multiprocessing computer system, the process scheduling facility of the operating system, together with the production control staff, fills this role.

11.5 Summary

Processors are the major building blocks of a system, since they execute a system's activities. They consist of four types of components: a sensory facility to receive stimuli from events, a central component to select and build a response, a memory, and an external interface component to send information to the outside world. Processors are imperfect, and their limitations give rise to the nonessential activities in a system.

In this chapter, we concentrated on the characteristics of a single processor/single essential activity system. The processor in such a system would carry out two kinds of essential activities, central activities and external interface activities. Its nonessential activities would be mainly approval activities, which check the processor's work for errors, and internal transportation activities, which move data to and from internal storage. Internal transportation activities are also known as the intra-processor infrastructure, and approval activities belong to a class of activities known as administrative activities. Other nonessential features are added to improve a system's performance; these are often batch data stores.

This discussion of single processor systems completes the first, and simplest, of the many activity patterns in existing systems. By knowing this pattern, along with those described in the next two chapters, you will be able to study large, complex systems and to identify the essential activities and memory quickly and accurately.

Chapter 12

The Anatomy of Multiple Processor Systems

Why can't all systems use only a single processor? If you try to implement many systems with only one processor, one or both of two problems will arise. First, one processor may not be talented enough for all phases of the job, or a processor with all the right skills may be unjustifiably expensive. The other possible problem is that a single processor may not be able to work fast enough or long enough, or if it can, it would simply cost too much compared to a set of less powerful processors that satisfy the requirement by working together. We call the first problem the skill vs. cost factor, and the second problem the capacity vs. cost factor.

To avoid these problems, many systems must use multiple processors, and unfortunately as a result take on additional physical characteristics that camouflage the essence of the system. In this chapter, we describe the anatomical patterns of these multiple processor systems.

There is an amazing variety of multiple processor incarnations. Like single processor systems, each processor has an intra-processor infrastructure to transport data and a set of internal administrative activities to check for mistakes and to schedule tasks. Each processor may also have its own external interface activities. However, some important features occur only in multiple processor incarnations:

- the fragmentation of essential activities among the processors

- an interprocessor infrastructure that allows the processors to communicate with each other

- an interprocessor administration consisting of activities that monitor and coordinate the operation of a set of processors

Each of these multiprocessor features is described in the following sections, so that you can see a definite pattern emerge by the end of the chapter.

12.1 The fragmentation of essential activities among processors

Essential activities are usually fragmented and the fragments spread among two or more processors. It is a rare processor whose combination of skills, work capacity, and cost make it an appropriate choice to carry out an entire essential activity by itself. When you analyze the demands placed upon a system by any one of its essential activities, you typically discover that the most effective and efficient way to carry out the essential activity is to use a team of specialized processors, each of which is suited to a portion of the activity.

So when you study an existing system, you actually are studying division of labor through specialization. The labor entailed in each essential activity is divided among several processors according to the skill vs. cost factor and the capacity vs. cost factor.

12.1.1 Fragmentation according to processor skill and cost

The skill vs. cost factor comes into play when it is too expensive to purchase, rent, or hire a single processor that can perform the entire essential activity. Consider the essential activity shown in the DFD in Figure 12.1. This activity issues tickets for an air shuttle that flies between New York and Washington, D.C. When presented with a ticket request and a credit card, the activity generates a ticket and a receipt. This is a very simple system, and we have taken some liberties to make it even easier to present. For example, any problem that might arise from lack of seats is scrupulously ignored. (Maybe this is a realistic example after all!)

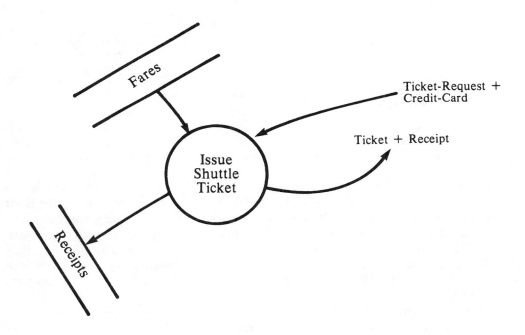

Figure 12.1. An essential activity for a ticketing system.

Now, to implement this activity, a company could use any number of incarnations. One possibility is to use a single processor, such as the automated ticketing machine shown in Figure 12.2, which is driven by a microcomputer. The company that obtained this incarnation clearly had ample funds, because it was able to develop a system that is both friendly to passengers and efficient at bookkeeping.

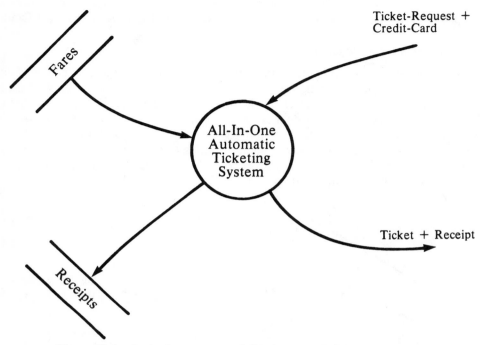

Figure 12.2. A single processor, fully automated ticketing system.

A less wealthy company chose to implement the activity with a computer and a human clerk, as shown in Figure 12.3. The company decided to divide the labor between two processors, because it was too expensive to buy or build a computer system that could communicate courteously yet effectively with the human passengers. On the other hand, although human clerks could talk to the passengers easily, they lacked the speed and accuracy needed for looking up fares, validating credit cards, and typing out tickets. So in this case, the skill vs. cost factor forced the company to create a multiple processor incarnation.

A third company lacked enough money to develop either the super computer ticket clerk or the average computer-and-clerk combination. So, the company hired an obstreperous, out-of-work son-in-law to write the tickets, and it supplied two mechanical processors to assist him: an adding machine and a credit card imprinter, as shown in Figure 12.4. Of course, during peak periods, this company isn't able to do as much business as the other two companies.

The point of these examples is, The less money you have to acquire versatile processors, the more you will have to make do with a host of more narrowly skilled workers. And the greater the division of labor, the more fragmented and disintegrated the essential activity becomes, further camouflaging the essence of the system.

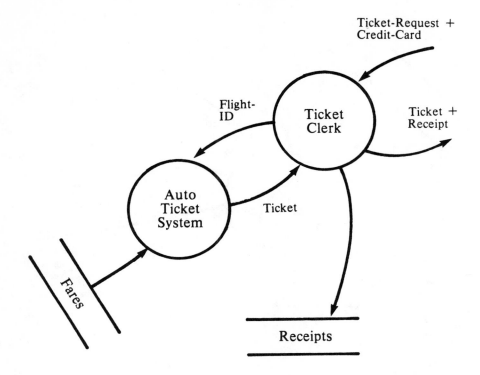

Figure 12.3. A two-processor ticketing system using clerk and computer.

12.1.2 *Fragmentation according to processor capacity and cost*

Fragmentation by capacity and cost results when it is too expensive to employ a single processor with sufficient workload capacity to handle an entire essential activity. One processor might be too slow to carry out the entire activity in a reasonable amount of time, while another processor may not be able to stay in service long enough without relief. So, many systems employ multiple processors to provide additional speed and endurance.

In our shuttle example, three relatively inexpensive processors performed the essential activities for the third company. As we said before, this incarnation may not be able to handle the load that the others can, but presumably the people who established the incarnation accept that. But what would happen if they got more money? They could create a system with increased capacity to keep up with a greater workload. Of the many possible ways to do this, they settle on the same clerk-and-computer option that the second company used for reasons of skill vs. cost. The same fragmentation of the essential activity takes place, but this time the motive is to increase the capacity of the system at an acceptable increase in total cost and a possible decrease in processing cost for each event.

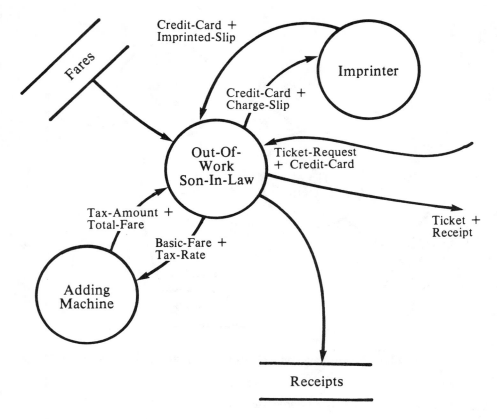

Figure 12.4. A three-processor ticketing system using son-in-law and machines.

Either the skill vs. cost factor or the capacity vs. cost factor could justify the fragmentation of the Issue Shuttle Ticket essential activity. In reality, the fragmentation of essential activities usually results from a combination of these two factors.

The processors in our second example, shown in Figure 12.3, have the characteristics of single processor/single essential activity systems. Each processor remembers and carries out a fragment of the essential activity Issue Shuttle Ticket. The ticket clerk also carries out external interface activities that allow him or her to communicate with the world outside the system — the customer — *and* with the other processor in the system — the automated ticketing system. In the same way, the automated ticketing system edits data coming from the ticket clerk and formats responses for presentation to the clerk. The automated ticketing system maintains the system's essential memory of fares and taxes, while the ticket clerk keeps track of the receipts. Each processor checks the validity of its own work before sending it on. So within this multiple processor system, each processor behaves very much as if it were an independent single processor system. Despite these similarities to single processor systems, multiple processor systems have two additional characteristics: They need facilities for communication between processors and for coordination of processors working together, as discussed in turn in the next two sections.

12.2 The interprocessor infrastructure

All systems that employ two or more processors to carry out one essential activity must provide a way for the processors to communicate with one another. This mechanism is called the *interprocessor infrastructure*. In addition, each processor has its own intra-processor infrastructure, the activities that transport data to a processor from data stores connected to only that processor. Yet, for processors to communicate with each other, additional transportation facilities are needed.

The interprocessor infrastructure exists for either one or both of two purposes. First, it provides access to a piece of essential memory used by two or more processors. Second, when two or more processors implement different fragments of the same activity, it sends the intermediate products of these fragments from one processor to another.

Let's assume for the moment that we have a multiple processor incarnation in which each processor carries out one entire essential activity. With this assumption, we can concentrate on the features of the infrastructure that provide shared access to stored data.

12.2.1 Sharing stored data

All of the many possible ways to share stored data among a group of processors are variations of two basic options: establishing a data container to be shared among the processors, or creating a private data container for each processor.

12.2.1.1 Shared stored data containers

Figure 12.5 illustrates the first option, in which the system's essential memory resides in a single data container that is used by a number of processors. The container is typically divided into subcontainers, because it is easier to transport an individual subcontainer than the entire set of stored data. The data elements that make up essential memory are divided among the subcontainers in a way that conserves storage and retrieval resources.

Figures 12.6 through 12.8 show three typical examples of this phenomenon. Figure 12.6 depicts part of a clerical system in which three processors use a common data container, in this case, a metal filing cabinet. Inside the filing cabinet are manila folders as the subcontainers of data. The cabinet's contents have been organized so that each folder contains a commonly used subset of essential memory, such as the data about one customer, one project, or one account. When a clerk needs to access a portion of the system's essential memory, he or she can remove those particular folders.

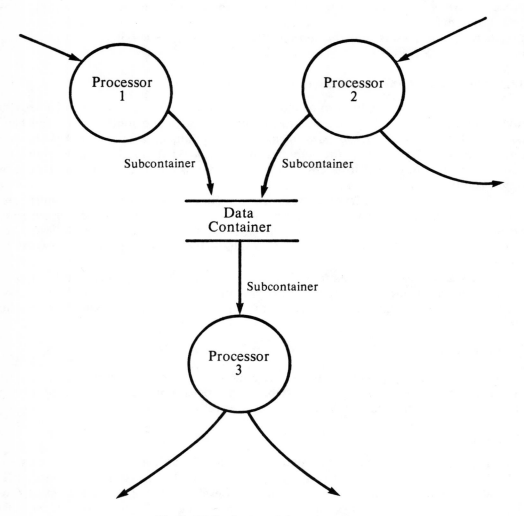

Figure 12.5. A shared data container.

Figure 12.7 shows the same strategy used with a different technology. Here the common data container is an online database, and the subcontainers are the segments into which stored data has been partitioned. A processor — in this case, a computer program — can access the segment that contains the data it needs, leaving the other segments undisturbed and available to the other processors.

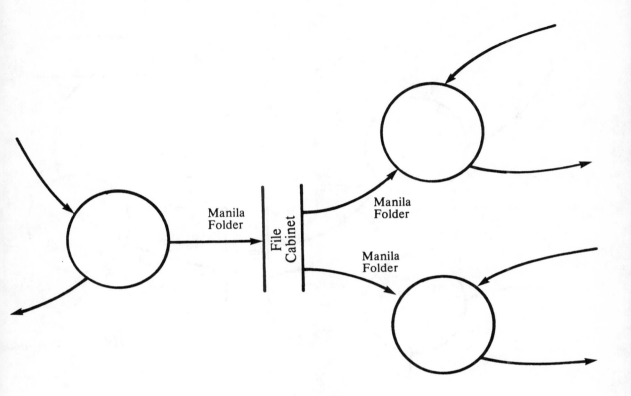

Figure 12.6. A filing cabinet shared among processors.

Finally, Figure 12.8 shows yet another example of a data container shared by processors, in this case, an automated batch system. The shared data container is a flat master file, and the subcontainers are the records that make up the file.* The primary difference between these three examples is the sophistication of the mechanism by which the processors access the data containers. The clerical system relies upon human action, the batch system uses sequential file processing techniques, and the online system enjoys a powerful direct access facility. Beyond these differences, however, all three systems exhibit two characteristics of multiple processor systems that are not present in single processor systems: The subcontainer may not contain *exactly* the stored data elements that a given processor requires, and the organization of the subcontainers may not be convenient for the purposes of a particular processor.

---

*A flat master file is a data store that does not employ database technology and typically has limited access capabilities.

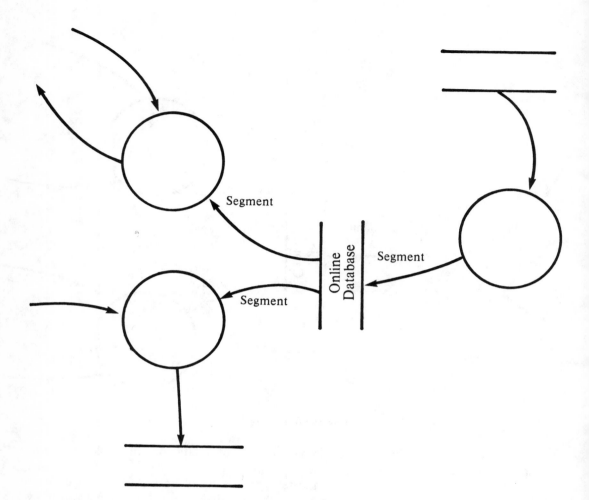

Figure 12.7. An online database shared among processors.

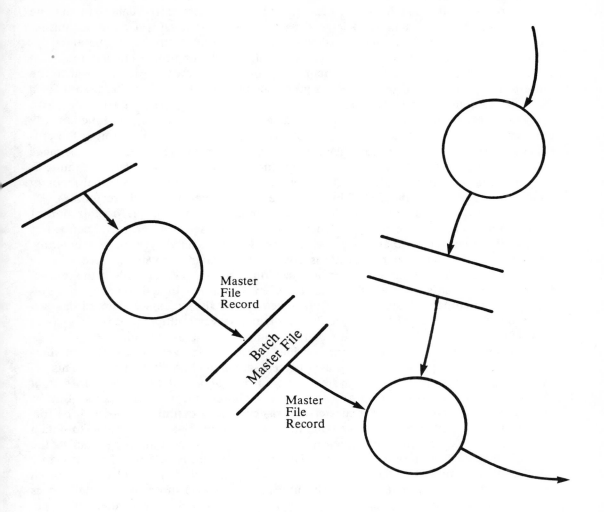

Figure 12.8. A batch master file shared among processors.

Since the shared data container must satisfy the essential memory requirements of all the processors that use it, it not surprisingly contains more stored data than any one processor needs. For example, a large department store maintains a customer master file that is used by the computer programs to carry out many of the store's essential activities. The customer master file is composed of records, one for each customer, that share a common format. They all contain the same data elements, which appear in the same position on each record. When a processor retrieves the master file record for a particular customer, it receives the entire record, including all the data associated with that customer, regardless of whether it needs all the data to perform its function. By providing a uniform interface to the physical data store, the system now forces the processors to handle more data than they require, all in the name of efficiency. This tactic makes discovering the essence of the system more difficult, because it is not immediately obvious what stored data *is* required by the processor. The true essential memory requirement has been camouflaged by the universal customer master file record.

The second characteristic of shared data containers concerns the organization of the containers and the means by which they can be accessed. Physical data containers are usually organized so that the processors can get the data they want conveniently. For example, the customer master file is organized to provide convenient access to the data stored about one customer. This organization is good, however, only if the processor in question wants to access the information by customer. In a single processor system, the data container can always be organized to suit the processor's stored data access requirements. But when a single data container serves multiple processors — each of which may desire a different access path — compromises are often made.

In the department store system, consider a processor that produces a report summarizing sales by type of credit card as a percentage of total credit card sales. This activity does not require all of the system's stored data about all customers. In fact, it does not require any stored data about customers who do not use credit cards. But, as the organization of the customer master file was chosen to optimize the overall performance of the multiple processor system, this particular processor is not able to obtain the stored data it requires conveniently. Instead, the processor must reorganize the system's essential memory to its needs. This reorganization is accomplished through a set of "extract and sort" processes, shown in Figure 12.9, that are familiar to all batch system designers and programmers. The first process selects those master file records that include credit card payments, the second process sorts these by credit card type, thereby grouping together all sales made with the same type of credit card, and the third process summarizes these detailed records to produce the desired report. None of these contortions would be necessary if the implementation technology provided a convenient way to access a desired portion of stored data directly.

You have seen two physical characteristics of shared data containers: standard data subcontainers that do not contain exactly the data elements required by the processes that use them, and file organizations that limit access capabilities. Both of these characteristics result from the use of imperfect data storage and retrieval technology to implement a data container that is shared by processors with different essential memory requirements. The most important consequence of such technological limitations is that they mask each processor's true requirements for stored data, making identification of the system's essence more difficult.

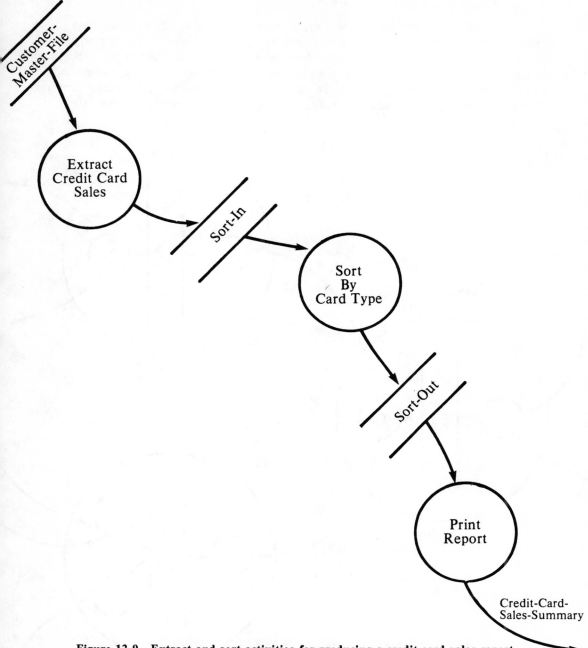

Figure 12.9. Extract and sort activities for producing a credit card sales report.

12.2.1.2 Private stored data containers

Figure 12.10 depicts another scheme by which processors share stored data. Each of the three processors has a private data container, inaccessible to the other processors, that provides the required stored data. An advantage of this scheme is that since each container serves just one processor, the content and organization can be chosen to conform to the needs of that processor. So each processor is less likely to receive unwanted data and less likely to have to reorganize an inconveniently structured data store.

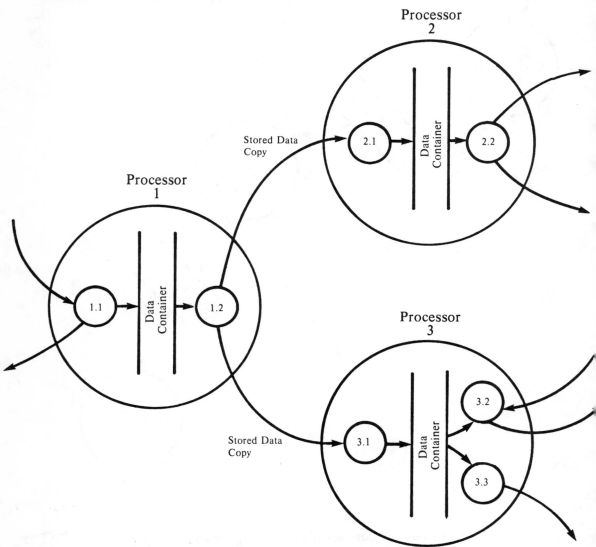

Figure 12.10. Private data containers.

A disadvantage is that establishing and maintaining two or more copies of the same stored data creates a different set of physical characteristics that, like those of shared data containers, camouflage the essence of the system. These physical characteristics result because the system must transport stored data elements from one processor's private data container to those owned by other processors in the system. In its simplest form, data transportation among private data containers adds three physical features to a system: transmission activities in the processor that originates the data, redundant custodial activities in the processor receiving the data, and dataflows between the two processors.

Figure 12.11 shows the transportation of stored data about newly hired employees from a personnel clerk's private data store to the payroll file controlled by a payroll clerk. The activity called Report New Employees does nothing more than transmit stored data across the boundary between one processor and another. (This boundary exists because of internal technological limitations, specifically the need to employ more than one processor to carry out the essential activity.) The activity Update Master File establishes and maintains the essential memory required by the payroll clerk; it receives data transmitted from other processors in the system, such as the personnel clerk, as well as data from outside the system. So, Update Master File is to some extent a duplicate of the custodial activity that establishes and maintains the same information in the personnel master data store. This type of "custodial clone" activity results from sharing the same stored data in private containers. The dataflow New-Hire-List is the physical data channel through which the two clerks communicate. If it were not for the internal technological imperfections, neither the internal transmission activity, nor the custodial clone, nor the data channel would be necessary. A single data store and a single custodial activity would suffice.

Sometimes, communicating stored data among processors involves more than simple data transportation. If the processors are sufficiently different from one another, they may use incompatible technologies to represent information. To pick an obvious example, we know that transmitting data from a human processor to a computer processor almost always requires *translation,* as well as transportation, since humans generally store data in different forms than do computers. Figure 12.12 shows a simple example of data translation. The personnel clerk communicates information to the automated personnel system via an online terminal. In addition to moving the data from the clerk's desk to the computer room, the interprocessor infrastructure also translates the data from a form understandable to the clerk into a form acceptable to the automated system. The translation activity itself is not visible on the diagram but takes place within the transmission process and the custodial clone activity, Build Update Transaction. Having to translate the data complicates the process descriptions and the dataflow definitions, thus making the underlying essence of this system more difficult to identify.

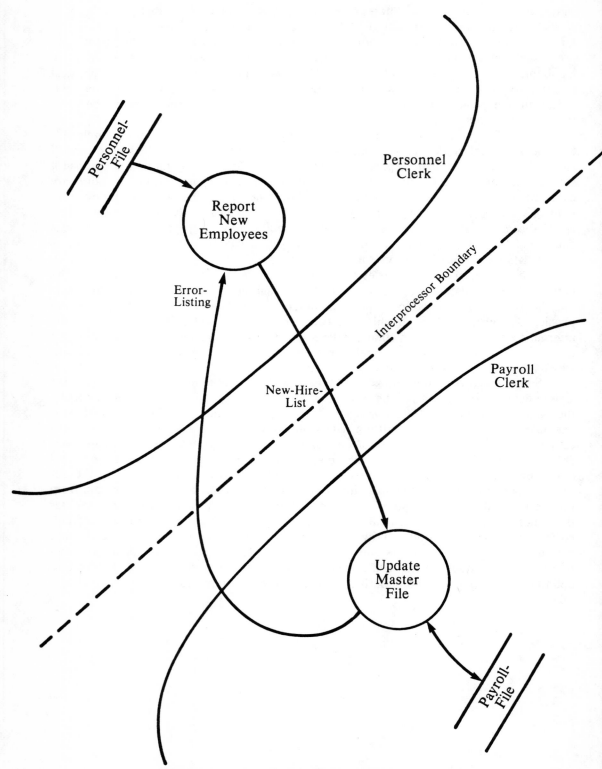

Figure 12.11. Transportation of stored data between two private data containers in a payroll system.

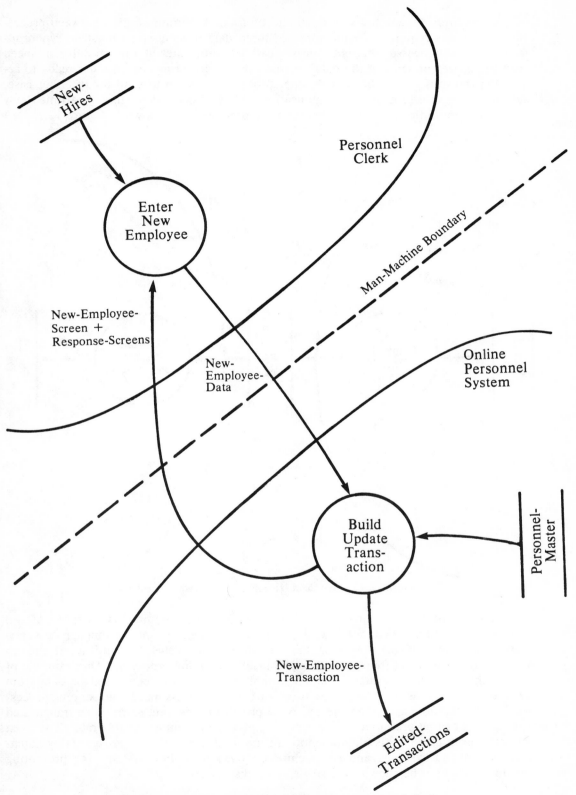

**Figure 12.12. Data translation between a personnel clerk
and an automated personnel system.**

In addition to data transportation and translation, communication between processors sometimes requires *synchronization*. Stored data transmissions must be synchronized when the processor that is to receive data communicates at a much different speed than the processor sending the data, or when the receiving processor is not ready to accept the transmission at the time another processor is ready to send it. In either case, synchronizing interprocessor communication adds a physical data store — a time delay — to the communication channel between the processors, as shown in Figure 12.13.

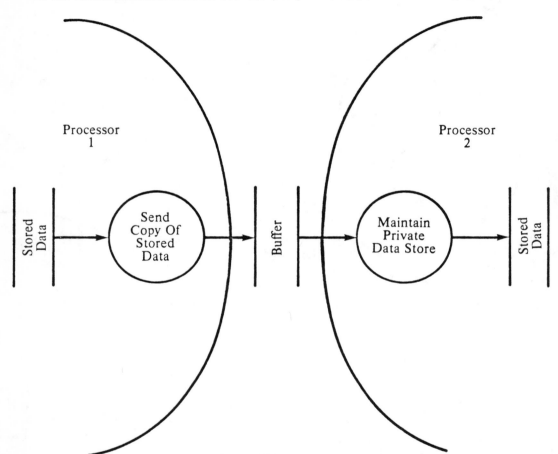

Figure 12.13. Synchronization between two processors.

Synchronization to adjust for the differences in the communication speed of two processors is often called *batching, buffering,* or *spooling.* If the sending processor is slower than the receiving processor, the data store accumulates data until it has enough to form an economical transmission to the faster, receiving processor. One example of this is the way online systems build a complete transaction, composed of data from several input screens, before passing it on for further processing. If the sending processor is considerably faster than the receiving processor, the data store holds transmitted data until it can be presented to the receiving processor at an acceptable rate. The most common examples of this phenomenon are the production of printed reports by batch-style automated systems, and the techniques used by online systems for presenting results that cannot be shown on a single video display screen.

Synchronizing interprocessor communication when one processor isn't ready to receive data requires nothing more than scheduling transmissions between processors. Data to be transmitted is saved in a data store by the sending processor until the receiving processor is ready to accept it. At that time, it is simply removed from the data store and sent on its way. Figure 12.14 depicts two examples of this type of synchronization. The transactions prepared by the payroll clerk for submission to the automated payroll system are held in the Daily-Transactions file until the overnight batch run. Output from the payroll system is held in the Daily-Reports store until the clerk returns to work the following morning. In each case, the data store serves as a kind of mailbox by holding a message until the addressee is available to receive it.

You have now seen that using private data containers to share stored data among processors adds a number of physical characteristics to the system: activities and data channels that transport and translate transmitted data, and data containers that synchronize interprocessor communication. Add to these the physical characteristics arising from sharing data containers among processors — unnecessary stored data and awkward data store organization — and you can understand why it is often so difficult to discern a system's essence.

12.2.2 Communicating intermediate products

The second major function of the interprocessor infrastructure is to send intermediate products from one processor to another. This is necessary when a processor carries out a fragment of an essential activity and needs to transmit the product of that fragment to another processor that performs another fragment of the same essential activity. In Figure 12.15, you see a piece of an incarnation set up to send intermediate products between processors that perform one essential activity, Satisfy Book Order. The intermediate products are the book orders and the book invoices. A book order is accepted as the result of a positive supply check; at this point, an invoice and a book shipment still need to be constructed for the book order. The next fragment produces book invoices after receiving an accepted book order and creating a receivable record, but the books still haven't been shipped and the book Inventory file needs to be updated. Both of these intermediate products, results of essential activity fragments, must be communicated along the chain of processors that are responding to the event "customer orders book."

The components that transport intermediate products are the very same ones that communicate updates to common stored data elements from one processor to another. The containers, files, and activities are the same; they are merely being used to carry out a different function.

The interprocessor infrastructure is itself partly responsible for the second major feature of multiple processor systems: the interprocessor administration.

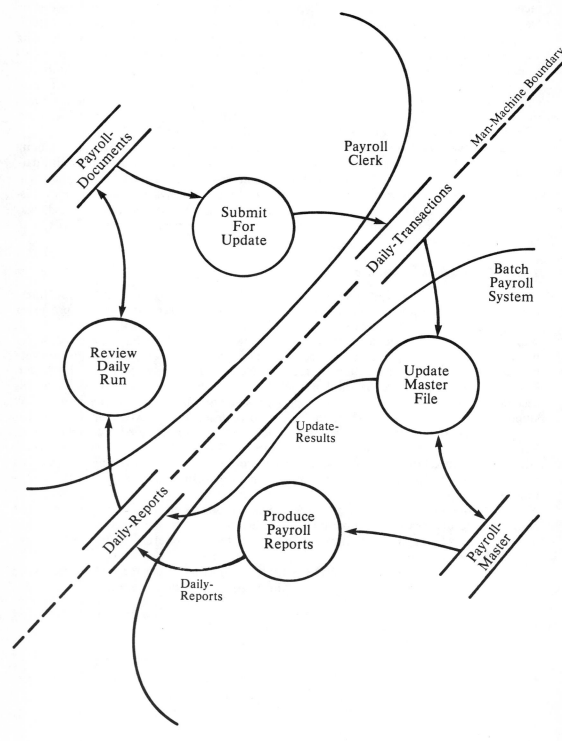

Figure 12.14. Synchronization between two processors when one isn't ready to receive data.

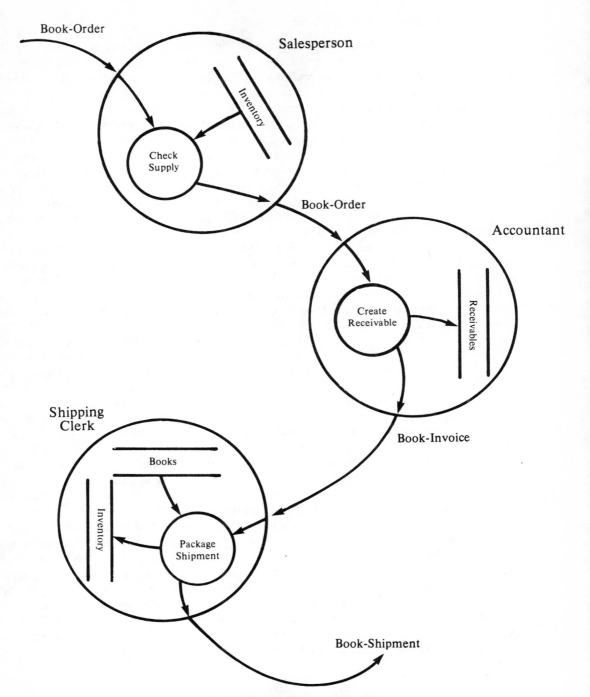

**Figure 12.15. Intermediate product communication
for the activity Satisfy Book Order.**

12.3 The interprocessor administration

The fallibility of both the environment that the system interacts with and the system's processors are by now well known to you. Now, we must deal with the equally fallible multiple processor incarnations. Although the intra-processor administration is already in place, each processor now needs additional administrative activities to detect and correct errors. Errors in multiple processor systems arise from two additional sources: the intra-processor administration and the interprocessor infrastructure. Think about it: If the intra-processor administration were perfect, if it always detected and corrected errors that an individual processor made, the system would not need extra checks for correctness. But imperfect technology puts the double whammy on the system, because a processor monitoring its own execution of an essential activity is no less likely to make a mistake than when executing the essential activity. Furthermore, components of the infrastructure can fail as well: Information can be destroyed or distorted while in transit from one processor to another.

To counter this exposure to new sources of error, a system turns processors against each other. Each processor becomes responsible not only for executing its own activities correctly, but also for detecting any problems produced by an "upstream" processor. All processors carry out additional editing activities that check for errors introduced not only by other processors, but also by the infrastructure. You saw examples of these interprocessor administrative activities in Figures 12.11 and 12.12. Both Update Master File and Build Update Transaction inspect the data they receive and report errors to the originating processor. In this way, they protect themselves against erroneous data produced by another processor or damaged by the infrastructure. The DFD in Figure 12.16 highlights the interprocessor administration. The large circles represent processors; the circles labeled EAF are essential activity fragments; and the small circles labeled E are the new administrative activities that edit material coming from other processors. The circles with an A represent the approval activities carried out by each processor to check its own work.

Whether a processor implements whole essential activities or fragments, or whether it is connected to another processor to receive the latest update of common stored data or to receive an intermediate product, every processor must always check another processor's work and verify that the infrastructure communicated it accurately.

Other parts of the interprocessor administration are related to the editing of information from other processors. The first of these is the dataflow that results when one processor notifies another processor of the acceptability of data that it has received. If the data is acceptable, a confirmation may be sent back. If the data contains an error, the rejected data along with an explanation of the problem may be sent back. The administration also may need additional stored data. A processor may keep filed copies of the information sent on to other processors so that it can trace the problem if any of the data is sent back. Keeping the copies is also a good idea because if the transmission never gets to the processor for which it was intended, a backup copy is available for retransmission.

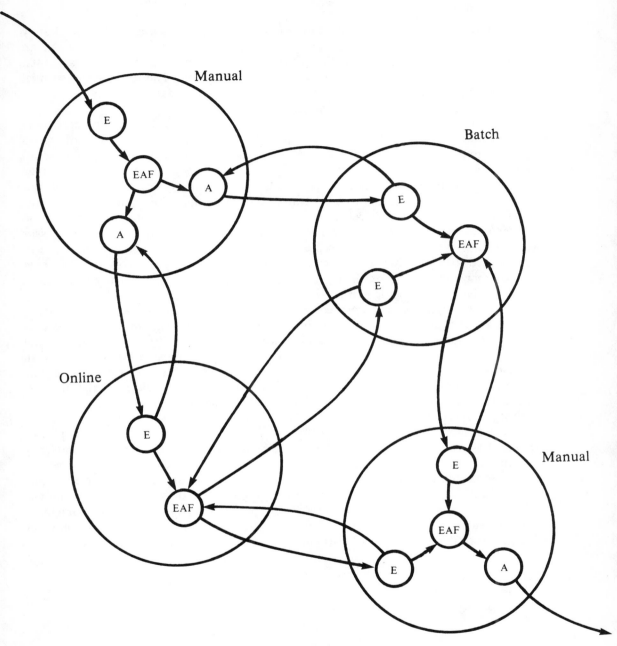

Figure 12.16. Interprocessor administration.

12.4 Summary

Real systems employ multiple processors to carry out essential activities. Rarely does a single processor have sufficient skills and workload capacity to carry out an essential activity alone. The essential activity is consequently divided into fragments, each of which is allocated to the most appropriate processor. The division of essential labor camouflages the system's essence simply because each essential activity loses its identity when it is fragmented into seemingly separate tasks.

Multiple processor systems include an infrastructure that allows the processors to share essential memory and to communicate intermediate products. Processors share essential memory either by using one common data container or by storing data redundantly in containers that are privately owned by each processor. Each of these methods of sharing data introduces physical characteristics into the system. Shared data containers may require processors to accept more data than they need and may force processors to use awkward data retrieval techniques. Private data containers create the need for new infrastructural activities to transmit shared data items from the processor that originates them to the other processors that need them. Activities that receive shared stored data and channels for that data further complicate the system. The same interprocessor communication facility is used to transmit the intermediate results of an essential activity that is being carried out by a team of processors.

The fallibility of both the processors and the infrastructure create the need for activities that monitor interprocessor communication. These administrative activities protect each processor from the mistakes of other processors and from the failure of the infrastructure to communicate information accurately. The most common form of an administrative activity is an edit that detects errors in data as it enters a processor. The interprocessor infrastructure and most administrative activities would not exist in a system that enjoyed perfect internal technology; they camouflage a system's essence by being entirely nonessential.

Real systems do not always exhibit all of the characteristics discussed in this chapter. For example, many automated systems lack adequate checks on the integrity of incoming data. The next system you study may not have a complete and correct interprocessor administration, or you may discover processors that combine infrastructural and administrative activities. But regardless of the type of systems you encounter, you will observe many of the patterns we describe. The effect of all these characteristics is to render the essence of the system virtually unrecognizable among the activities and data stores that make up its incarnation.

Chapter 13

The Consolidation
of System Activities

In the multiple processor systems discussed so far, a processor executes either one whole essential activity or a fragment, plus the requisite administrative and communication activities. In a real system, however, a processor often performs a set of multiple essential activities or essential activity fragments. We call this set a *conglomerate.* In this chapter, we examine how a conglomerate originates and its effect on the infrastructure, the administrative activities, and the essential memory. We also introduce the concept of specialized processors that carry out nonessential system activities, such as editing, auditing, and processor synchronization. Finally, we complete our model of a real system as a nested set of processors.

13.1 Consolidation of essential activity fragments

Most conglomerates are formed because one or more processors have excess processing capacity. What this means is that a processor is idle either because it has the skills and capacity to complete its assigned work more quickly than needed, or another processor is late in delivering critical input.

Excess capacity is expensive since the cost of a processor depends at least partially upon its capacity. To save money, most incarnations fire, furlough, sell, or lease their extra processors and divide the work among the processors that remain, thus forming a conglomerate.

For an illustration of how consolidation occurs, consider the tasks the accountant performs in our book selling system, shown in Figure 12.15. Given the daily volume of book orders, the accountant can create all the necessary receivable records in less than two hours. Obviously, she is not very busy. What would you do about it? You could get the accountant to accept part-time employment, or you could give her more tasks to help justify her salary. As commonly happens, you decide to give the accountant more responsibilities, taking them from other essential activities within the system. It is natural to allocate all the financial bookkeeping tasks in the activities Accept Book Payment, Receive Wholesaler Delivery, and Pay Book Wholesaler to this underutilized processor. The effect is to create a *packed processor,* one that carries out two or more essential activity fragments, typically from different essential activities. In our example, shown in Figure 13.1, the ground for consolidating essential fragments is that each fragment involves the same skill. Ignored is the fact that this skill is being applied to completely different essential activities.

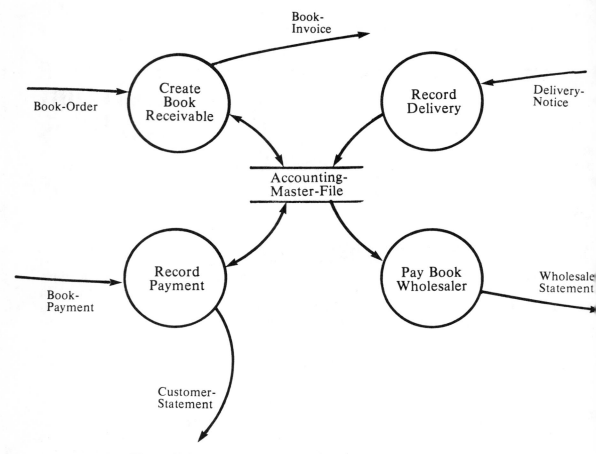

Figure 13.1. Activities of the packed processor, Accountant.

Packed processors are extremely common. They are a rational response to the limitations of a system's implementation technology. Because they conserve resources that might otherwise be wasted, they are one of the many ways to reduce the cost of the system.

Including packed processors in our anatomy of existing systems, we arrive at the updated view of a single processor shown in Figure 13.2. The figure is an idealized version and shows no communication between the fragments; each has its own private infrastructure and memory.

13.2 Consolidation of nonessential system activities

Now, moving from the idealized to the complex, we consider the effect of consolidation of nonessential system activities on the incarnation. Specifically, in the following subsections, we examine consolidation of the infrastructure, of data stores, of administrative activities, and of activities in a multiple processor system.

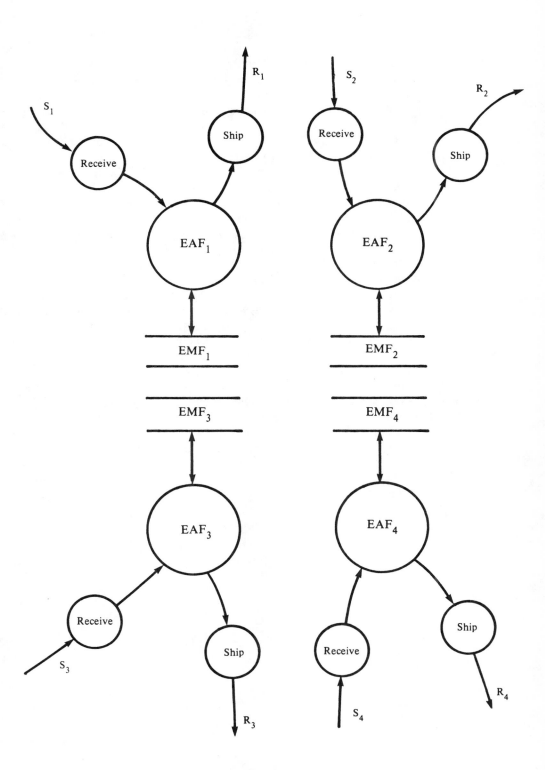

Figure 13.2. Anatomy of a packed processor.

13.2.1 Consolidation of the infrastructure

Looking at Figures 13.1 and 13.2, you might suppose that the separate inputs and outputs for each essential activity fragment reach the processor by different routes. Figure 13.2 goes even further by implying that a processor has a separate component to receive each input and to ship each output. In other words, according to these diagrams, every input and output travels on its own dedicated channel. This is rarely the case. It isn't likely, for example, that each processor communicating with the accountant in our example would have a private line. Supposing the accountant receives all messages by phone, she wouldn't have three telephones on her desk, one for each input dataflow, unless she lived in a world of free telephone service. The accountant probably has one phone that is used to carry four types of messages.

Systems frequently consolidate several logical dataflows into a single physical data channel to conserve scarce resources. We call such a multi-message dataflow a *packed channel.* Figure 13.3 shows the accountant — a packed processor — as part of a network of packed processors connected by the infrastructure. The dataflows, Accounting-Memo and Accounting-Statement, are packed channels, as shown by their data dictionary definitions accompanying the figure.

Packed channels lead to one of the most common infrastructural activities: A *transaction center* is a process that receives a message from a packed channel and routes it according to message type. This kind of "traffic cop" process is shown in Figure 13.4 as the bubble labeled "Receive And Route." The stimulus dataflow is a packed channel; it can carry four types of messages. Each message type comes from a different event, and hence goes to a different essential activity fragment. Receive And Route identifies each incoming message and passes it along to the appropriate essential activity fragment.

Transaction centers are everywhere. The most obvious example is a computer system that processes business transactions, such as an airline reservation system. On a smaller scale, your bank's automated teller machine uses a transaction center process to distinguish between deposits, withdrawals, transfers, and inquiries. For that matter, a human teller performs exactly the same transaction identification process using speech and paper as the technology, instead of a video display screen and keyboard. Regardless of the form, a transaction center almost always marks the connection between a packed channel and a packed processor.

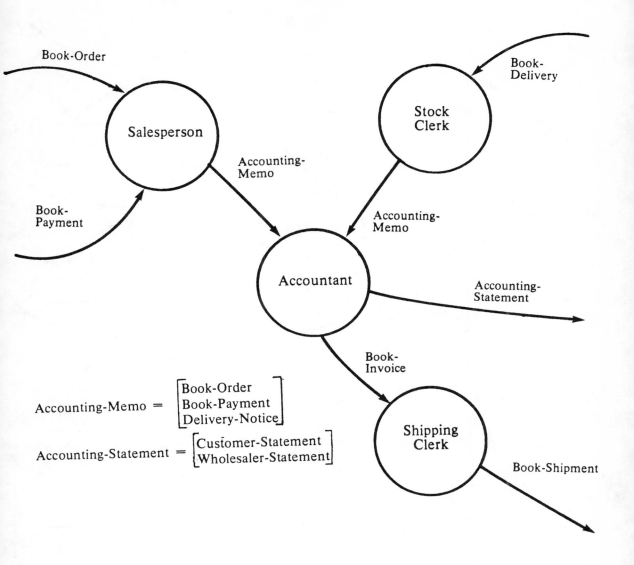

Figure 13.3. A set of packed processors in the book selling system.

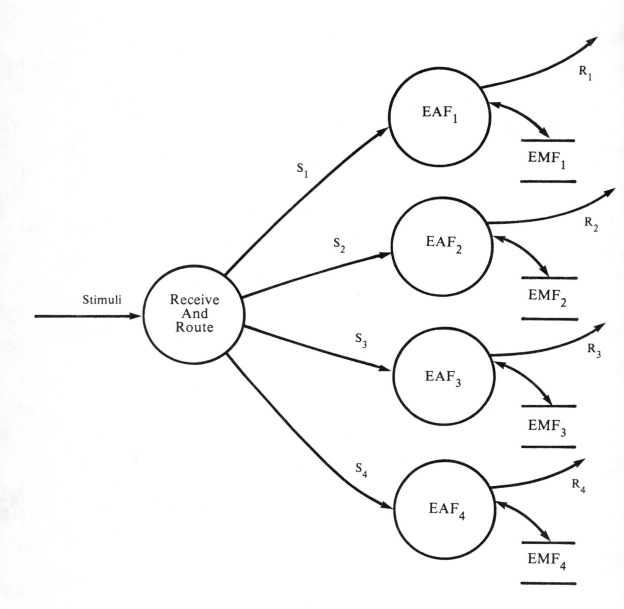

Figure 13.4. A transaction center.

The communication of results from the packed processor is organized in much the same way. Again, to avoid the expense of a dedicated channel to each processor, we consolidate all outputs into a single channel, as depicted in Figure 13.5. The Pack And Ship activity pulls all of the responses together, tags them with some means of identification so that a transaction center on the other side of the packed channel can distinguish them, and sends them off through the common channel.

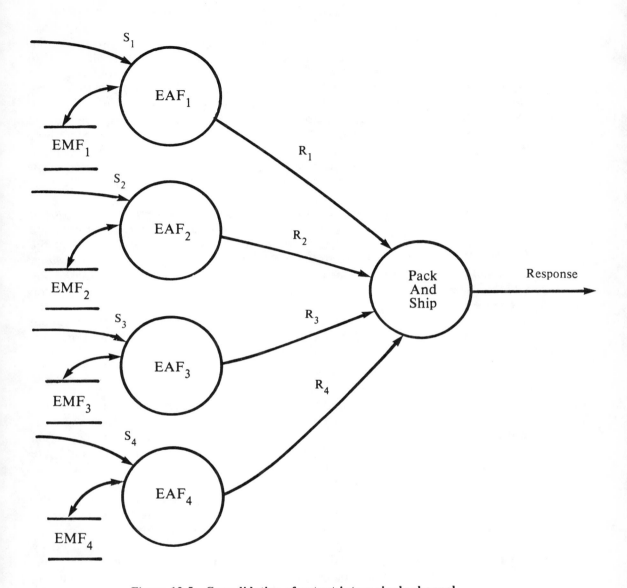

Figure 13.5. Consolidation of output into a single channel.

13.2.2 Consolidation of data stores

Where the essential activity fragments go, the essential memory they need must follow. For example, in order to handle the fragments shown in Figure 13.1, the accountant needs the memory accessed by those activities. Specifically, she must have stored data about the books sold to customers and about those ordered from wholesalers.

Because we haven't said much about the organization of an essential activity's memory requirements, you might think that each fragment's requirement becomes a separate data store, as in Figure 13.2. (In DeMarco's work, this is called a private component file [11].) By contrast, in Figure 13.1, all of the stored data is allocated to a single data store called Accounting-Master-File. Why did this happen? Although the accountant must carry out four separate tasks, and must provide these tasks with four potentially different packets of stored data, she did not set up four physically separate files. In this case, the accountant realized that the data required by the four fragments overlap to some extent. Some of the information stored by the Create Book Receivable fragment is the same as the data required by the Record Payment fragment. Similarly, both Record Delivery and Pay Book Wholesaler need to know what books were ordered from a given wholesaler. If separate files were set up with redundant data in each, they would be expensive to maintain, since updating a data element would mean having to access and change data in multiple data stores. To avoid the expense, the accountant consolidated the stored data for the four essential activity fragments into a single accounting master file.

Again, the reason for creating the fewest data stores stems from the limitations of the stored data technology of the system. Maintaining separate files is expensive, and the cost is reduced by creating a single data store that satisfies the needs of all the essential activity fragments allocated to the processor. We call this kind of file a *variety show* data store because it reminds us of the programming strategy used for *The Ed Sullivan Show*. The success of that long-running television favorite was based upon offering a mix of acts that satisfied most viewers, even though they had widely differing tastes. You use that same strategy when you create consolidated data stores, resulting in a file that satisfies the stored data requirements of every essential activity fragment, even though any one fragment may require only a small portion of the data in the file.

13.2.3 Consolidation of administrative activities

The consolidation of administrative activities within a packed processor is similar to the consolidation of the infrastructure. The individual edits that protect the fragments carried out by the processor are factored out and grouped with the transaction center process; so, in addition to routing the messages that come in, the edit and route activity can now protect all of the essential activity fragments from erroneous data.

In much the same way, the approval activities may be grouped with the activity that collects and tags responses before sending them to the single output channel. Before the message is loaded into the common output channel, it is checked again to make sure the essential activity fragment that produced it was executed correctly.

Finally, the data stores that contain backup copies of information produced by each fragment are likely to be merged into a variety show file. That single file can contain the backup information from many essential activity fragments. As before, putting everything into one file often saves money. Figure 13.6 shows what happens to the typical processor after the consolidation of nonessential incarnation components; the backup file has been left out for clarity's sake.

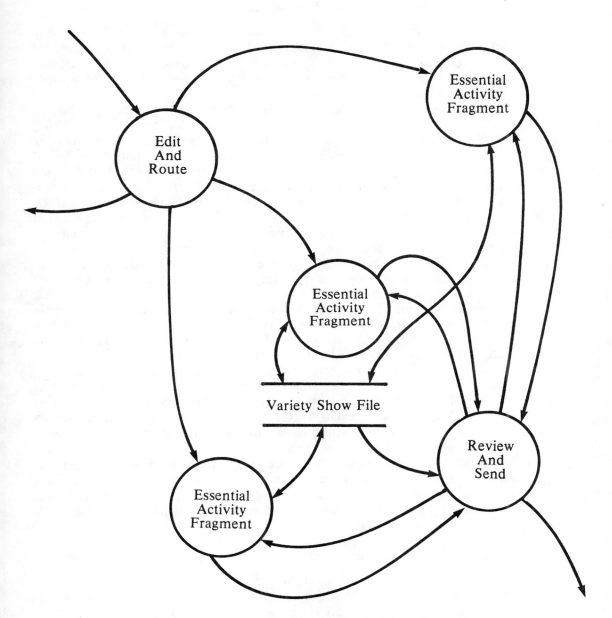

Figure 13.6. A processor with consolidated administrative activities.

13.2.4 Consolidation of activities in a multiple processor system

Some activities carried out by separate processors can be joined and assigned to one special processor. When multiple processors share a physical file, for example, the file access activities carried out by the processors are the same. If there are enough uses of the file, it is usually cost-effective to factor out these activities and consolidate them within a specialized processor.

The administrative activities in a multiple processor system are also consolidated. Managers in an incarnation often execute three kinds of administrative activities: They make sure that all processors are executing their activities correctly, they allocate resources in a cost-effective manner to the processors, and they coordinate the individual processors working together on some phase of an activity. All of these tasks could theoretically be carried out by the individual processors. However, because these processors are imperfect, they are probably not able to administer themselves. If a single processor is responsible for much of the quality assurance and resource management of the multiple incarnation, these activities are more likely to be carried out adequately.

Auditors are another good example of specialized administration processors that execute a conglomerate of activities. Many of the quality assurance tasks to be executed by each processor can be more cost-effectively performed if they are assigned to a processor specializing in those tasks.

It is especially obvious in the last two cases, but no less true over all, that these specialized processors often have particular skills that make the consolidation of activities much more cost-effective. Managers are skilled at planning resource allocation, and auditors are skilled at spotting subtle errors. The specialized processor can achieve economies of scale that an implementation without specialized processors cannot.

A typical multiple processor incarnation with both packed and specialized processors is shown in Figure 13.7. The essential activities in this system are carried out by the processors labeled Worker. Each Worker may perform several essential activity fragments, much like the accountant in Figure 13.1. The other processors perform infrastructural and administrative activities that would otherwise fall to the workers. Reception and Distribution provide a uniform interface to the outside world. Supervisor and Auditor check on the worker processors to detect and correct their inevitable mistakes. Supervisor also serves as a transaction center, routing incoming messages to an appropriate worker. These are examples of division of labor among processors specially skilled to perform nonessential tasks.

13.3 Nested processors

Figures 13.6, 13.7, and 13.8 illustrate why real systems can be viewed as nested sets of processors. By comparing the figures, you can contrast the activities of a multiple processor/multiple essential activity system with those of a single processor carrying out just one essential activity. The diagrams, while not identical, have the same basic shape. The systems shown perform many of the same activities: receiving and editing data from the outside world, carrying out planned responses with the aid of essential memory, and reviewing each response before sending it out. The key difference is that several of the activities carried out by the single processor (in Figure 13.8) have become the work of separate specialized processors in the multiple processor system (in Figure 13.7).

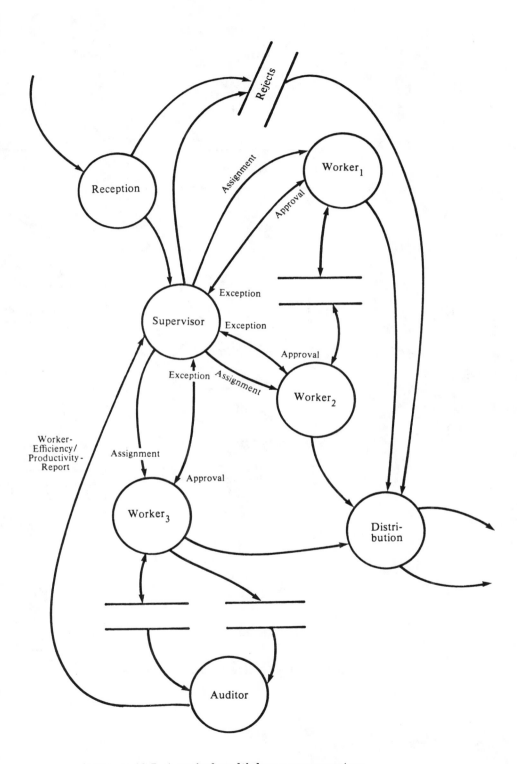

Figure 13.7. A typical multiple processor system.

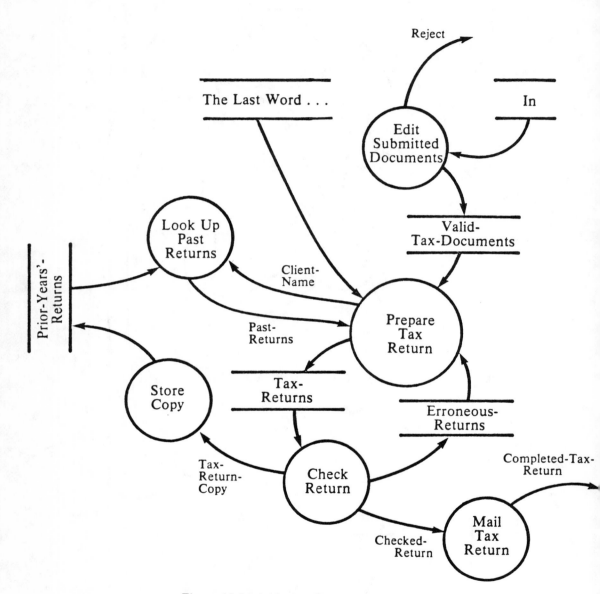

Figure 13.8. A single processor system.

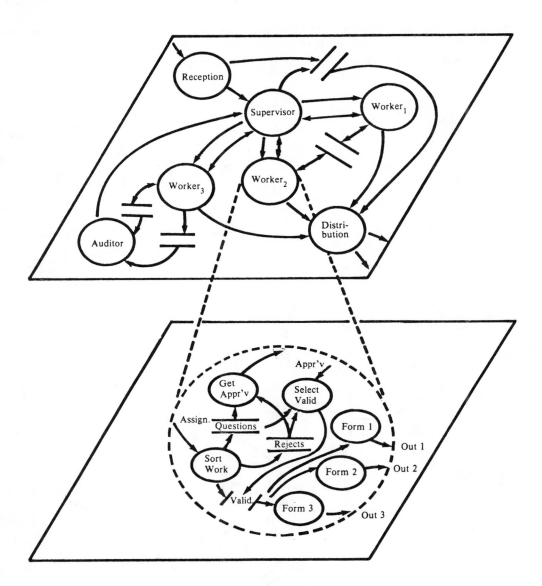

Figure 13.9. A system of levels of processors.

The typical processor is like the whole system of processors in microcosm. You can see this clearly in Figure 13.9, where one processor at the system level carries out a pattern of activities that is similar to the pattern of a system as a whole. So, an implemented system with all its packed processors, specialized processors, files, and channels is composed of processors, each of which is a complete incarnation in its own right. Just as the incarnation as a whole has an interprocessor infrastructure and administration, so an individual processor has an intra-processor infrastructure and administration. Just as processors share common variety show files, a processor will have private variety show files. Just as the incarnation as a whole performs a set of essential activities, so a processor carries out a set of essential activity fragments. Thus, a system is like a set of puzzle boxes such that one box (the incarnation as a whole) contains a set of other boxes (processors), each of which contains a set of boxes (activities carried out by processors). This concept is portrayed in Figure 13.10.

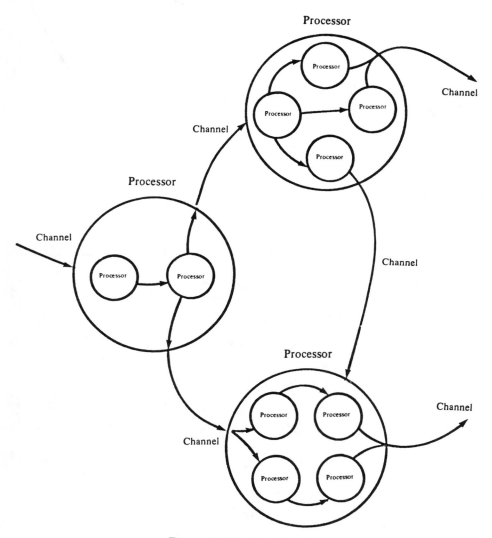

Figure 13.10. Nested processors.

This concept of nesting is not new. Everyone who understands the structure of atoms can imagine that within our planet are atoms that in turn are planets in a much smaller universe made up of atoms that are planets within an even smaller universe, and so on. The concept is even popular with Madison Avenue. Witness the Frosted Flakes box that shows Tony the Tiger holding a box of Frosted Flakes that also has a picture of Tony the Tiger holding a box of Frosted Flakes . . . and so on. However obvious this concept may seem, it is no less important, because the anatomical patterns of a single processor are the same as those exhibited on a larger scale by a group of processors working together as a superprocessor, as discussed below.

13.4 Superprocessor incarnations

Systems can be very large, sometimes consisting of hundreds, even thousands, of processors. All these processors are necessary to handle the workload, the number of essential activities, and the complexity of the essential activities within a large system. In such an incarnation, a single interprocessor infrastructure and especially a single administration may not be able to keep up with the load. It is close to impossible with today's technology for either the individual processors or a specialized administrator to monitor the performance of a huge system of people and machines.

The most popular solution to this problem is to organize the processors into manageable groups of several processors that together act as one processor. We call such a set of processors a *superprocessor*. It usually maintains a common interface to the external infrastructure and operates under the control of one set of administrative activities. Superprocessors are organized just like other multiple processor systems. Forming superprocessors allows you to extend the capacity of a multiprocessor incarnation without losing control over the administration of the processors.

Figure 13.11 shows how superprocessors interact with one another in a large system, in this case a book selling company. Each bubble represents a superprocessor; it is a component of the company, often called a department, that itself is made up of processing components — often, people and machines. Each superprocessor, though it contains many processors, has all the behavioral characteristics of a single processor. That is, it carries out essential activity fragments and has an infrastructure and administration.

Activities are allocated among superprocessors in much the same way as activities are allocated among processors in a multiple processor incarnation. Essential activities are fragmented and allocated according to the comparative skill, capacity, and cost of the superprocessors, leading to superprocessor specialists and packed superprocessors.

The fragmentation of essential activities across superprocessor boundaries leads to yet another level of infrastructure: an infrastructure that carries information across the boundaries between processors that belong to different superprocessors. However, in form, it has the same characteristics as the interprocessor infrastructure in a simple multiple processor system.

Because every superprocessor makes mistakes, consumes scarce resources, and can't supervise itself properly, yet another level of administration is needed. The inter-superprocessor administration is responsible for ensuring the quality of response and for allocating resources among two or more superprocessors. This higher level of administration tends to be assigned to a person or group independent from each superprocessor. This independent unit and the superprocessors under its control can in turn be considered one unit, an even more super superprocessor. If there are enough superprocessors, they may be grouped into several super-superprocessors. So, once there is more work than the superprocessor incarnations can carry out, you can organize an even larger superprocessor system. Regardless of the scope of a system, however, you will always observe the same pattern of activity as discussed in the past three chapters and as illustrated in Figure 13.9. On a very small scale, all of the activities may be performed by a single processor. More typically, the labor is divided among specialized processors. In a very large system, entire groups of specialized processors may be employed.

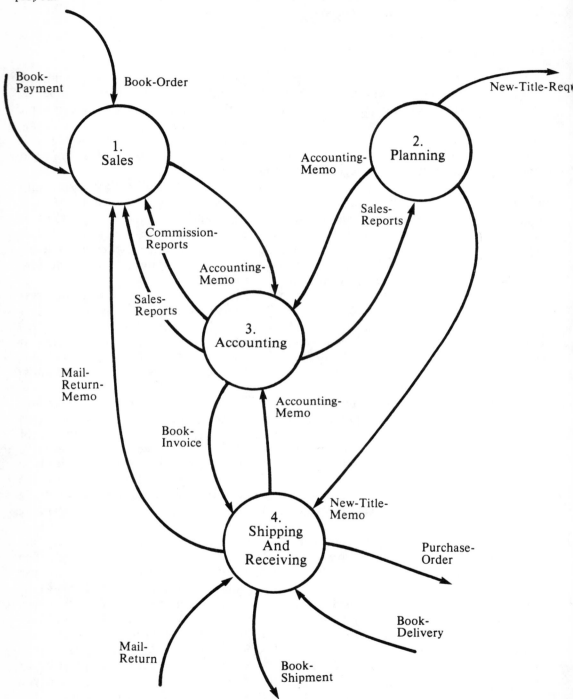

Figure 13.11. Superprocessors in a book selling company.

13.5 Summary

As this chapter illustrates, incarnations can be understood in terms of nested processors. Each processor has its own intra-processor infrastructure and administration, and all processors share an interprocessor infrastructure and administration. This pattern is also present in superprocessors and groups of superprocessors.

You should also understand that essence is camouflaged by the effects of imperfect technology. Essential activities are fragmented among processors according to processor skill, cost, and capacity to do work. The essential memory required by the processors is allocated to accessible data containers. The interprocessor infrastructure enables processors to share essential memory and to communicate intermediate results. Administrative activities check for errors in data arriving from the outside world and for errors committed by the processors or the infrastructure. Fragments from different essential activities are consolidated into one processor, and dataflows are grouped into packed data channels. Finally, nonessential activities and data are consolidated into specialized infrastructural and administrative processors. All together, these phenomena camouflage the essence of a system in two ways: by introducing nonessential components that can confuse the untrained observer and by contorting the essential activities and memory in a way that is also confusing.

Consider one last analogy to illustrate the effect of imperfect technology on a system's essence: The object of a puzzle called Rubik's Cube® is to scramble the small colored squares around the face of the cube and then rearrange them so that each face of the cube has squares of a single color. Suppose the essence of a system is a Rubik's Cube. If for each level of processor in the incarnation the cube is rearranged, the resulting cube, which now represents the incarnation, is an almost unsolvable mess. It is just as difficult to find the essence in a real-world incarnation as it is to unscramble the cube.

Chapter 14

Discovering the
Essential System
Through System Archaeology

In this chapter, we give an overview of our strategy for deriving the essence of an existing system from a model of its current incarnation. To derive the essence, you first must have a physical model of one or more existing systems. This model is composed of data flow diagrams, data dictionary definitions, and minispecifications. It does not matter how physical the model is. The procedure will work whether the model includes a lot of technological detail, or whether the model contains only the fragments of essential activities and selected aspects of the infrastructure and administration.

The amount of physical information in the model of the existing system will determine in what detail you perform each of the activities we describe in the following chapters. For example, if your current physical model is partitioned to show the existing boundaries between the high-level processors, such as departments and computer systems, you will have to perform a significant portion of the expansion activities at the start. But if you have partitioned your current physical model using some other, less physical method, such as by drawing boundaries around parts that respond to certain events and groups of events, you will find that you have already accomplished a good deal of the expansion phase.

This procedure is *not* foolproof; you cannot perform it as if it were a magical spell, without thinking. Use the steps only as guidelines, because your next systems analysis effort always includes factors that affect how you apply the steps.

Specifically, you must assess the managerial, technical, and political traits of your project and derive your own tactical procedure based upon how these traits affect our strategy. Applying this procedure blindly does not produce the best possible results, and in fact, if you ignore critical political factors, you may not finish the project at all. In Part Eight, we recommend how to adapt this technical procedure into a managerial project plan. Because of the dangers of ignoring political and managerial factors, if this procedure were a consumer product, we would want every copy to display this warning label: *This product should be used only with the brain in the ON position.*

You also must carry out the procedure with care because it seems complicated, especially at first. To help you remember all the steps, we compare our strategy to the process used by an archaeologist who unearths a set of artifacts in order to reconstruct an ancient civilization.

14.1 The discovery strategy

Our procedure for discovering the essence of a system reminds us of how archaeologists discover facts about older civilizations. In each case, knowledge of either the civilization or the system's essence has been lost or is unavailable. Both archaeologists and analysts must recover this knowledge without adding distortion through incorrect deduction or personal bias. To find artifacts in an apparently promising area, archaeologists must remove the earth and debris that hide them. In their search, they usually find artifacts in fragments, which must be cleaned and classified. All fragments seeming to be part of a shattered plate are put in one pile, while fragments of a vase go into another. At this point, an archaeologist may try to reconstruct the object by piecing the fragments together. With glue and clamps, he rebuilds the plate and he then moves on to the vase.

After all of the objects have been rebuilt from the classified sets of fragments, the archaeologists may finish the job by organizing the objects into a scene typical of how they were used originally.

You follow a similar process when you set out to discover the essence of an existing system. You can unearth the essence of a system by following these three steps:

1. Find the fragments of essential activities. As in archaeology, the essential fragments are typically buried deep within the incarnation. Remove the physical characteristics of the system and classify the fragments that you find.

2. Reconstruct and model each essential activity.

3. Integrate the individual essential activity models into a single essential model.

Figure 14.1 contains a data flow diagram of these activities and their products. To see where these activities fit into the systems analysis process, look back at Figure 10.5. The three activities above compose the activity Derive Essence in that diagram.

Each of these steps is discussed in detail in the following chapters. In Part Four (Chapters 15 to 17), we discuss the process of finding and classifying the essential activity fragments of an existing system. Part Five (Chapters 18 to 21) covers methods for reconstructing and modeling each essential activity. The discovery process is completed in Part Six (Chapters 22 to 24), where we discuss how to integrate the individual essential activity models to produce a global model of the system's essence.

The next three sections briefly explain each of the three steps. If you have a clear overview of the procedure while reading the next few chapters, its details will make more sense to you.

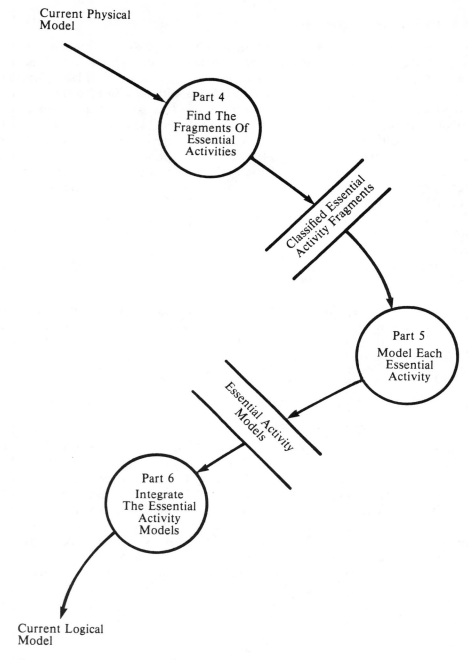

Current Physical
Model

Part 4

Find The
Fragments Of
Essential
Activities

Classified Essential
Activity Fragments

Part 5

Model Each
Essential
Activity

Essential Activity
Models

Part 6

Integrate
The Essential
Activity
Models

Current Logical
Model

Figure 14.1. The process of deriving the existing system's essence.

14.2 Finding and classifying the essential activity fragments

Your first task is to discover which parts of the system are the essential fragments. In an incarnation, these fragments are obscured by the system's infrastructure and administration. Once they are isolated from the infrastructure and administrative activities, you identify the fragments that belong together so that you can later reassemble them into whole essential activities.

To that end, you perform the following steps:

1. Expand the current physical model by removing the boundaries between processors and superprocessors. This exposes the essential fragments, infrastructure, and administration.

2. Remove the infrastructure and administration activities to reveal the essential fragments.

3. Classify these essential fragments according to what event they respond to.

14.3 Modeling an essential activity

Once you have a classified set of essential fragments, remove any remaining physical characteristics and build a model of the essential activity. You can control the complexity of the task by working on one set of classified essential fragments at a time. You perform these steps:

1. Remove the remaining physical characteristics belonging to each classified set of fragments.

2. Synthesize a model of each essential activity model from what remains.

14.4 Integrating the essential activity models

To produce a complete model of a system's essence, you join the local essential activity models that were developed in the previous step. Although each essential activity model is logical by itself, you must eliminate any physical characteristics that linger in the connections between activities. Most of these characteristics are in the system's essential memory, since activities are connected through data stores. To finish the process, review the global model to insure that it is understandable and free from physical constraint.

To these ends, you perform the following steps:

1. Build an integrated model of the system's essential memory.

2. Eliminate unused data and data that the system is not required to remember.

3. Assess the quality of the model and improve it as needed.

Having completed our overview of the process of discovering the essence of an existing system, we now plunge into the details of the process.

14.5 Summary

In this chapter, we presented an overview of our strategy for deriving the essence of an existing system from that system's present incarnation. To begin the procedure for deriving a model of essence, you need a current physical model composed of data flow diagrams, data dictionary definitions, and minispecifications. With this model, you carry out the following three steps: Find fragments of essential activities and group those that are part of the same activity; reconstruct and model each essential activity; and combine the essential activities. The result is a single essential model of the existing system.

Part Four

Finding the
Essential Activities

The three chapters in Part Four describe the critical first step in the derivation of an essential model from the current physical model: the identification of the fragmented essential activities and memory. By removing the nonessential components of the existing system, this step begins to reveal the essential shape of the system.

Chapter 15 shows you how to expand the current physical model by clearing away the physical upper levels of processes, dataflows, and data stores. The model that results enables you to distinguish components of the system's essence from its physical characteristics.

Chapter 16 describes the removal of all but the essential activities, dataflows, and data stores. The interprocessor infrastructure and intra-processor administration are removed, and the remaining fragments are reconnected.

Chapter 17 provides a method for organizing the many fragments of essential activities and essential memory into a leveled set of data flow diagrams. Grouping the activity fragments according to the event to which they respond yields a model composed of pieces that can be independently studied. While the essential fragments retain some physical characteristics, the upper-level diagrams produced at the end of this phase offer a preliminary view of how the completed current logical model will be organized.

Chapter 15

Expanding the Current Physical Model

In this chapter, we discuss the process of expanding a current physical model to create an interim product called an *expanded physical model*. We call the model *expanded* because we remove the upper-level DFDs and data dictionary definitions, thus dramatically increasing the amount of detail in the current incarnation model. When you completely expand the current physical model, your new model will display fragments of essential activities and essential memory, along with the detailed activities, dataflows, and memory that make up the infrastructure and administration of the system. Because both the essential and the physical details are now visible, it becomes easy to distinguish the technology-dependent aspects of the system and to eliminate them from the model.

There are two reasons for removing the upper levels of the current physical model. First, the upper levels often are physical, with each grouping in an upper level corresponding to a component of the incarnation (such as a processor, processor group, physical file, physical channel, or physical data packet). As a result, the lower-level groupings of activities and data reflect the consolidation that exists in the current incarnation. Second, even if the upper levels aren't too physical, they typically don't use the budget for complexity effectively. Analysts who don't use physical components to partition the upper levels generally choose to partition the model according to their preliminary notions of system functions. Carving the model in this fashion usually produces pieces that are either too complicated or too simple, and the subsequent unbalanced distribution of activities and data on the many separate diagrams of each lower level makes the model harder to understand.

Even after you have removed the upper levels of the DFD, some parts of the model will need further expansion. To find them, you examine the whole model, looking for clues that tell you which activities, dataflows, and data stores need to be divided and their pieces modeled separately. This process exposes still more details and separates still more essential fragments from physical ones.

In summary, the benefit of expansion is that it untangles the essential activity fragments, essential memory fragments, and essential dataflows from groupings that are either too physical or too complex. So, your objective when expanding a model is to liberate as many essential fragments as possible from the nonessential groupings. To accomplish this, you examine each modeling tool in turn, starting with the data flow diagrams, followed by the data dictionary and minispecs. Then, you need to consider the effects of one tool or another, as we suggest at the end of the chapter.

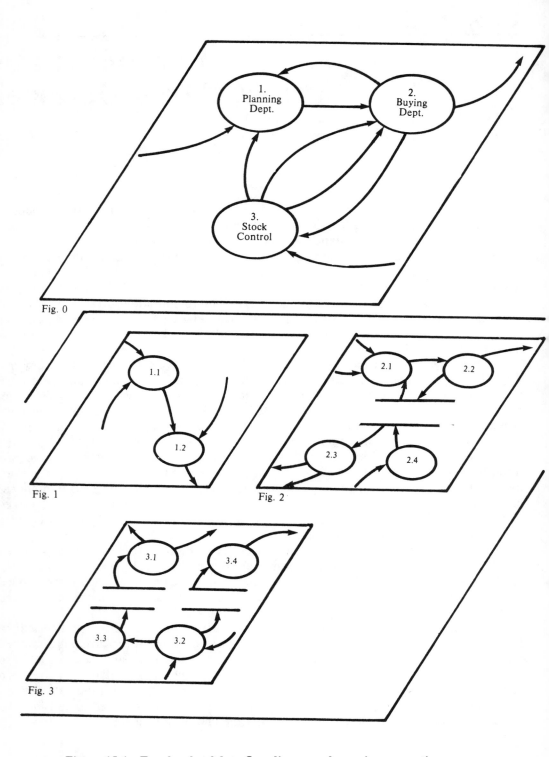

Figure 15.1. Two levels of data flow diagrams drawn in perspective.

15.1 Expanding the data flow diagrams

The expansion of the data flow diagrams as an essential modeling activity was introduced by DeMarco in his work on structured analysis [11]. You begin with a leveled set of data flow diagrams, such as the simple set shown in Figure 15.1. You remove all upper-level DFDs, leaving only the lowest-level DFDs in the set, shown as Figures 15.2a, b, and c.

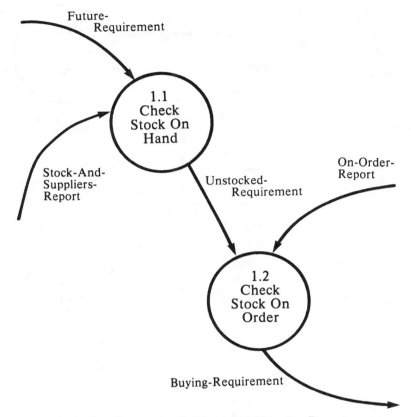

Figure 15.2a. Lowest-level diagram of Planning Department.

The last step is to combine the separate lowest-level data flow diagrams into one diagram. To connect these DFDs, you find the dataflows and data stores that appear on two or more diagrams, then join the diagrams at the points where the identical dataflow or data store meets. Notice, for example, that the same dataflow that leaves bubble 1.2, Buying-Requirement, goes into bubble 2.1. Therefore, these two dataflows are connected, joining the two diagrams. Figure 15.3 shows the result of carrying out this procedure on the three diagrams in Figures 15.2a, b, and c.

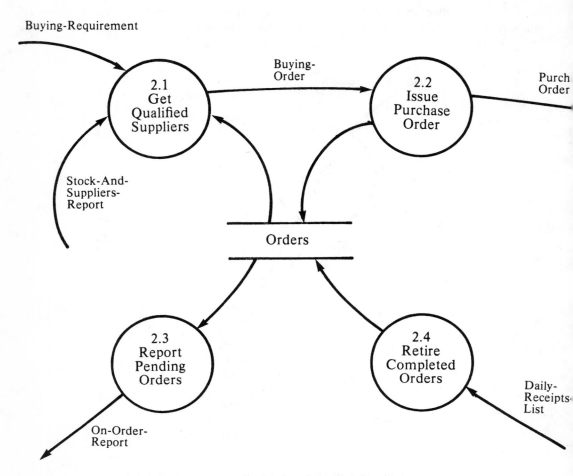

Figure 15.2b. Lowest-level diagram of Buying Department.

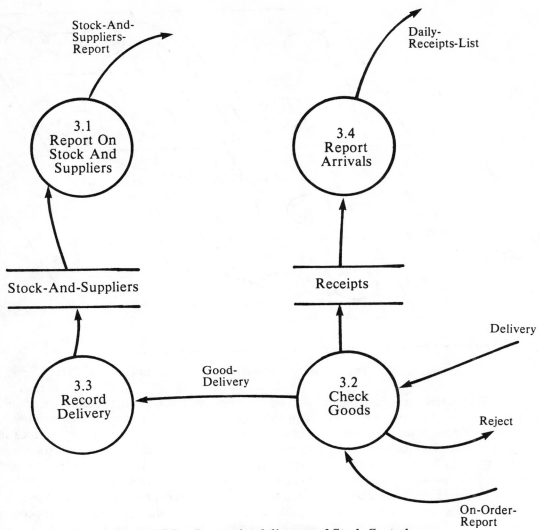

Figure 15.2c. Lowest-level diagram of Stock Control.

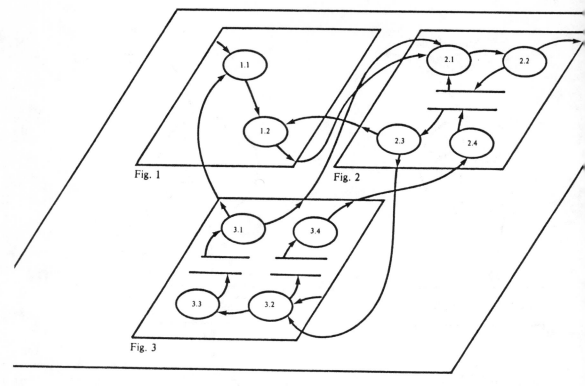

Figure 15.3. Three lowest-level diagrams connected to each other.

This simple manipulation of the DFDs removes most of the boundaries around essential features that are based upon processors, groups of processors, or groups of groups. These boundaries formed conglomerates of unrelated fragments of essence in many cases. Expanding the DFDs also removes any ineffective functional partitions that may have been used in the current physical model. Finally, it separates the dataflows in packed channels, which carry many different essential dataflows.

However, as beneficial as the expanded DFD is, you probably will not have expanded the current physical model as far as you should. Many lowest-level processes may still represent clusters of essential fragments that should be expanded further. Dataflows to these clusters may still be packed channels, and some data stores may contain clusters of fragments from different objects. The reason that you missed these clusters is that the DFD you start with is not detailed enough to expose all essential fragments.

To find these insufficiently partitioned portions of the current physical model, you search the expanded DFD for clues to indicate that an activity, dataflow, or data store consists of unrelated essential fragments. The clues are

- a physical name or an overly general name on an activity, dataflow, or data store

- a data store that connects a large number of activities, or several arrows that have the same name and so indicate that one dataflow connects many activities

- an activity that has many more dataflows going into or out of it than do the other activities, or one that accesses many more data stores than do the others

Finding any of these features tells you to consider further partitioning the activity, dataflow, or data store. To do this, you must consult the data dictionary or the mini-specs. The following sections tell you how; they also tell you how to find clues in both the data dictionary and the minispec that indicate that more partitioning is needed.

15.2 Expanding the data dictionary

At this point, your data dictionary is far from technologically neutral; in many projects, the data dictionary during early analysis is nothing more than a collection of the input forms, record layouts, and reports depicted by printer and CRT layout sheets. Keeping this documentation physical is beneficial at this point, for it prevents you from wasting time constructing logical data dictionary entries that you will only discard later. Yet, these same physical packets of information contain hidden parts of the system's essence. By expanding the contents of these packets, you often find significant pieces of essence.

Especially well camouflaged in the data dictionary are the external stimuli to the system, the system's responses to a given event, and the object data stores that form essential memory. In the cases of stimuli and responses, what hides essential features is the consolidation of many individual packets of data into a bundle. The essential object data stores are camouflaged because, like the essential activities, they have been fragmented and then bundled into unrecognizable packets.

In addition to these bundles, there are extraneous pieces of data mandated by the limited capacity of the processors, infrastructure, and administration. Batch data stores exist throughout the system to store intermediate products until another processor is ready to process them. Other data stores hold onto information until a channel is available to carry a transmission. Still others hold information for approval by the administration. Bundled in with dataflow packets are audit trail and approval information from the administration and routing information from the infrastructure.

You have the same objectives for expanding the data dictionary as for expanding the data flow diagram. You want to find as many pieces of essence within the physical flows and files as you can and at the same time expose elements of the infrastructure and administration. To achieve these objectives, you examine each dataflow and data store in the data dictionary — pay special attention to those that look suspicious on the DFD — and employ two tactics on them: unbundling dataflows and removing repeating groups.

Unbundling dataflows is usually easy. By inspecting either an input or output data channel, you can see that several kinds of packets flow through the channel. In the case of stimuli, the presence of a transaction code typically indicates that more than one kind of packet uses a given channel. For example, in Figure 15.4, Mortgage-Transaction is defined by a transaction code and one of three payments, as indicated by the brackets or the "or" construct. For the system's response, the different report names making up Mortgage-Reports indicate that the channel is shared. If you are working from sample forms or record layouts, rather than from data dictionary definitions, look for alternate formats or alternate ways to fill out the form. Each alternative may turn out to be one type of message that is carried by this container.

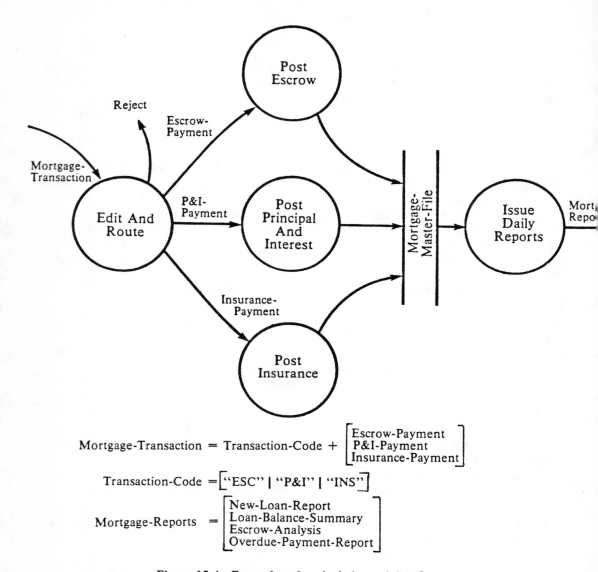

$$\text{Mortgage-Transaction} = \text{Transaction-Code} + \begin{bmatrix} \text{Escrow-Payment} \\ \text{P\&I-Payment} \\ \text{Insurance-Payment} \end{bmatrix}$$

$$\text{Transaction-Code} = \begin{bmatrix} \text{``ESC'' | ``P\&I'' | ``INS''} \end{bmatrix}$$

$$\text{Mortgage-Reports} = \begin{bmatrix} \text{New-Loan-Report} \\ \text{Loan-Balance-Summary} \\ \text{Escrow-Analysis} \\ \text{Overdue-Payment-Report} \end{bmatrix}$$

Figure 15.4. Examples of packed channel dataflows.

Figure 15.5 shows the result of unbundling a dataflow. As you unbundle dataflows, revise the data flow diagrams and data dictionary definitions. But don't make your changes either formal or especially neat. You may very well find yourself *deleting* some of these processes and flows in the next step of our procedure, so you don't want to spend too much time documenting them now. Just draw the unbundled flows right on top of the existing DFD. As for the data dictionary, you can often delete the dataflow definition of the packed channel — in this case, Mortgage-Transaction and Mortgage-Reports.

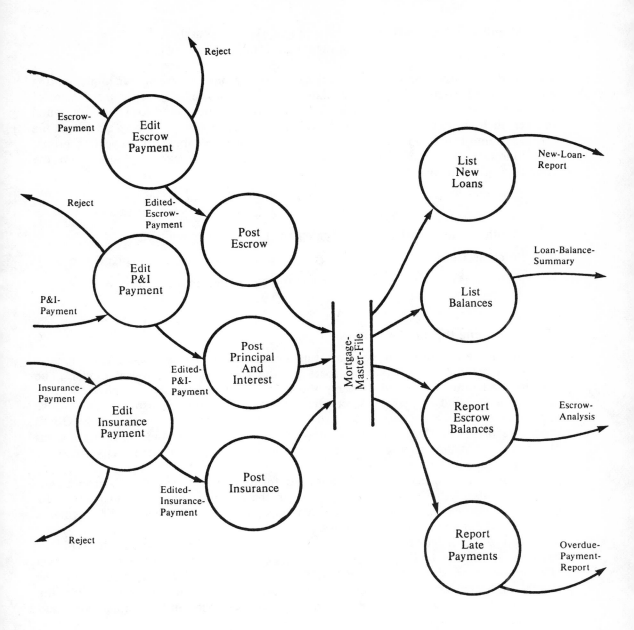

Figure 15.5. Examples of expanded dataflows.

In order to isolate fragments of essential memory, you simply remove repeating groups. Examining a file layout in an existing system may reveal that cohesive groups of data occur along with or within another cohesive group, as in the following data dictionary definitions:

Customer-Master = {Account-No. + Customer-Name + Customer-Address + {Monthly-Statement-Data} + {Payment-Data}}

Monthly-Statement-Data   =   {Statement-Date + {Purchase-Detail} + Statement-Total}

This nesting of repeating groups, as it is called, is one physical means for implementing a relationship between two groups of data. In many cases, each group is a fragment of an object data store that you want to build into your essential model. Therefore, when you see such internal repeating groups, you will remove them and create a separate file for each removed group and for the original file that contained the repeating group. The following definitions show the result of removing repeating groups from the definitions given above:

Customer-Master   =   {Account-No. + Customer-Name + Customer-Address}

Monthly-Statement-Data   =   {Statement-Date + Statement-Total}

Purchase-Detail   =   {Item-No. + Item-Description + Quantity + . . .}

Payment-Data   =   {Payment-Date + Payment-Amount + Payment-Type + . . .}

15.3 Expanding the minispecifications

Just as we do not recommend creating a complete current physical data dictionary, we do not recommend writing minispecs for the entire current physical model. The reason is the same: You will probably throw away most of the current physical mini-specs once you find the essential activities. Instead of writing minispecs, you can use some of the existing documentation, such as policy manuals, run books, and COBOL procedure divisions. Only consider writing minispecs when you want to clarify a complex activity that the existing documentation cannot provide. For many activities, your memory is sufficient at this point.

Your minispecs may take just about any form, provided they allow you to examine the details of the activities that appear on the expanded data flow diagram. The purpose of this examination is to find activities that are conglomerates of fragments from several different essential activities.

You discover conglomerates not only by looking for clues on the DFD as described in Section 15.1, but also by classifying these activities according to their level of cohesion. Cohesion is a concept borrowed from structured design; it refers to how closely connected are the different functions carried out by the activities within the conglomerate [33]. You will tend to find four levels of cohesion between activity fragments in the conglomerates:

- *sequential cohesion:* The output from one activity becomes the input to the next.

- *communicational cohesion:* The activities use the same data, but perform completely different procedures on the data.

- *logical cohesion:* All the component activities perform a similar type of action on completely unrelated pieces of data.

- *coincidental cohesion:* Neither similarity of procedure nor any common data ties the activities together.

Certain levels of cohesion imply conglomeration, especially logical and coincidental cohesion. Communicational and sequential cohesion also indicate conglomeration, but of more strongly related activities. The different levels of cohesion in conglomerates reflect compromises made when the incarnation of the existing system was selected. This is especially true for every level but sequential. The three levels often result from attempts to optimize either the skill or the processing time of the processor or to isolate data stores for security reasons.

You needn't expand activities that exhibit sequential cohesion at this point, since they aren't a major source of technological bias. You would do better to focus your attention on expanding conglomerates that exhibit the other forms of cohesion. If the conglomerates are left intact, you imply that you want them to be implemented that way. That would be a problem, since the specific cause for merging the activities is now technologically obsolete. Even if there were no problem with technological bias, you wouldn't be able to produce as concise a model with the conglomerates intact. Essential activities that should be independent in all regards except for essential memory would be involuntarily joined through these conglomerates, making it harder to tell just what the essential activities are.

It is relatively easy to spot the various forms of cohesion when you examine a complex activity's minispec. In general, conglomerates contain activities that could operate independently from the others. If you had the technology to allocate each independent activity to a microprocessor, then all the activities could be performed at the same time. Activities that show logical or coincidental cohesion levels can also be recognized because, as described by the minispec, the data they transform are totally unrelated. Activities that show communicational cohesion use common data, but significant time lags usually separate the execution of each activity from the others. In many cases, an unexposed data store lies at the center of the conglomerate. Some of these characteristics are illustrated in the minispec below, which is accompanied by the DFD in Figure 15.6.

> For each Accounting-Memo:
>   Do the following based on the type of Accounting-Memo received:
>   Type 1:   a Book-Order
>   :
>   Type 2:   a Delivery-Notice
>   :
>   Type 3:   a Book-Payment
>   :
>
> Once each week check each Account-Payable:
>   For each Account-Payable for 30 or more days:
>   Issue a Wholesaler-Statement . . .

The Books Accountant processor carries out four essential activity fragments, as you saw in Figure 13.1. Since they all require financial data about books purchased from wholesalers and sold to retail customers, these four activities exhibit communicational cohesion. To a lesser extent, they also have logical cohesion, because the tasks performed share a requirement for bookkeeping skills. Since both logical and communicational cohesion suggest that these fragments belong to separate essential activities, the Books Accountant bubble should be expanded.

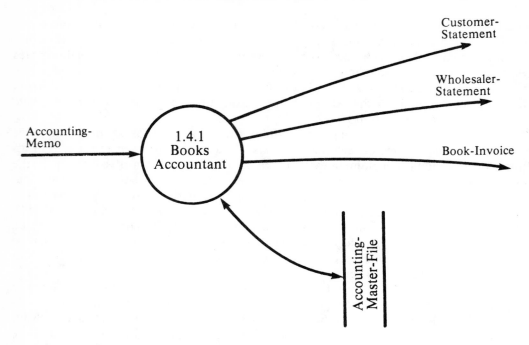

Figure 15.6. A process to be expanded.

Once you have identified a minispec that manifests these telltale levels of cohesion, you expand it by transforming it into a DFD. That requires two steps. On the DFD, first establish an activity for each group of sequentially cohesive activities within the minispec, and second establish any data stores that connect these activities. Repeat this process for any other conglomerate minispecs that you find in searching through the expanded DFD.

15.4 Feedback effects

In sum, there are three different ways to find areas of the expanded DFD that should be expanded even more: by examining the labels on the expanded DFD, by looking for packed dataflow or data store definitions in the data dictionary, and by looking for packed activities within a minispec. Each time you expand a component contained in one modeling tool, you must revise the other two so that all three correspond. In this way, changing one model produces feedback that affects the other models.

When you find components on the DFD with vague or physical names and you repartition them, you must rewrite the data dictionary definitions and the minispecs for the dataflows, data stores, and activities that are changed or added. When you unpack dataflows and data stores, you need to replace the dataflows and data stores on the expanded DFD. You also need to consider splitting the appropriate minispecs. When you find unrelated essential fragments within a minispec, you must replace each affected bubble in the expanded DFD with several bubbles representing the fragments. Finally, the new data stores need to be added to the data dictionary. By applying to all tools the results of investigating one tool, you maximize your chances of liberating essential fragments from the constraints of the current incarnation.

15.5   A tranquilizer for expandophobes

All through our discussion, you may have been thinking to yourself, "Expansion is nice in theory, but how would I ever expand the information I obtain about a really large existing system?" We sympathize with your concerns. Fortunately, there are ways to make expanding a large system easier.

When deriving the essence of a large system, you can divide the system into portions that each can be reasonably expanded. Once each area is expanded, you might then try to join them into an expanded model of the entire system. If the areas are too big to be put back together easily, you can go on to the next step of the process, which reduces each area, and then try to fit the reduced areas into a global model. If the system is really large, go much further in the derivation process before attempting to unify the model.

Once you've studied a system for a time, you will be able to understand a much larger expanded model of it than you expect. In some of our past projects, we have produced expanded diagrams containing nearly two hundred activities. Project members had no real problem in understanding the diagrams; however, taping the paper together or finding enough whiteboard space to contain the entire expanded DFD was a significant difficulty.

At the end of the book, we describe an approach called blitzing that will help you avoid the problems of deriving the essence of a large system. But for now, we continue our detailed presentation of deriving a system's essence without taking into account all practical realities; the shortcuts we give later will make sense only if you already understand the full process.

15.6   Summary

In this chapter, we discuss how to create an expanded physical model: First, remove the upper-level DFDs; this leaves only the bottom-level diagrams, which you connect to form one diagram. Packed channels are separated into constituent dataflows by unbundling dataflows in the data dictionary and removing repeating groups. Last, you search the lowest-level activities for conglomerates of essential fragments. Each conglomerate is expanded by turning it into a DFD wherein the separate fragments are connected by data stores. The final result is a model that displays most of the essential fragments intermixed with activities, dataflows, and data stores from the existing system's infrastructure and administration. Since the overall objective in this phase is to obtain a classified set of essential activity fragments, our next step is to remove the obvious technology-dependent aspects of the expanded physical model. This removal or reduction process is the subject of Chapter 16.

Chapter 16

Reducing the
Expanded Physical Model

From the previous chapter, you know that the purpose of the expanded physical model is to reveal the model's physical characteristics. Now, you want to eliminate those physical features, producing what we call the *reduced physical model*. In this new interim model, all but the unclassified essential fragments are removed, thus moving you significantly closer to the overall objective of deriving the essential model.

Figures 16.1 and 16.2 show an abstract expanded DFD and a reduced DFD, respectively. Whereas expansion liberates essential fragments from physical or functional conglomerates, reduction removes the extraneous dataflows, data stores, and activities that are obviously part of the infrastructure and administration. Because the remaining essential fragments are more visible, they are now much easier to classify and therefore much easier to reassemble into whole essential activities.

The reduction process consists of three steps:

1. Remove the components of the interprocessor infrastructure.

2. Remove the entire administration, both intra-processor and interprocessor.

3. Reconnect the remaining fragments of essential activities.

Each of these steps is detailed in the following sections.

16.1 Removing the interprocessor infrastructure

In order to remove the infrastructure's components, you naturally must know how to recognize them. As depicted in the current physical model, the infrastructure consists of

- *processors* or *processes* that pack, ship, receive, and unpack data traveling across processor or superprocessor boundaries, known as transporter, translator, and batching activities

- *dataflows* that represent actual channels of data; these are the physical medium by which data is carried from one processor to another

- *data stores,* commonly known as batch data stores, that represent the actual physical storage devices, or containers of data in order to synchronize communication between processors

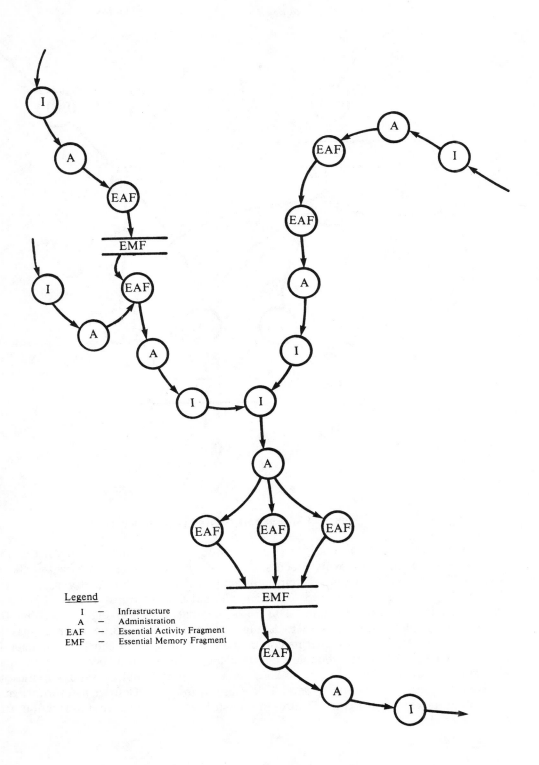

Legend

I   —   Infrastructure
A   —   Administration
EAF —   Essential Activity Fragment
EMF —   Essential Memory Fragment

Figure 16.1. Abstract expanded physical model.

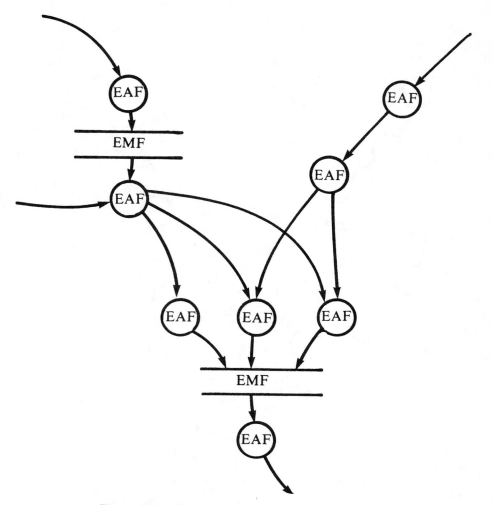

Figure 16.2. Abstract reduced physical model.

At this point in the process, you search through the expanded DFD for the obvious components, checking every process, dataflow, and data store. In your search, you typically find transporter activities, translator activities, and batch data stores.

A transporter activity or bubble carries data but doesn't transform it. It is easy to spot on a DFD, because the same data flows out that flows in. Figure 16.3 has four transporter bubbles. Although each dataflow acquires a new name when it leaves a transporter bubble, you know that its contents are the same from checking the data dictionary. All the transporter bubble does is change the physical location of the data. Figure 16.3 also shows the two basic kinds of transporter bubbles: send activities and receive activities. For example, Send Telex and Warehouse Courier send information to another place, while Route Telex and Distribute Mail receive information from somewhere else.

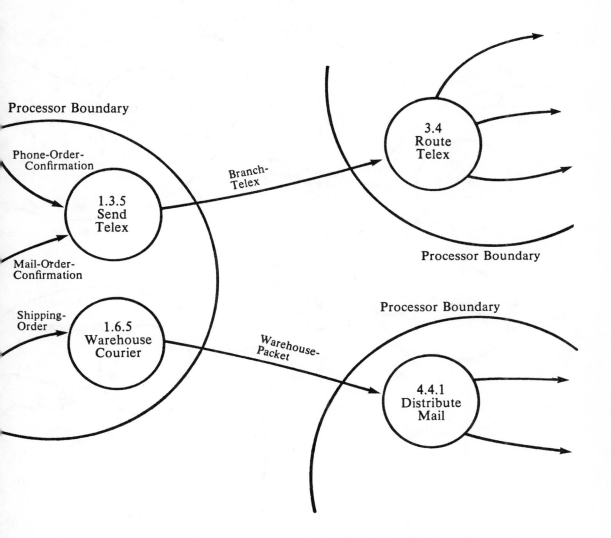

Figure 16.3. Four transporter activities.

A translator activity or bubble may appear to change data, but in reality changes only the physical data container or the way in which a particular data value is represented. The top two activities in Figure 16.4 merely change the physical package in which a dataflow is transported. Specifically, Mail Copies To Warehouse packs data into a physical container by bundling envelopes into a labeled bag, and Distribute Mail unpacks data from that physical container.

The third bubble in Figure 16.4, Enter Orders, is an example of a translator activity that changes the means of representing the data elements. Activities of this kind often translate dataflows between human and computer processors, or between two human processors, as when an activity translates the codes used to represent logical data values from one departmental standard to another. That activity could be very important to people communicating with each other from those different departments.

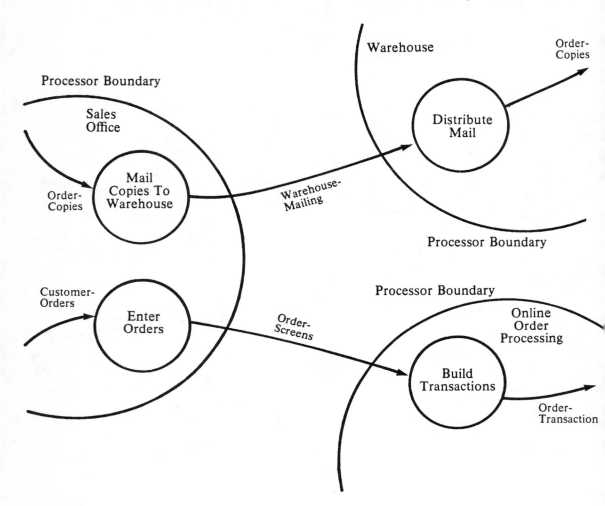

Figure 16.4. Four translator activities.

Translator activities are often combined with transporter activities, and both are usually found where a processor boundary used to be. On one side of the processor boundary, the activities pack the data in a form acceptable to another processor and send the data to that processor. On the other side are activities that receive the data and unpack it from the physical form it was sent in. On either side, there may be activities to change the physical representation of logical data values.

Once you find infrastructural activities, you simply remove them from the data flow diagrams and toss out their minispecs. As you remove infrastructural activities, you also remove the physical dataflows that connect them. For example, you would remove the dataflows Branch-Telex and Warehouse-Packet from Figure 16.3, and the dataflows Warehouse-Mailing and Order-Screens from Figure 16.4. Figure 16.5 shows the effect of removing the infrastructure from the abstract expanded physical model in Figure 16.1.

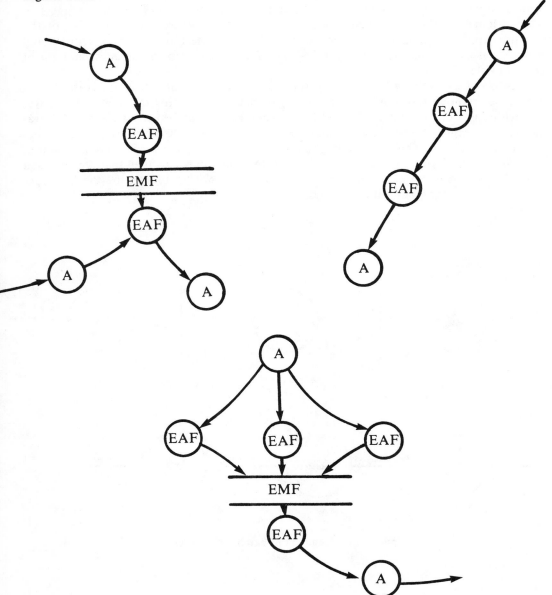

Figure 16.5. Abstract expanded physical model after the infrastructure is removed.

The final components of the infrastructure that you must remove are the physical storage devices called batch data stores. Many of the data stores on the expanded physical model are of this type. They hold data produced by one processor until a data channel or some other processor is available. The two parts of Figures 16.6 and 16.7 show data flow diagrams before removing batch data stores and afterward.

You will usually have no trouble finding lots of activities, dataflows, and stores that are 100 percent infrastructure, but you will also frequently find activities that are from 70 to 99 percent infrastructure. In other words, there may be tiny nuggets of essence lodged within a given component of the infrastructure. Since this is not the best time to do the detailed work necessary to isolate these bits of essence, you should keep any aspect of the expanded physical model that has even the smallest amount of essence within it.

A major exception to what we've said about removing infrastructure is this: You should not remove transporters, translators, or batch data stores that connect the system with the outside world. Figure 16.8 shows two examples of this kind of infrastructure. Keeping these activities and files is consistent with the principle of perfect internal technology. That is, you are free to remove the influence of technology within the boundaries of your system, but you must keep activities and files that are imposed upon the system by an external technology that would not be affected by the introduction of perfect technology within the system.

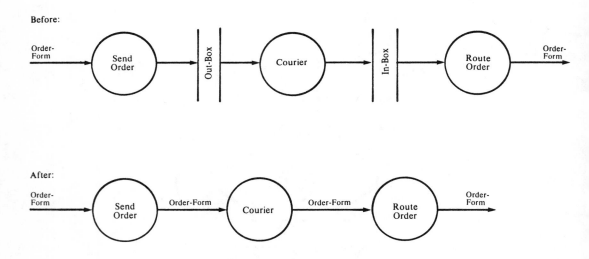

Figure 16.6. Removal of batch data stores.

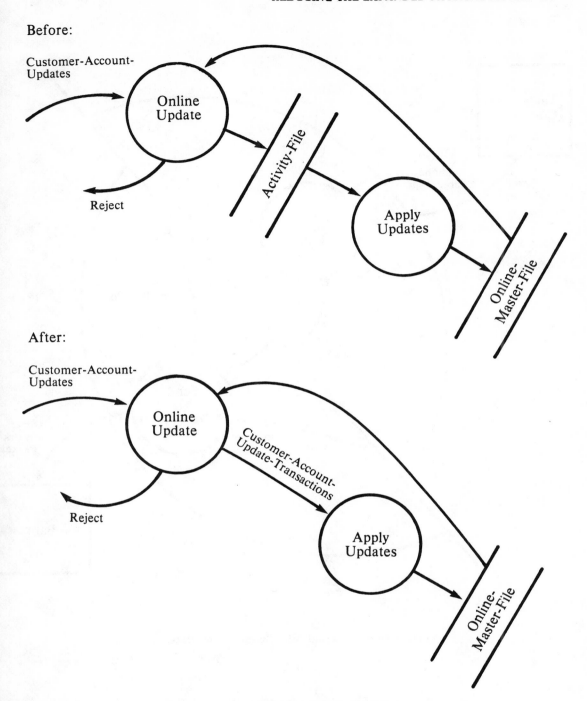

Figure 16.7. Removal of batch data store.

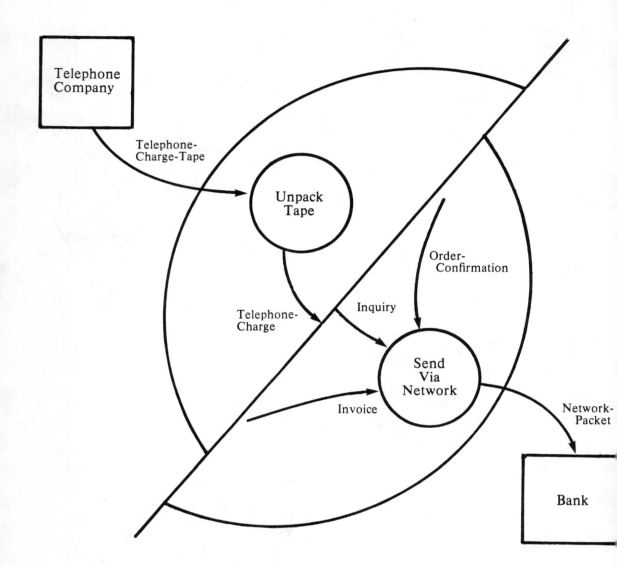

Figure 16.8. External infrastructural interfaces.

16.2 Removing the administration

The second step in creating the reduced physical model is eliminating the entire administration. In the current physical model, the administrative components are

- *edits* that protect the essential fragments from erroneous data produced by another processor or some part of the infrastructure

- *approvals* executed by a processor to detect its own errors before they are passed on to other processors

- *audits* that detect errors overlooked by both the edits and approvals

- *backups* that preserve a current version of the work being done by the system; backups can be used to recover from a system error

Figure 16.9 shows the arrangement of administrative features around the core of essential activity fragments. Since this abstract pattern exists in systems of all sizes and types, learning to recognize it will help you to identify administrative activities in real systems. You remove the administration in much the same way that you remove the infrastructure: by examining all of the remaining components of the expanded physical model for administrative activities, dataflows, and data stores. Figure 16.10 shows that there isn't much left of our abstract physical model once the infrastructure and the administration are removed.

Again, you must worry about the impact of imperfect *external* technology when you remove the administration, since the technology outside the system may be responsible for creating bad or bogus data. Procedures to protect the system from these external errors will probably be mixed in with the administration. So, as you remove administrative activities, you want to verify that they are solely the result of imperfect internal technology.

If you find edits that check the work of imperfect technology outside the system, you must retain these activities. After all, these edits are part of the essence of the system. But sometimes, it's not easy to distinguish the edits that check for external errors from those that protect against internal imperfections. For one reason, external edits do not always occur at the front-end of the system; they may be scattered throughout the current physical model, and may be commingled with internal edits. For this reason, remove *only* those edits that check exclusively for errors produced by the system itself. In this way, you insure that edits for outside errors will be a part of the essential model.

Another approach is to remove *all* edits at this time, put them aside, and reintroduce them later in the derivation process. There is an advantage to removing the entire administration. Since the edits in the reduced physical model occur near the boundaries of existing processors, keeping the edits means the essential model may retain physical characteristics of the existing system. By removing them all, if only temporarily, you avoid this risk.

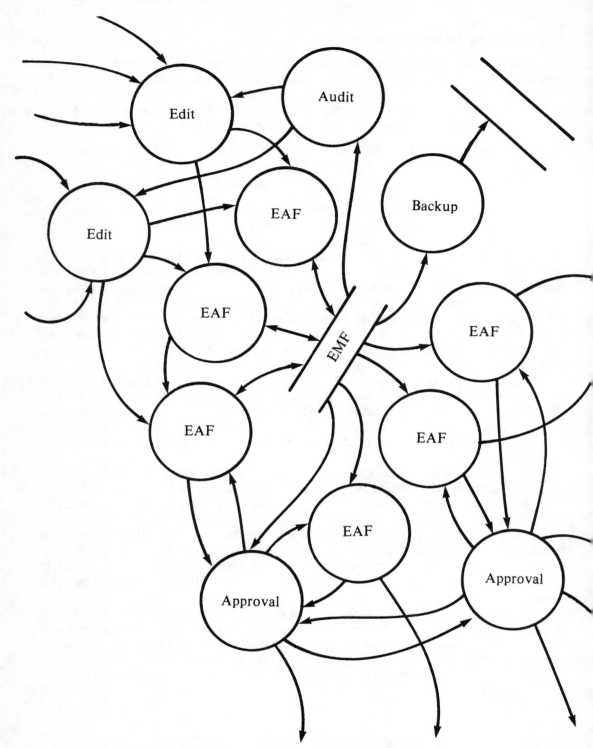

Figure 16.9. Abstract arrangement of administrative activities around a logical core.

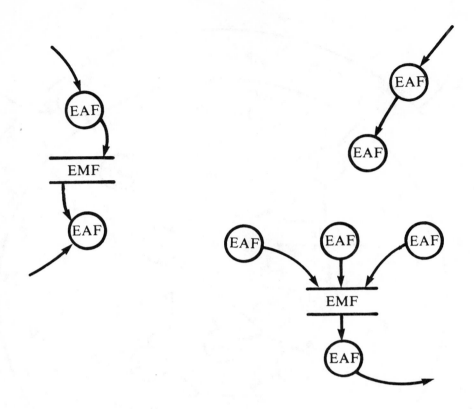

**Figure 16.10. Abstract expanded physical model
after the administration has been removed.**

16.3 Reconnecting the essential fragments

Although the fragments of the system's essence are uncovered and most of the physical characteristics are removed from the system model, the job of reducing the expanded physical model is not yet done. When you eliminate the infrastructure and administration, you leave gaping holes in the expanded physical model. Most of the data connections between essential activity fragments are severed, leaving isolated fragments scattered across the face of the model. However, in order to perform the next step of classifying the essential activity fragments, you are going to need to see the data connections between essential fragments. So, before you finish the reduction step, you have to reconnect the disjointed pieces of essence into a coherent system model.

Reconnecting the essential fragments is not difficult, because it is simply a matter of sending a dataflow or an essential memory access directly to an activity, instead of passing the data through a series of infrastructure and administrative activities. You can even refer to the expanded physical model to see how the pieces were linked by the infrastructure. The result of reconnecting the fragments is shown in Figure 16.2.

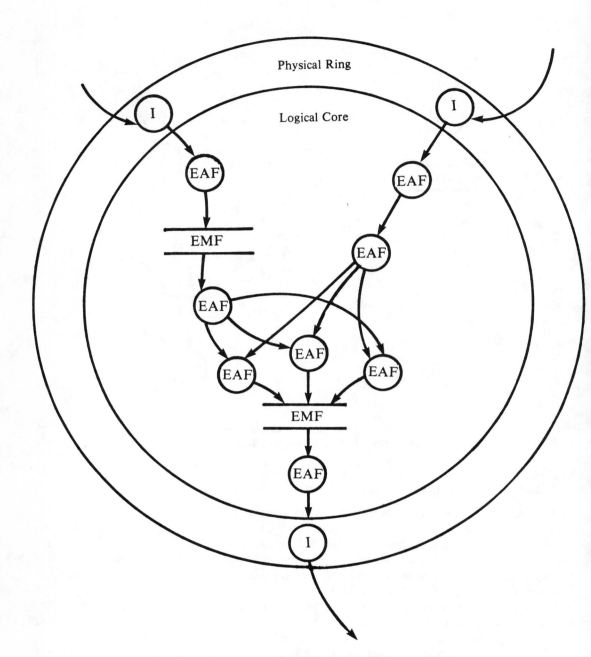

Figure 16.11. A ring of physical interface activities.

16.4 The results of the reduction process

The parts of the reduced physical model are almost always arranged in a particular pattern (see Figure 16.11). Around the edge of the model are the activities, dataflows, and data stores that are totally physical. These features, called the *physical ring* of the system, are the system's interface to the imperfect technology of the environment surrounding the system. Further inside, you find activities, dataflows, and data stores that are mostly essential. Although these features still need to be scrubbed clean of subtle physical characteristics, they are at least a beginning for what will ultimately be the logical core of the system.

The characteristic shape of the reduced physical model results from applying the principle of perfect internal technology. You eliminate any activities, dataflows, and data stores that exist because of the specific technology used to implement the system. However, you must respect the technological impositions of the system's environment. Although the technology of the environment could be changed, you probably have no direct authority to do so. So until you do, you must abide by the environmental constraints imposed on the system and reflect these constraints in your model.

The reduced physical model allows you to see the fragmented essence of the system without the mental, clerical, and financial burden of modeling the technological characteristics of the existing system. The reduction process significantly removes the impact of imperfect technology from your interim model of the system. As a result, the reduced physical model assumes that the system uses perfect technology. Since the model shows no infrastructure, it assumes perfect, telepathic communication between activities. The disappearance of most of the batch data stores from the model implies perfect processors, whose speed is such that synchronization is unnecessary. Now that most approvals and internal edits have been removed, the model reflects an assumption that processors are infallible. The assumption of perfect technology also eliminates the need for the other function of the administration: resource allocation. Perfect processors work at such a capacity that no administration of their work is necessary. Infinite resources make resource management irrelevant.

It is in the reduction step that the most dramatic part of the discovery process occurs. Even though perhaps 40 to 70 percent of the model disappears at this stage, you aren't close to being done. Many subtle physical characteristics are still embedded in the reduced model. Furthermore, the model is an eyesore, since the model does little to budget complexity. We resolve these problems in the following chapters.

16.5 Summary

During the reduction process, you remove the obvious physical characteristics from the expanded physical model. First, you eliminate the infrastructure's transporter activities, translator activities, and batch data stores. Next, you remove the edits, approvals, audits, and backup files and activities that make up the administration. The final step of the reduction process is to reconnect the essential fragments. The resulting model has a core of mostly essential fragments surrounded by a ring of physical activities that interface with the system's environment.

Chapter 17

Classifying the Essential Fragments

Upon finishing the first two steps of the process to discover an existing system's essence, you have removed the majority of physical characteristics from the current physical model. Thus, you might conclude that the remaining activities and data stores in the reduced physical model are primarily essential.

On the contrary, the model still exhibits characteristics of the system's implementation technology. The actions themselves are fragments of essential activities, but the fashion in which they are carried out, together with the language used to describe them, reeks of technological limitations. For example, the actions take place in sequences that are dictated by the nature and quantity of the processors in the existing system, and the data containers used to store essential memory are variety show files chosen to optimize the performance of the existing system. Moreover, the minispecifications from the current physical model still refer to data stores that you deleted in the previous step. Although you may think that your next task is to remove these remaining physical footprints, it isn't. Before you can thoroughly examine each fragment of essential activity and essential memory found in the reduced physical model, you must guard against the risks that arise from working with complexity.

Most systems that you will study are relatively large and complex. Even after you have removed the infrastructural and administrative processes and data stores, there will probably be many dozens, if not hundreds, of remaining activity and memory fragments. Whether the exact quantity is fifty or one hundred or five hundred, it is far too large for human beings to deal with efficiently. The abstract reduced physical model in Figure 17.1 gives you some idea of the uncontrolled complexity of the model at this point. The eighteen essential activity fragments, presented as they are, are more than most model reviewers can comfortably handle.

Of course, you would never attempt to logicalize two hundred essential activity fragments *simultaneously;* you would work on perhaps several at a time, until you had removed the remaining physical characteristics from all of them. This is a natural way to avoid the difficulty, frustration, and error caused by dealing with too much at one time.

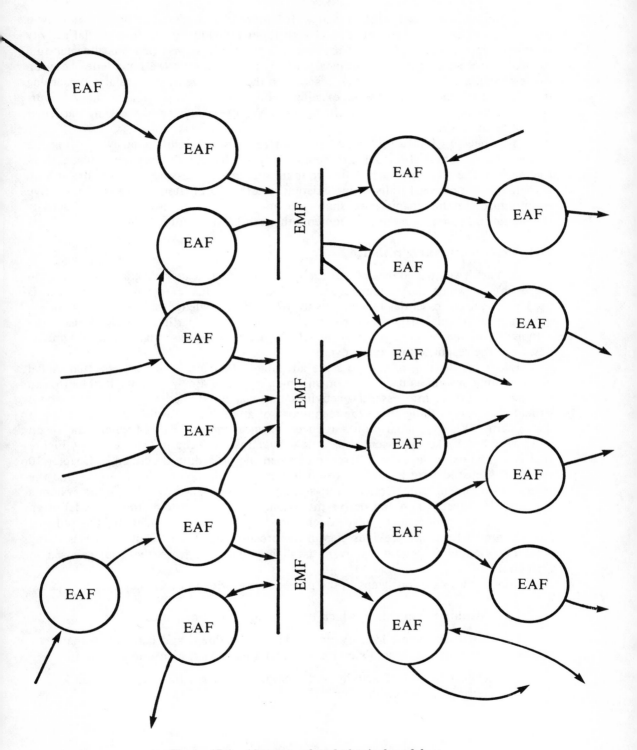

Figure 17.1. Abstract reduced physical model.

Although you don't want to deal with hundreds of essential activity fragments at one time, neither do you want to deal with them individually. Each essential activity fragment represents the portion of the system's planned response that was allocated to a particular processor in the present incarnation. If you preserve the boundaries between two essential activity fragments that respond to the same event, you are also preserving the division of labor caused by the existing technology. To do so violates the very purpose of deriving an essential model: to transcend the characteristics of any particular type of technology.

You need to find a midground somewhere between simultaneously studying the entire reduced physical model and sequentially studying individual essential activity fragments. The first approach defies your information processing abilities, the second perpetuates the physical traits of the existing system. Fortunately, there is an attractive compromise. In this chapter, we show you how to create groups of essential activity fragments and essential memory fragments that can be studied together.

17.1 Event partitioning revisited

In Part Two, we advocated partitioning essential data flow diagrams according to the events to which the system responds. Chapter 10 illustrated how this approach is used to create an essential model directly from the knowledge of a set of system requirements. In this chapter, we show how to use event partitioning to decrease the complexity of the reduced physical model: You group the essential activity fragments that form the response to a single event.

Event partitioning offers the same advantages for partitioning an essential model of an existing system as it does for organizing a model of a new system: It helps you to organize the model into essential activities that can be studied separately, but whose structure reveals nothing about the technology of a present or future implementation of the system. When you partition a reduced physical model in this fashion, the result looks very much like the essential data flow diagrams produced in Chapter 10 — Figure 10.3 is a good example. The difference between the procedure described in Chapter 10 and that discussed here is not the result but the starting point. When you define the essence of a system from scratch, your starting point is your knowledge of the system's purpose. However, when you derive the essence of a new system from a model of an existing system, you usually have much more to work with — namely, the reduced physical model from the previous step in the process. So, the procedure for classifying a set of essential fragments by event is slightly different from the one we used in Chapter 10.

Briefly, there are four steps in the procedure described in this chapter:

1.   Identify the events to which the existing system responds.

2.   For each event, identify the activities, dataflows, and data stores that make up the system's entire immediate planned response to the event.

3.   Depict the set of activity and memory fragments for each event on a separate data flow diagram.

4.   Draw an upper-level data flow diagram on which each response to an event — each essential activity — is represented by a single process. If there are too many essential activities to put on one data flow diagram, create even higher-level DFDs to group essential activities that respond to related events.

During this procedure, you identify both the activities that respond to external events (occurring outside the context of study) and those that respond to temporal events (specific times when an essential activity is performed). The distinction is important because you apply the above procedure differently to each type of event. You identify external events by examining the external entities that act upon the system, while you usually find temporal events by studying the schedules that govern the system's automatic actions. Furthermore, external events can result both in updates to the system's essential memory and in responses to the outside world, while the results of temporal events are nearly always responses to the outside world. Because of the varied nature of external and temporal events, we discuss essential fragments that respond to external events first, and those that respond to temporal events afterward.

17.2 Identifying external events

You begin the classification process by identifying the events to which the system has a planned response. Although you will be inclined to identify both types of events at the same time, we advise that you concentrate first on the external events. Identifying true temporal events is tricky, because many seemingly reasonable temporal events turn out to be physical timing constraints caused by the technological limitations of the existing system. It is easier to distinguish true temporal events from bogus ones if you have at least a partial set of externally driven essential activities.

External events are found in the system's environment. In terms of modeling tools, you find these events by studying the context diagram from the current physical model, which shows the entire system as a single process and also shows the people, computer systems, and organizations with which the system interacts. On the context diagram, external events are nearly always heralded by the arrival of a dataflow from one of the external entities.

We discussed the identification of external events earlier in this book, but in quite a different context. In Chapter 10, we showed how to create a list of events to which you *believe* the system should respond. Here, however, armed with the reduced physical model, you can face the task more confidently. You can list the external events to which you *observe* the system responding, as documented in your data flow diagrams, data dictionary, and minispecifications. Of course, there's still room for error: You may have omitted a response from the model of the existing system, but generally you are much better equipped to identify external events when you can base your conclusions upon the system's documentation.

Consider the context diagram in Figure 17.2. Two dataflows enter the materials supply system: Material-Requisition and Delivery. Both are the manifestation of an external event to which this system has a planned response. Identifying external events is not always that simple. Sometimes, the reduced physical model does not show the original source of a dataflow, but just the last component of the external infrastructure that brings it into the system. Similarly, physical DFDs often depict dataflows that are physical data containers carried by the external infrastructural processors.

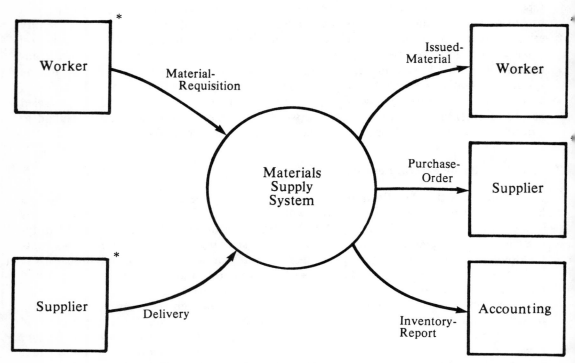

*An asterisk next to each of two or more identically named boxes indicates that the boxes refer to the same entity.

Figure 17.2. Context diagram of materials supply system.

Consider the exceedingly physical context diagram in Figure 17.3. Employees in this company routinely submit material requisitions via interdepartmental mail, but they also can do so by phone. Nevertheless, this system does not really respond to the phone and the mail courier, but to the employee on the other side of both of them — the original source of the requisition.

Figure 17.4 illustrates a similar problem. The incoming dataflow Telex is a physical data container employed by the external infrastructure. Of course, a telex may contain a variety of messages, each of which announces the occurrence of a different event, to which a different response must be generated.

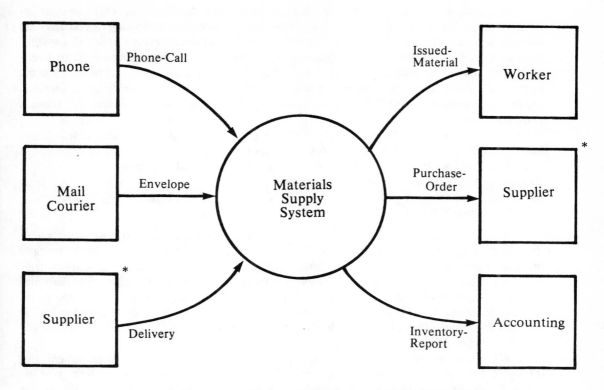

Figure 17.3. Context diagram showing the external infrastructure as the source of stimulus dataflows.

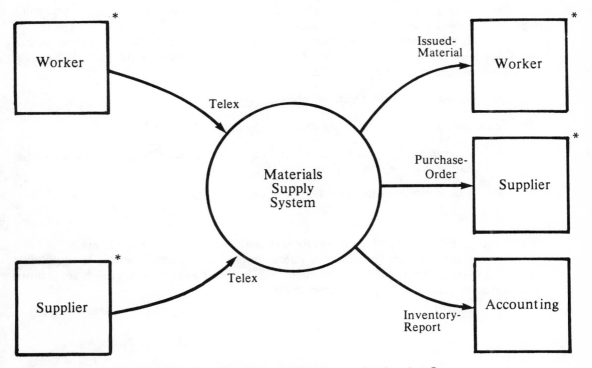

Figure 17.4. Physical data container as a stimulus dataflow.

In Chapter 15, we recommended that you expand a dataflow like Telex into its individual essential components. If your expansion is incomplete however, you may be tricked into declaring that the external event is "telex arrives," even though the system has not one, but several dozen or even several hundred, possible responses to an event with such a general name. We are not saying that "telex arrives" is *not* an external event; of course it is. But since the whole point in partitioning the reduced physical model by events is to divide it into manageable pieces, "telex arrives" does not make a good candidate.

As you study a physical context diagram or a reduced physical model, watch out for external entities that are merely transporting dataflows from another, more significant entity. Beware, too, of incoming dataflows that are physically motivated bundles carrying many independent messages.

17.2.1 Naming external events

Before you can study the system's response to a particular event, you must name the event, and sometimes that's the trickiest part of the whole process. After habitually thinking about systems in terms of functions, you are likely to name the activity triggered by the event, or part of that activity, rather than the event itself. For example, you might see the incoming Material-Requisition dataflow and declare the event to be Process Material Requisition. Process Material Requisition is not an external event at all; it is an activity that occurs inside the system under study. Therefore, it isn't an appropriate event name.

What's in a name? The names you use to describe a system both reveal and influence your thinking about the system. In this system modeling discipline, you attempt to discern the essence of a system by concentrating on its relationship to the world around it. Using the names of internal system processes to describe external events distracts you from studying the system's interactions with its environment. You run the risk of getting bogged down in the system's internal physical structure, which will make knowing its essence more difficult.

So, it's important to name external events properly, and while there are no formal rules for such names, we have found this informal guideline quite useful: External event names take the form

external entity + active verb + object

Some examples may help you understand how to apply this guideline:

- Customer (an entity external to the Good Skate Company) requests (an active verb) skates (an object).

- Police officer issues ticket.

- Bank returns check.

There are perfectly acceptable names for external events that do not fit this pattern exactly. "Student registers for class" is one example. The attitude behind the guideline is the important point: When you search for external events, you search for actions performed by *others,* not by the system itself.

Now, identify and name the events represented by the dataflows entering the materials supply system. We offer these two:

- Worker requisitions material.

- Supplier delivers stock.

As a final step, assign a number to each external event in your list. (The reason for doing this will become apparent in the next section.)

17.3 Identifying essential fragments for each external event

Once you have a list of well-named external events, the next step is to associate or map each event with one or more essential activity fragments from the reduced physical model. The following general procedure is performed for each external event:

1. Identify the essential activity fragment that first recognizes the occurrence of the event. Usually, this is the first activity within the system to receive the dataflow that signals the event. Mark this fragment with a number representing the event to which it responds, using the number from the event list produced in the previous step.

2. Follow the path of the system's response to the event from one essential activity fragment to the next by following the dataflows that connect the fragments. You can think of this step as watching a relay race where you follow the baton from runner to runner until the race is complete — in this case, until the system's entire immediate response is complete. Along the way, you mark each essential activity fragment that takes part in the response with the event number.

3. Draw a new data flow diagram that shows only the essential activity fragments marked with the event number, along with the dataflows and essential memory fragments connected to them.

We illustrate this procedure using the materials supply system. Figure 17.5 shows this system's response to the first of our two external events, "worker requisitions material." The incoming Material-Requisition is intercepted by process 3.1.1, which checks that the level of stock is sufficient to fill the request. If it is, an Issue-Ticket (the baton) is passed to process 3.3.1, which actually issues the requested material. While recording the removal of material from inventory, process 3.3.1 checks if the stock has fallen below the reorder point. If it has, process 3.3.1 sends a Restock-Memo to process 2.3. This activity fragment, the third leg of the relay race, refers to the Suppliers data store to identify those vendors who are able to provide the particular stock item. The fourth and final component of the system's response to the external event "worker requisitions material" is process 1.5.5, which issues a purchase order to a selected supplier to replenish the inventory. Having traced the system's immediate response to the first event on the list, you should tag each essential activity that takes part in the response with a "1," as shown in Figure 17.6.

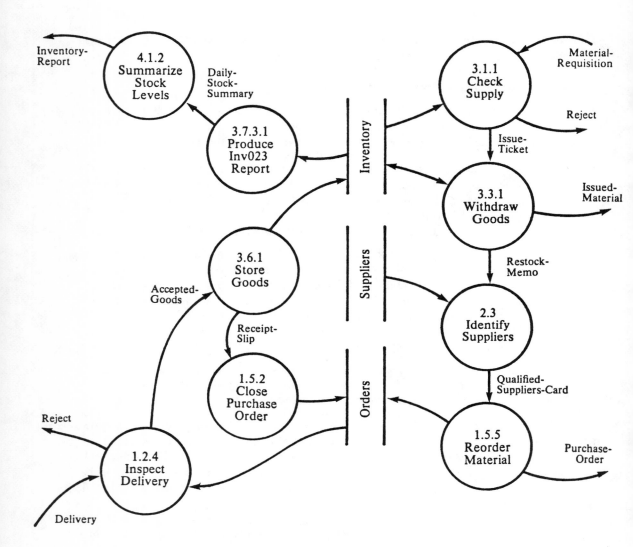

Figure 17.5. Reduced physical model for the materials supply system.

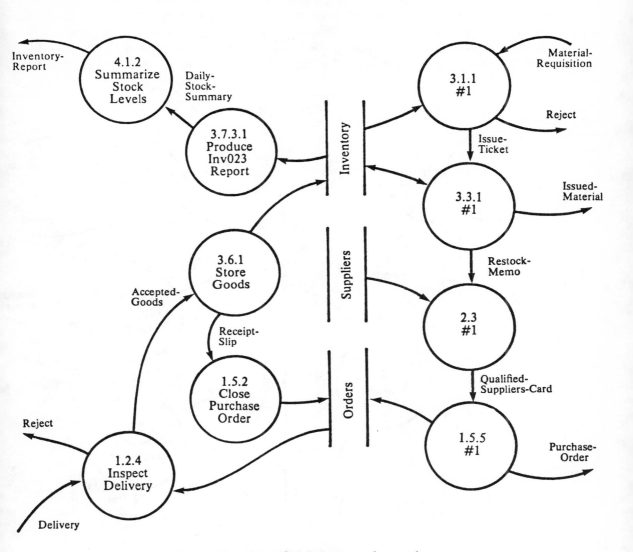

Figure 17.6. Identified fragments of event 1.

Let's proceed to the second external event, "supplier delivers stock." Process 1.2.4 receives the delivery and checks it against the Orders file to ensure that it is correct. The Accepted-Goods are passed to process 3.6.1 so that it can record their arrival in the Inventory data store. Finally, a Receipt-Slip is passed to process 1.5.2, which closes out the purchase order that originally initiated the delivery. All three of these essential activity fragments are labeled with a "2" to signify that they respond to the second event on the list.

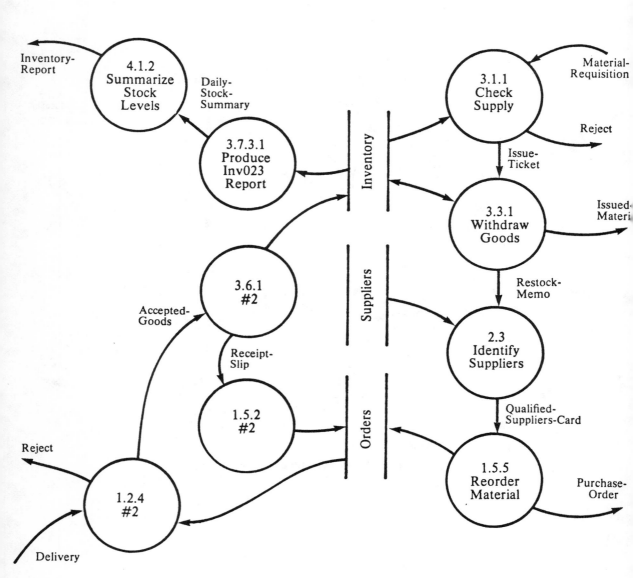

Figure 17.7. Identified fragments of event 2.

In addition to the relay race analogy, you may find it useful to think of the essential activity fragments as dominoes standing upright next to one another. Each external event is a hand that knocks over one of the dominoes — the first fragment to know of the event — which in turn knocks over adjacent fragments until the system's response is complete. The event "worker requisitions material" knocks over four essential activity fragments, while the event "supplier delivers stock" knocks over only three. Although both analogies are useful, they are somewhat oversimplified, because there are a number of issues that may complicate the mapping process. For example, an essential activity fragment is sometimes "knocked over" by more than one event, as shown in Figure 17.8, where bubble D participates in the response to both events 1 and 2. Process D is a conglomerate of two different essential activity fragments grouped for some technological reason. This situation can result from not fully expanding the current physical model. You merely complete the expansion by dividing it into its two constituent processes, as shown in Figure 17.9.

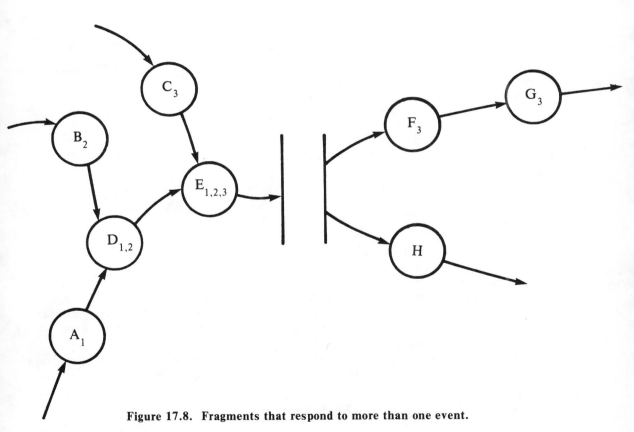

Figure 17.8. Fragments that respond to more than one event.

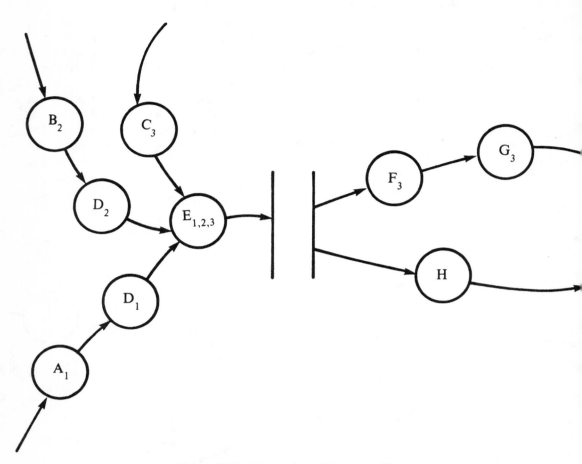

Figure 17.9. Expansion of fragment D.

The same complication — a fragment responding to more than one event — can also result when the fragment implements a single common function that is a part of the system's response to two or more events. Process E in Figure 17.9 is a single function used in common by all three events. To treat this case, you must make a copy of the shared activity for each event response that requires it, as depicted in Figure 17.10. Notice that in the resulting diagram you create redundancy so that you can see how the shared activity fits into each event response. If at the end of the logicalization procedure you find that the activity truly is shared by many responses, you can document it accordingly. We discuss this process in Part Six.

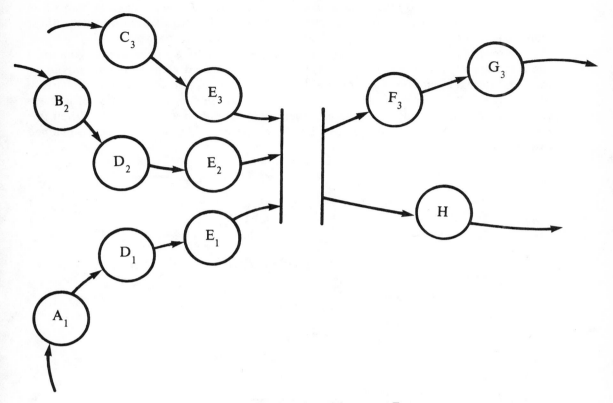

Figure 17.10. Duplication of fragment E.

17.3.1 Identifying the system's complete response

Throughout this chapter, we talk about following the flow of data among essential activity fragments until the system's response is complete. To understand this concept, you must realize that a system's planned response to an external event consists of either or both of two types of response: an *external* and an *internal* response. If the system responds to the event by involving the outside world, usually by sending data or other matter to one or more external entities, it makes an external response. In the materials supply system, both Issued-Material and Purchase-Order are external responses. If the system reacts to the external event by revising its essential memory, that is an internal response. Since internal responses are the purpose of custodial activities, you can also call them custodial responses. The essential activity fragments that respond to the second event, "supplier delivers stock," carry out internal responses when they update the Inventory and Orders data stores.

Essential activities are not limited to just one response or to just one type of response. In the course of accepting the delivery from the supplier, the materials supply system produces two internal responses, and when it processes a material requisition it produces up to four responses, two that are external and two that are internal. Many real-life essential activities produce a dozen or more distinct responses to a single event.

In order to discover the system's entire response to a given event, you must check that each path of essential activity fragments associated with the event ends in either an external response or an internal response. That is, you must follow the stream of activity until it either leaves the system (when a dataflow is sent to an external entity), or it terminates within the system (when essential memory is updated). When this condition is satisfied, you can be confident that you have identified the entire planned response to the event.

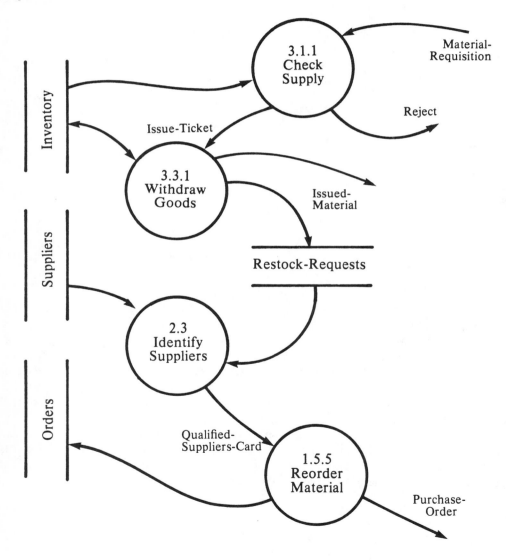

Figure 17.11. A physical time delay in the materials supply system.

While it is fairly easy to determine if a response has left the system entirely, it is not always obvious that a data store update is an essential memory update. You have to beware of data stores that are time delays resulting from internal technological limitations. These can fool you into thinking that a response is complete when it is not. Figure 17.11 illustrates this problem using a modified version of the reduced physical model for the materials supply system. The data store called Restock-Requests contains some Restock-Memos that have yet to be sent on to process 2.3 for further processing.

This batching might happen for a number of reasons. Perhaps the processor that issues material is not the same as the processor that reorders material from the suppliers. But if this were the only reason for the data store, you probably would have identified it as a part of the interprocessor infrastructure and eliminated it earlier. It is more likely that the existing system has a schedule for ordering material, and that the data store represents the time delay between the generation of the Restock-Memo and the next scheduled replenishment. Does this mean that the system's response to the event "worker requisitions material" is complete when the Restock-Memo is placed into the Restock-Requests data store? Is it really an internal response rather than an external response?

Whether the response is external or internal depends entirely upon the reason for replenishing stock only periodically. It seems simpler for the system to reorder as soon as it discovers the need. So why does the system wait for a seemingly arbitrary period of time to elapse before sending the purchase order to the supplier? Answering this question is important, because some of the reasons for batching documents and waiting for a specified time to perform an activity are purely physical, while others are essential to the system. If this particular system processes batches of Restock-Memos once a week because of its *internal* technological shortcomings, this means the data store is physical and should be deleted at this time. But if the delay arises from *external* technological constraints, the data store is essential to this system, and it should remain.

Be suspicious of all responses to external events that end in data store updates. Ask yourself why the response must stop at this point. What is the next step? Why can't that step happen immediately? If the limitation is external in nature, such as a schedule imposed upon the system by an external entity, then there is nothing you can do about it. But if the stream of activity is broken because "the first job runs daily and the second runs only on Thursdays," or because "it's too much trouble to reorder until I've got at least a couple of dozen requests," you should delete the physical data store and complete the system's immediate planned response to the external event.

To handle physically motivated stored data accesses, you need to modify the procedure for identifying external-event-related fragments. As you follow the path of the system's response, look for a situation where a data store is updated and yet, as soon as the data store is updated, some essential activity fragment could immediately process the data stored. When you find this situation, you must hop over the data store and continue following the flow of the response, marking each essential activity fragment you find on the other side that takes part in the response. You also want to transform the stored data update into a dataflow so that the diagram shows the immediate flow of the response from fragment to fragment.

Figure 17.12 shows the result of these additional steps. Upon close examination, we find that essential activity E_1 makes a physical stored data update. We hop over the data store and find F_3 and H. We eliminate F_3, having previously determined that it is exclusively part of the response to external event 3. The H fragment, on the other hand, could pick up the response to external event 1 where E_1 left off. So, we mark H with a "1" and de-couple E_1's update to stored data. We replace that update with the new dataflow shown in Figure 17.12.

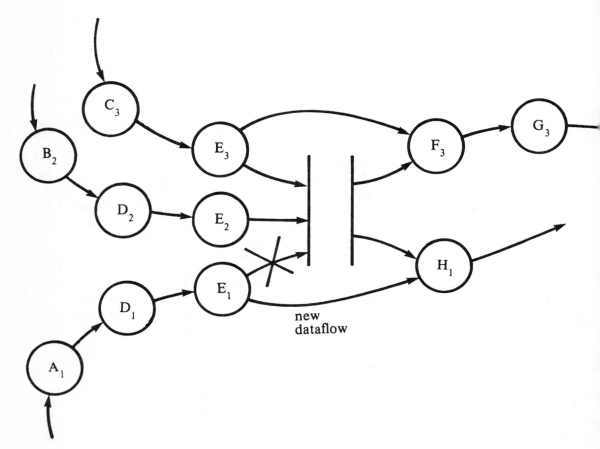

Figure 17.12. New dataflow added to bypass physical data store.

Figure 17.13 shows the same steps carried out on the example in Figure 17.11. Notice that the new dataflow assumes the name of the eliminated physical data store. This is only a temporary measure.

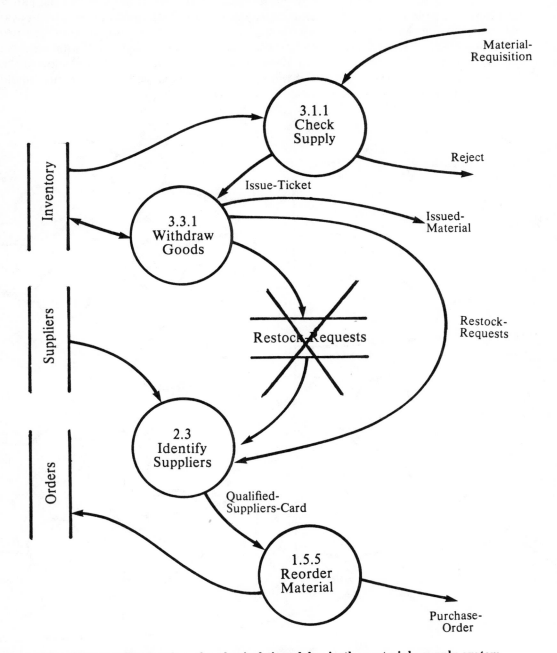

Figure 17.13. Elimination of a physical time delay in the materials supply system.

Once you have attributed one or more essential activity fragments to each external event and have resolved the special cases just discussed, the resulting reduced physical model should exhibit the pattern shown in Figure 17.14. In this diagram are two types of essential activity fragments: those that have been identified with external events, and those that have not been allocated at all. Provided you have followed the procedure correctly to this point, all of the remaining essential activity fragments should respond to temporal events, as discussed in the next section.

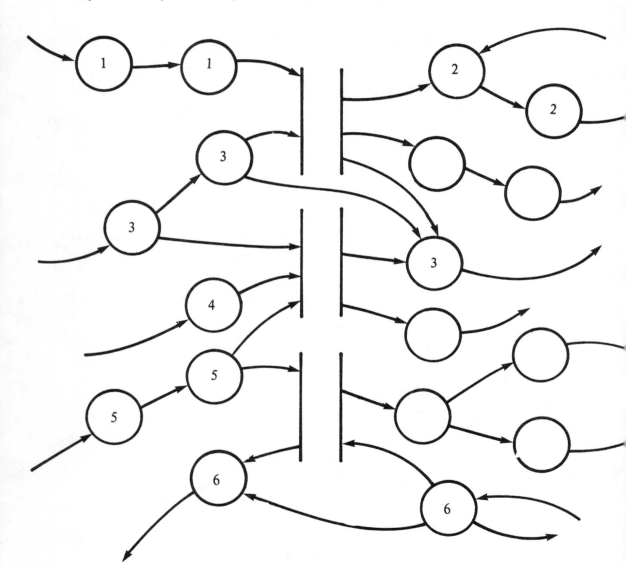

Figure 17.14. Reduced physical model after identification
of externally driven fragments.

17.4 Identifying temporal events

As we know, temporal events trigger essential activities that occur on schedule at a specific time, sometimes periodically, perhaps weekly, monthly, or annually. Other essential activities happen at a predetermined time and date. You classify essential activity fragments that respond to a temporal event in much the same way that you classify fragments associated with external events:

1. Identify the temporal events that drive the essential activity fragments from the reduced physical model.

2. For each temporal event, identify the fragment that first learns of the occurrence of the event, and follow the flow of processing from that fragment to all the connected fragments until the response is complete.

3. Draw data flow diagrams, each one showing the identified fragments for one temporal event.

Despite their similarities to external events, temporal events present a few unique problems: They are not triggered by incoming dataflows, so they are not as easy to find as external events, and since no external entity initiates temporal events, you need a new guideline for naming them. Most important, identifying true temporal events is tricky simply because you discover them by looking *inside* the system, where a variety of physical characteristics lie waiting to be mistaken for essential system features. In this section, we concentrate on the special challenges posed by using temporal events to classify essential activity fragments.

The important trait shared by all temporally driven essential activities is that the schedule they follow is based solely upon the relationship between the system and entities outside the system. The origin of many temporal events can be traced to contractual agreements between the system being studied and other systems, such as a company's agreement to pay its salaried employees biweekly or an individual's obligation to make a mortgage payment every month.

A true temporal event has nothing to do with limitations on the processing capacity of a company's data processing system, nor is it determined by the cost of performing an activity. Unfortunately, some purely physical characteristics may fool you into thinking you've spotted a temporal event. For example, a common tactic for improving the operating efficiency of a system is to make processors execute their activities on batches of data that have accumulated in physical data stores. That usually means that the processor works on another activity *until such time* as the batch is ready to be processed. That time can be when the amount of work being batched reaches a certain level or when a physical processing schedule states. This kind of physical activity trigger is a false temporal event, because the schedule is not determined by an entity outside the system. The Identify Suppliers activity fragment in Figure 17.11 is an example of a batch-executed activity. The batches are stored in the Restock-Requests data store, and the activity is triggered by a physical schedule also embedded in Restock-Requests. Such timing constraints, as conventional structured analysis wisdom teaches, are quite often physical. So, as you search for the temporal events that a reduced physical model responds to, it pays to be suspicious.

Having warned you of the dangers ahead, we offer an informal procedure for finding temporal events. You return again to the context diagram, where you found incoming dataflows that led back to external events. This time, however, you study the outgoing flows, since they are the only sign of temporally driven essential activities visible from outside the system. You don't need to study all of the outgoing flows because many of them represent responses to external events. So, make a copy of the context diagram or of the reduced physical model, and delete the output dataflows that you have already accounted for by studying the external events. For example, in our materials supply system, we deleted the flows Issued-Material and Purchase-Order, as seen in Figure 17.15.

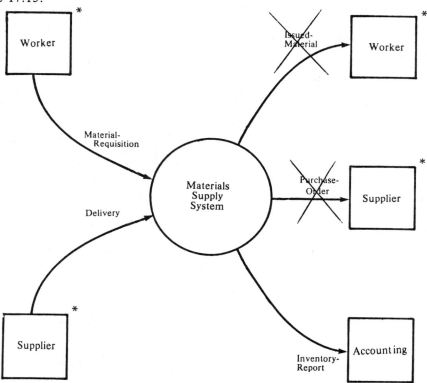

Figure 17.15. Remaining output dataflows in the materials supply system.

You can now narrow your focus to the remaining output dataflows, which you know are in response to either a true or false temporal event. To determine which, ask the following questions about each flow:

- For what reason does the system produce this output? If the answer is to improve the efficiency of the system, you probably have found a false temporal event. On the other hand, systems often use temporal events as prompts to entities outside the system when they can't count on an external event triggering the output. Mortgage systems, for example, don't wait for a payment from a mortgagor to inform that mortgagor that the payment is three pay periods late. Instead, the system uses a temporal event to prompt the mortgagor to pay. If the mortgagor never pays, even after receiving all the proper notices, it becomes time to foreclose the mortgage.

- What are the expectations of the entities in the outside world receiving the flow? Do they expect the flow to reflect events and results processed over a certain period of time, as in the case of the government getting a report of taxable interest earned in an account over a year's time? True temporal events are often the prompts to a system that allow it to meet obligations it undertakes to these entities.

These questions help direct you toward the outside world, where essential activity schedules originate, and discourage you from thinking of the scheduled outputs in terms of the schedules themselves. If you concentrate on the actual time at which the existing system performs an activity, you are liable to mislabel an event such as "02:52 — time to run PDS compress job" as a temporal event when it is really a physical constraint. Rather than saying that "April 15" is a temporal event, look to the reason behind the schedule — that it's time to submit a personal income tax return. So, try to discern the external expectation that is fulfilled by a scheduled activity, and express the temporal event in those terms.

17.4.1 Naming temporal events

The names of temporal events are not usually very revealing. Once you identify an outgoing dataflow that does not belong to the response to an external event, your first instinct is to name the event using the form

time to issue [dataflow name]

If you use this method in the materials supply example, the essential activity that produces Inventory-Report is triggered by the event "time to issue inventory report." Boring? You bet, especially when you are modeling a system with one hundred or so such temporal events. After a while, the phrase "time to issue" becomes so common that you abbreviate it to TTI. That is, if you don't delete it altogether and simply use the dataflow name as the event name, which we do not recommend.

Unfortunately, sometimes you just can't do much better than to use "time to issue." Sometimes, the system you are studying has entered into an agreement with an external entity that obligates it to produce a report according to a schedule. Once you have named the report, there isn't much more you can say about the event, except that it is "time to fulfill our obligation to [external entity] by issuing [report name]." There are many examples of this phenomenon: "time to report affirmative action compliance status," "time to report employee earnings," or "time to falsify project progress report."

The good news is that sometimes you *can* do better. You can find a better name in the cases where the temporally driven outgoing dataflow notifies the outside world of the occurrence of an event that is known to the system through its essential memory. Consider a magazine subscription system that notifies a customer that his or her subscription has expired. The system knows when to notify the customer because it remembers when the subscription started and what its duration is. Just because the outgoing dataflow is called Expiration-Notice doesn't mean you should settle for an event name like "time to issue expiration notice." You can name the event that really inspires this essential activity, namely "subscription expires."

Whenever possible, name temporal events in terms of the original, external motivation for the essential activity that produces an outgoing dataflow. But be prepared for the cases in which the name of the dataflow itself tells the whole story. Such is the case with Inventory-Report, so we declare the driving temporal event to be "time to report stock on hand."

17.5 Identifying essential fragments for each temporal event

Once you have named the temporal event, you identify the essential activity fragments associated with it. To find the essential activity fragment that first learns of the occurrence of a temporal event, you look for processes in the reduced physical data flow diagram that get all their input from data stores; that is, they are not triggered by dataflows from other processes or from the outside world. In Figure 17.16, processes A, S, and D fit this description.

To be sure that you have identified a temporally driven essential activity fragment, you must examine its minispecification. There, you should find a reference to a time interval or to a specific point in time, or a comparison of the present time or date with the system's essential memory. For instance, you might see one of these statements:

Once each hour at 15 minutes past the hour,

On July 1, do all the following things:

If today is the Subscription-Expiration-Date,

Just because you find a similar statement does not guarantee that you have found an activity that responds to a temporal event. You may still be dealing with a physical time delay like those discussed in Section 17.4. You've got to agonize over the reasons behind the schedule and convince yourself that the schedule would still exist if the system were implemented using perfect internal technology.

Once you are sure that the temporal event is valid, follow the flow of data from one essential activity fragment to the next until the response is complete, exactly as you do for the external events. There is only one difference worth mentioning: While externally driven essential activities have both internal and external results, the results of temporally driven essential activities are almost always entirely external. Temporally driven custodial activities are so rare, in fact, that we used to doubt their existence. In all our experience using this technique, we have encountered exactly two such essential activities. This observation provides you with an additional defense against falsely identifying physical timing constraints as temporal events. You can expect to find temporally driven custodial activities once in a great while, but if they become commonplace, you are probably dealing with essential activity fragments that respond to one of the previously identified external events.

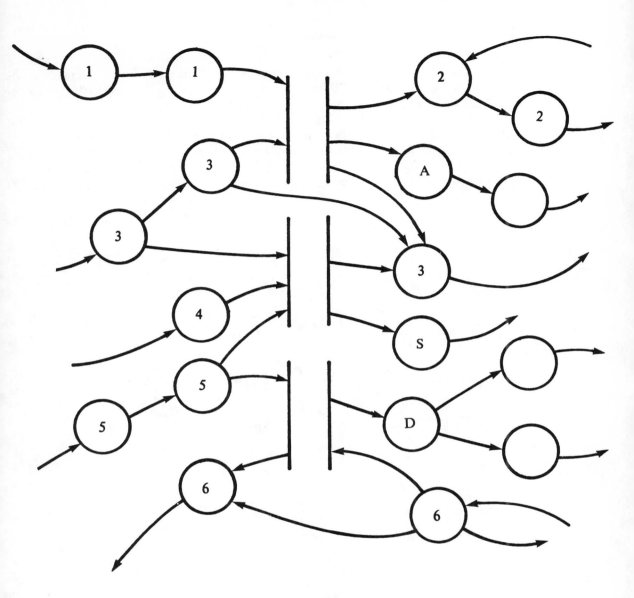

Figure 17.16. First essential fragments to learn of temporal event occurrence.

17.6 Building the initial essential model

Once you have attributed all of the essential activity fragments from the reduced physical model to the external and temporal events that trigger them, you are ready to create a first-cut essential model. You create it by repartitioning the reduced physical DFD into a set of data flow diagrams consisting of at least two levels.

At the most detailed level, you draw a separate data flow diagram for each of the essential activities that you have identified using event partitioning. Each diagram depicts the essential activity fragments that make up the system's immediate response to a single event, and the associated data stores and dataflows. In Figures 17.17a, b, and c, each diagram shows one of the three essential activities of the materials supply system.

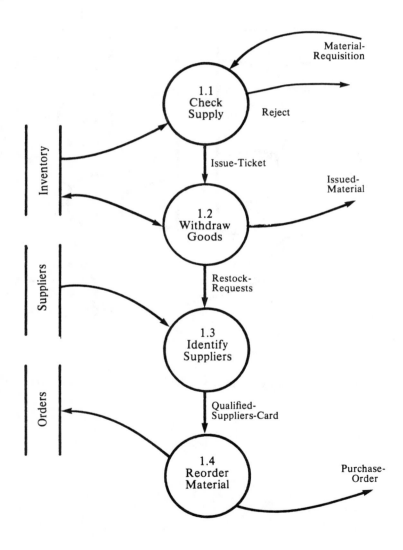

Figure 17.17a. Lower-level DFD for Issue Stock.

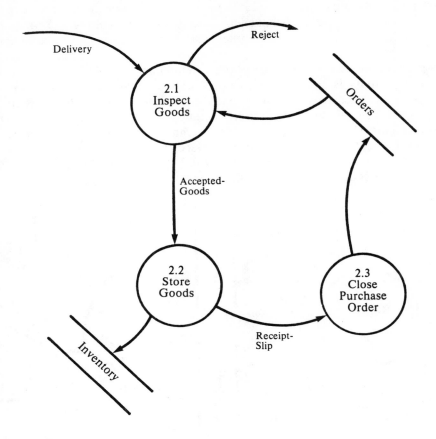

Figure 17.17b. Lower-level DFD for Accept New Stock.

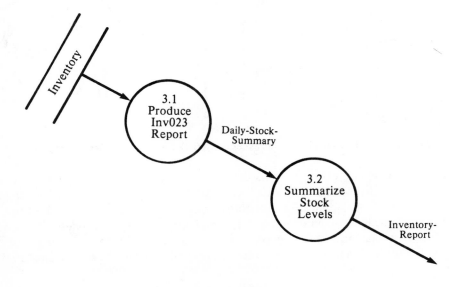

Figure 17.17c. Lower-level DFD for Report On Stock.

These diagrams serve as the primary input to the next step in the logicalization procedure, in which you study the fragments of each essential activity in detail. But you need more than this one lower-level view of the system if you are to produce a consistent and integrated essential model. You need to be able to see easily how the essential activities interact with one another, and how they share essential memory data stores.

To get this broader perspective, create a temporary upper-level diagram in which each essential activity is shown as a single process, as in Figure 17.18. This diagram coordinates the detailed diagrams of fragments attributed to individual essential activities. The activities in Figure 17.18 are the parent activities for the individual fragment diagrams in Figures 17.17a through c. This diagram is only temporary because you will have paid little attention to the data store interfaces between essential activities. As the result of later work on defining essential memory (described in Chapter 20), you will revise the high-level diagrams drawn at this point.

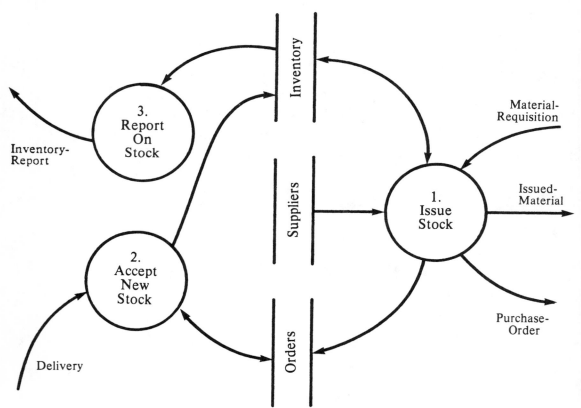

Figure 17.18. Temporary global essential DFD for materials supply system.

The only drawback of an essential-activity-level data flow diagram for a real system is the complexity of the diagram itself. Most real systems respond to dozens of events, and some large applications systems respond to several hundred events. Because of the importance of controlling the complexity of system models, it is not wise to build data flow diagrams that contain hundreds, or even dozens, of processes. So in real life, you should consider creating additional levels of data flow diagrams above the essential activity level. These diagrams contain processes that represent groups of essential activities, and each of the diagrams should of course be limited to a reasonable level of complexity.

Having partitioned the reduced physical model so that it is effective yet not physical, you now can work on a reasonable portion of the essential activity fragments without becoming entangled in all of them. In addition, since you have placed the fragments in groups that are related only through data stores, groups of analysts can work simultaneously, each on a different essential activity. This will speed the analysis effort, without excessive effort being devoted to coordinating the groups.

17.7 Summary

This chapter discusses a procedure for classifying essential fragments in order to partition the system into pieces simple enough to work with. Each fragment is assigned to an essential activity; you discover which fragment belongs with which activity by using event partitioning.

The trickiest part of the procedure is the first step, when you must identify the events to which the system responds. External events are easier to find than temporal events, since they appear as boxes on the context diagram of the current physical model. Temporal events do not appear on the context diagram. To find them, you locate those outputs from the system that are not the result of an external event. You trace these outputs back to their origin, verifying that the origin is a temporal event.

The next step is to identify which fragments are part of the response to each event. You identify the fragment that first learns of the occurrence of an event, and then follow the path of the system's response from fragment to fragment. This means matching the dataflows that enter and leave the fragments. You draw a new DFD for the fragments belonging to each event — each DFD corresponds to one essential activity. After you have each model, you verify that each response is complete.

Your last step is to build the preliminary essential model. This consists of two levels of data flow diagrams: a lower level that contains a separate DFD for each essential activity and an upper level that shows each essential activity as a single process. If there are too many essential activities, an even higher-level diagram is needed. Each bubble on the higher-level diagram represents a group of essential activities.

Part Five

Defining an
Essential Activity

Part Five contains the central part of the derivation process and of the book. The steps described in the next four chapters accomplish most of the work of deriving the essence of an existing system from a current physical model. Because the rationale for these steps becomes clear only if you understand the whole process, Chapter 18 gives a preview of the remainder of the derivation procedure: It summarizes both the derivation of a single activity model, covered in this part, and the integration of the individual essential activity models into a single essential model, discussed in Part Six.

Chapter 19 reviews the physical characteristics that are likely to remain in the classified essential activity and memory fragments that were produced by the previous steps. These are a subset of the physical characteristics presented in Part Three — namely, those that would not have been eliminated through expansion, reduction, and classification.

Chapter 20 sets out the most important procedure in the book: transforming a set of fragments that respond to a single event into a logical model of that one essential activity. The result of this step is a definition of the activity's essential response, its required essential memory, and the information it needs from the outside world in order to carry out the response.

Chapter 21 shows you how to use the tools of structured analysis to model the essential activity defined in Chapter 20. At the conclusion of this part of the derivation process, the result is a set of essential activity models, each of which is a totally logical expression of the requirements for that activity.

Chapter 18

Completing the Derivation Process: An Overview

We now approach the second stage, or the heart, of the logicalization process. The remaining steps in the process are a distinct departure from the kind of work you performed to expand, reduce, and classify. These steps do not transform a model so much as they re-create it. Whereas the model that resulted from each of the previous steps consisted of the same components in the same form as in the working model used as input, now in the next steps, the model's physical organization becomes totally different from that of the existing system. Although the detailed activities and the data elements remain the same throughout the next steps, they are assigned to completely new data stores and higher-level activities. These data stores and activities compose what will be an entirely new model of the essential features of the system. The model must adhere to all of the essential modeling principles introduced in Chapter 6: It should contain no technological bias, it should be easy to understand, and it should state the minimum requirement for essential activities and memory.

Because of the creative leap required to derive an essential model from the classified activity and memory fragments derived in the previous step, the derivation procedure from now on appears in many ways like a strategy for creating an essential system model from scratch. This change in strategy and the complexity of the procedure necessitates our first presenting an overview in this chapter of the procedure for re-creating the essence of a system based on a classified set of essential fragments. We also explain the benefits and disadvantages of the strategy, so that the next chapters can deal exclusively with the procedure's several parts.

18.1 Partitioning the remaining work

The key to analyzing a complex system is to limit your exposure to its complexity. You must plan your work so that at any one stage you are making only a certain number of changes to a limited portion of the system being studied.

The desire to gain control over complexity is the driving force behind our choice of strategy for the remainder of the derivation process. You face two types of complexity as you derive essence: First, the process of deriving the essence of an existing system is itself difficult. You have to make many changes to the model, and you must base these changes often on judgments that are uncomfortably subjective. Second, the system under study may very well be both large and complicated. Systems analysts usually study systems composed of dozens of human and automated processors manipulating hundreds of items of data. Taken together, these two facets of complexity promise a formidable task ahead.

To make this task easier, we partition the work to be done in two ways: We limit the types of tasks performed at any one time, and we limit the portion of the system to be worked on at any one time. To achieve the dual partitioning of the derivation process, the key is to make two separate passes through the entire system model.

18.1.1 *The two-pass approach*

In the two-pass approach, you first focus your attention on one individual essential activity at a time. This first pass limits your exposure to complexity since you concentrate on only a small portion of the system at a time. Only when you are finished building each essential activity model do you combine these models and proceed to develop the model of essence of the whole system. During this second pass, you can make sure the interfaces are correct. Because different transformations are made to the system during each pass, the two-pass approach also limits the types of tasks performed at any one time. Figure 18.1 is a data flow diagram of the two-pass approach.

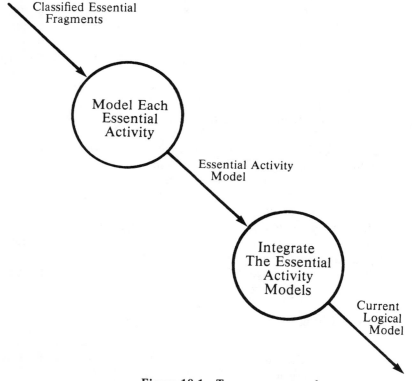

Figure 18.1. Two-pass approach.

Working on each essential activity separately is different from your focus so far in the derivation process. In the previous steps, you considered the entire existing system or the entire current physical model. You expanded all of it, reduced all of it, and used event partitioning to group essential activity fragments from throughout the model. Now, however, you focus initially only on the classified activity and memory fragments that make up a single essential activity, and you model each essential activity separately before considering the total model. With few exceptions, this modeling tactic applies to nearly all system development efforts. In this section, we describe the reason for our choice, and its advantages and disadvantages.

In this part of the logicalization procedure particularly, you study the existing system in great detail, and you can't afford to be overwhelmed by a large or complex chunk of the system. That is why you employed event partitioning in the previous step to divide the system into easily digested sets of physical fragments. Further simplifying your work is the probability that one such set of fragments is not too highly connected to the others.

Making the derivation and modeling of each essential activity a separate task also makes it easier for more than one person to contribute to this phase without excessive loss of productivity. For example, if there are a dozen sets of essential activity fragments and a dozen people on the project team, each person can build a single essential activity model, and all can work simultaneously. Although there will probably be several times as many essential activities as there are team members, our point still holds: Partitioning the project team in the same way as the logical model lessens the need for communication among team members, and so allows them to work in parallel.

You get an additional benefit from this approach during the second pass when you integrate the essential activities. The interactions between essential activities through essential memory are very complex. One essential activity may produce essential memory for many other essential activities, just as that essential activity may consume essential memory produced by many activities. With so many interactions to verify and physical features in these interfaces to remove, you will be grateful that you can concentrate on integrating the essential activities.

Despite its significant advantages, this scheme has a drawback: By focusing your attention, even temporarily, on the fragments of just one essential activity, you are condemned to producing an imperfect essential model. The imperfections come from two sources: First, some of the actions carried out by an essential activity may seem essential when in fact they are physical. Sometimes, you must be able to see the rest of the system before you can accurately determine that an activity is nonessential. Second, since the system's essential memory connects multiple essential activities, you cannot derive an accurate model of the essential memory for one essential activity without considering other essential activities simultaneously: namely, the fundamental activities that need the remembered information and the custodial activities that supply it. As you might guess, the benefits offered by the two-pass approach — control of complexity and ease of work allocation — outweigh the disadvantages cited here.

The first pass in the two-pass approach creates a set of essential activity models that are *locally* logical, but that may retain a hint of technological prejudice despite your best efforts. Because of the imperfections, you are forced to clean up the model in the second pass. That means you have to rework individual essential activities once you can see the entire, global view of the system's essence. In Part Five, we present the first pass through the derive-and-model cycle: the derivation of essence for an individual essential activity. Then, in Part Six, we discuss the second pass, the derivation of essence for the integrated set of essential activities.

To further simplify the tasks ahead, we partition each pass into two steps. In the first step, you *define* the essence of the portion of the system you are studying; in the second step, you *model* the essence using the tools of structured analysis. Figure 18.2 illustrates these subcomponents of the two-pass approach.

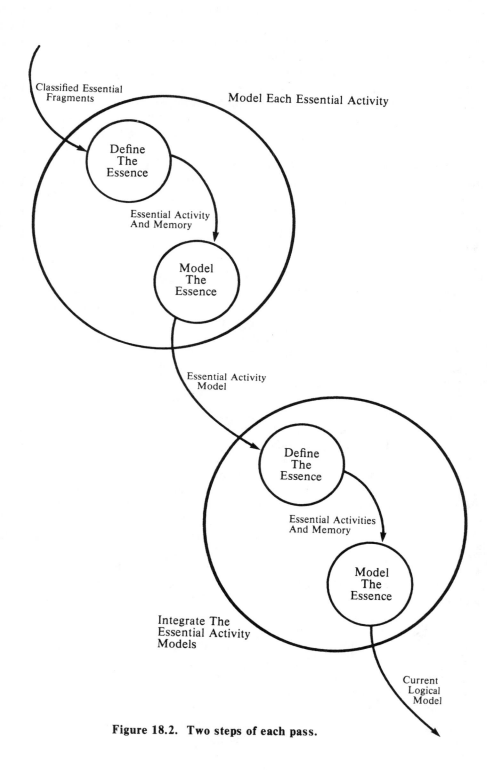

Figure 18.2. Two steps of each pass.

Starting with the fragments of a single essential activity, you first identify its remaining physical traits and recast its essence without them. Then you model the essential planned response using data flow diagrams, data dictionary definitions, and minispecifications. Once you have a set of local essential activity models, you apply the same two steps to the set: You define the essence of the entire system by removing the remaining physical characteristics of the stored data, and you construct an integrated model of the system's essence using the same modeling tools.

18.2 The difference between creating and deriving

There are two ways that the next part of the derivation of essence resembles the creation of the essence from scratch. In both tasks, you start with a blank piece of paper. Rather than marking up the DFD that contains the classified essential fragments, you examine the fragments, re-create their essence in your head, and commit it to a fresh sheet of paper. The second similarity concerns your point of view during this process. When you create the essence from scratch, you focus on the external interfaces because there are no system internals to study. Similarly, to derive the essence, you view each essential activity as a set of interactions with the systems, people, and organizations that lie outside the system being studied.

Despite the similarities between creating essence and deriving essential activities from the classified fragments, there is a big difference in the application of essential systems analysis to each situation. When you derive essence, you have evidence from the current physical model that can help you determine what should be in the essential model. When you create essence, you have no such help.

The existence of the current physical model means both good news and bad news for you at this point in the derivation process. The good news is that you can feel more secure that you are deriving an appropriate essence. The existing essential responses have been tested for appropriateness in the system, so if there are any problems with the essential responses, you are likely to know about them. It would be much riskier in the typical case to try to create a new essence that hasn't been used before. As good as the creation process may be, you may not be able to make up for the years of evaluation and refinement of essential response that take place in an existing system.

The bad news is that you have to capture faithfully the existing essence even though it is still camouflaged within the classified fragments. These fragments still contain substantial physical bias, and they are modeled in a way that inhibits easy understanding of the essence of the essential activity. So, even though you have a basic plan of attack and evidence of the essence of the existing system, deriving essence from this point on remains difficult.

18.3 Summary

In order to make the task of deriving essence less complex, you use a two-pass approach. The goal of the first pass is to model each essential activity separately. Then, in the second pass, you combine these models to develop the model of essence for the whole system. The result is a complete reorganization of the classified essential fragments. The procedure requires some of the creativity you would need to derive the essence from scratch, but it is less risky, since the classified fragments come from the model of an existing system whose responses have been tested and refined in use.

Chapter 19

Recognizing Remaining Physical Characteristics

Despite the many physical features already removed from the model at this point, the essential activity fragments and essential memory fragments still contain substantial technological bias. Although each fragment contains activities and memory that are needed no matter how the system is implemented, how these activities and memory are realized is physical. Sometimes, in a previous step, for example, the modelers merely chose a label for an activity that betrays a prejudice toward the existing implementation technology. More frequently, the remaining technological bias in the essential fragments stems from their heritage as being formerly a part of an existing processor-based incarnation. Despite your work to eliminate processor boundaries, communication facilities, and administrative activities, they all have left their mark on the essential fragments. All such bias must be completely removed before you can build a truly essential model.

In this chapter, we present a catalog of the most common physical traits found in the essential fragments. Our discussion is in two parts: We discuss first the subtle physical features remaining in the fragments that are part of a given essential activity, and, second, those features remaining in data containers that hold the essential memory produced or consumed by a given essential activity.

19.1 Physical characteristics of essential activity fragments

The fragments classified as part of an essential activity exhibit specifically five physical characteristics: fragmentation, extraneousness, redundancy, convolution, and nonessential sequence. With the exception of the last characteristic, these are among the qualities of technologically imperfect systems described in Chapter 4. The additional two qualities discussed there, conglomeration and vastness, have been eliminated from the model by reduction and by classification of fragments. In the following subsections, we explain how to recognize the five remaining physical characteristics, so that in the next steps you can eliminate them as well.

19.1.1 Fragmentation

In the current model, essential activities are fragmented to accommodate the skills and capacities of the available imperfect processors. Although in previous steps you removed all indication of which processors carry out what fragments, and you identified which fragments make up the response to what event, you did not change the initial fragmentation from that in the current physical model. Hence, the boundaries between individual fragments are still physical.

Figure 19.1 illustrates how the division between fragments is determined by the processors that carry out the activities. In the diagram, the string of essential activity fragments constitutes the response to the event "customer orders books." The large circles in the figure represent processor boundaries sketched in temporarily, although they were removed earlier in the process. Each processor boundary corresponds to an essential activity fragment.

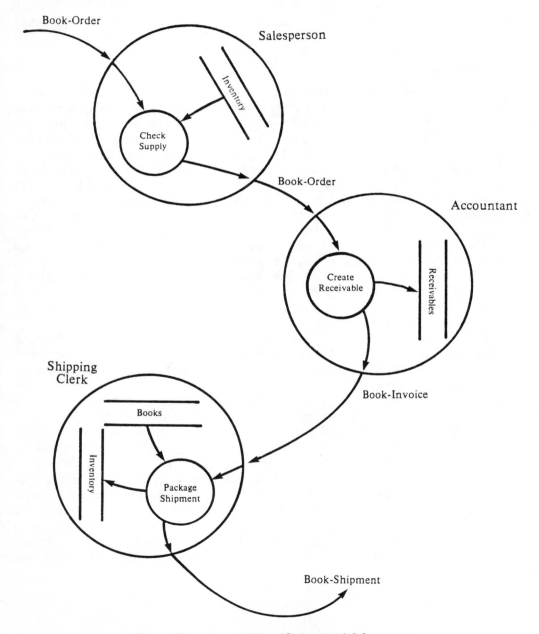

Figure 19.1. A set of classified essential fragments.

When this system was built, the fragments chosen were those that best suited the imperfect technology used to implement the essential activity. An accountant handles the creation of a receivable record, since he or she has the necessary accounting skills to carry out that part of the job. A burly shipping clerk takes care of packing the cartons of books to be shipped out. The salesperson checks the inventory, because only he or she can inform the customer immediately if there is insufficient stock to fill an order. The dataflows that connect the activity fragments are also features of physical fragmentation. Later, you will probably choose a different lower-level partitioning for each activity or perhaps no lower-level partitioning at all. If a lower-level partitioning is created, it will usually show different dataflows connecting the new lower-level activities.

19.1.2 Extraneousness

The many detailed infrastructural activities that remain in the fragments are extraneous, since a perfect technological environment has no need for any kind of infrastructure. They remain in the model because they aren't obvious enough to be removed in the reduction process. However, you find them now, because this is the first time you search the details of the essential fragments for subtle physical features.

The most common extraneous activities are those that transport data, unmodified, through an activity fragment. In the example above, assume that the activity Create Receivable receives the shipping order that informs the shipping clerk of the composition and destination of an order. Create Receivable does nothing with this data except pass it on to the Package Shipment fragment. Such a service would be unnecessary in a technologically perfect environment, because data would flow only to the point where it is needed. In the real world, however, the scarcity of resources for the infrastructure may dictate that the system route a dataflow through a processor that doesn't need it in order to avoid creating a separate physical dataflow. As a result, the dataflows entering and leaving a fragment that passes along transient data will contain extraneous data elements.

Another kind of extraneous activity is best explained through a new example. Figure 19.2 shows the two fragments for the essential activity Accept New Subscription. As is often the case, these fragments fall on either side of a data store. The activity Record New Subscription initializes the status flag to "new" and puts the name and address of the new subscriber, the type of subscription, and the expiration date into a data store, Subscription-Master-File. Later in the day, the fragment Print Notices goes through the master file, finds all of the new subscriptions, and prints the appropriate notifications. The new subscription notifications are batched together in this way to optimize the process of printing them. This batching adds three extraneous activities to the fragments of the essential activity: one to set the status flag, one to search the master file, and one to test the value of the status flag. With perfect technology, the system would be able to print new subscription notifications immediately upon receipt of an acceptable application. Although Subscription-Master-File may be extraneous since Accept New Subscription does not need it, remember that some other essential activity might — you won't be able to know this until you compare essential activities during the second pass.

19.1.3 Redundancy

Imperfect technology leads to infrastructures that store multiple copies of the same data element or set of data elements. This stored data redundancy, in turn, leads to the repetition of the activity that updates the redundant data. Figure 19.3 shows the fragments that make up the essential activity, Accept Mortgage Payment. Payment data is recorded in three places: a file of payment documents in the bank's branch, a master file within an automated mortgage system, and an online extract file of payment information available to the collections department.

Implementing this essential activity with perfect technology, you could consolidate the essential memory containing information about payments and thus remove the activities that updated the redundant copies of payment information. Like fragmentation and extraneousness, redundancy of activities leads to physical features in dataflows. Specifically, many dataflows carry data elements used by the activities that update redundant files. Once the redundant updates are removed, these data elements and perhaps the dataflow they are part of will be removed as well.

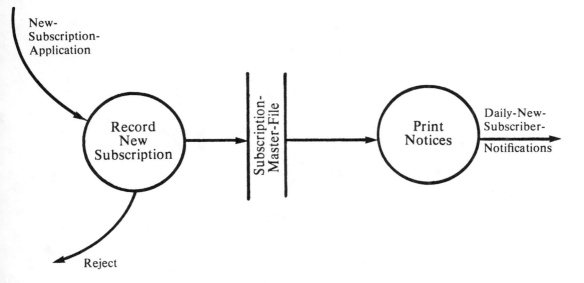

Figure 19.2. Fragments of the activity Accept New Subscription.

19.1.4 Convolution

Imperfect technology encourages efficiency in processing but unfortunately discourages customized access to stored data. In our model, activities get their data from variety show data stores. They are made up of data structures, which are nothing more than particular arrangements of information. (Some common physical data structures are forms, batch transaction records, online screens, and network data packets.) When a single data structure is used to convey different data elements to many activities, invariably that structure is convenient for use by some activities, while others must perform extra activities to get the data they need. As a result, some activities are convoluted in their approach to obtaining data. You want to examine the fragments for evidence that they must deal with awkwardly organized data structures.

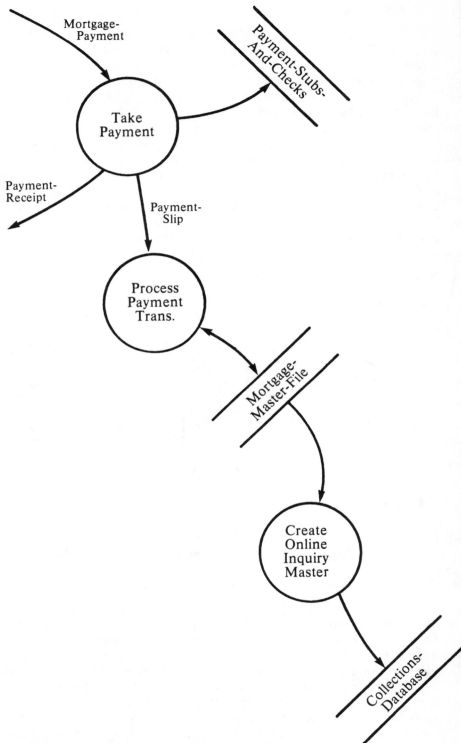

Figure 19.3. Fragments of the activity Accept Mortgage Payment.

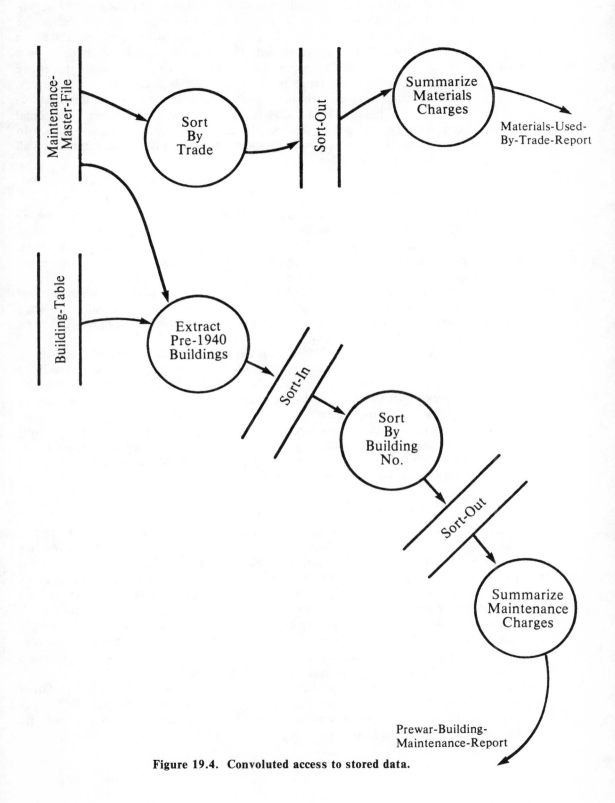

Figure 19.4. Convoluted access to stored data.

Figure 19.4 shows two examples of convoluted activities. In each of them, unnecessary sorts and extracts must be performed so the activity can select or derive the data it needs. In the first example, the Maintenance-Master-File is not organized so that the activity can easily refer to the records that describe the materials used by a given trade. Hence, the Sort By Trade process is necessary to sort through the entire master file. Similarly, the data structure in Building-Table is such that two processes, Extract Pre-1940 Buildings and Sort By Building No., are needed to get the data required for the activity. These extra physical features violate the principle of minimal modeling. Even though a perfect processor could execute the convoluted activities with ease, you must streamline these activities to specify the minimum requirement for all technologies, perfect or otherwise.

19.1.5 Nonessential sequence

One of the most obvious characteristics you observe about fragments making up an essential activity is that they occur in a certain order. In the DFD, the activity fragments are often linked to reflect that one activity fragment is executed after another. Sometimes, the DFD indicates that two or more activities can happen at the same time, and so no sequence is specified in this case.

There are three factors or limitations that influence the order in which actions take place in a planned response system: the activity's essence, the existing system's technological limitations, and the limitations introduced, often unwittingly, simply by the way you model the existing system.

These factors dictate either an essential order or an artificial order. Of course, in an essential activity model, the only valid justification for ordering the actions within an activity is that the order arises from the system's essence. Therefore, only the first type of sequence, the essential order, is allowable in an essential activity model. In order to produce such a model, you must identify the signs of artificial sequences created by one of the other two kinds of limitations: technological limitations and modeling tool limitations. Artificial sequence introduced by modeling tools was treated in Chapter 8. In the following subsections, we discuss artificial sequences caused by current technology, after first defining the essential order more carefully.

19.1.5.1 Essential order

Certain parts of a system's planned response to a given event must happen in a sequence, simply because some actions cannot be carried out unless another action has already been completed, as when an activity needs information that results from a prior computation. The sequence of activity dictated by such dependencies is the *essential order* of a planned response. Ultimately, the essential order must be built into the essential activity model so that the model is independent of an imperfect technology.

Consider the essential activity, Pay Hourly Workers, discussed earlier. Turning a time card into a paycheck is not simple, for the system must carry out several calculations and record the resulting data about earnings. Many of these activities can be done at the same time; for example, the system can calculate state tax and federal tax simultaneously. Figure 19.5 shows fragments of a similar essential activity, Pay Salaried Workers, that can be executed in parallel. This DFD shows that neither Update YTD Earnings nor Issue Check can be executed until Calculate Earnings has produced its output, but Update YTD Earnings and Issue Check can be executed simultaneously.

In a few cases, the system has no choice but to execute activities in a rigid sequence. For instance, it cannot compute the worker's net pay until the gross pay is known. It cannot compute net pay first, and it cannot compute net pay and gross pay simultaneously, *even if it has perfect computation technology,* simply because the computation of net pay requires the amount of gross pay as an ingredient.

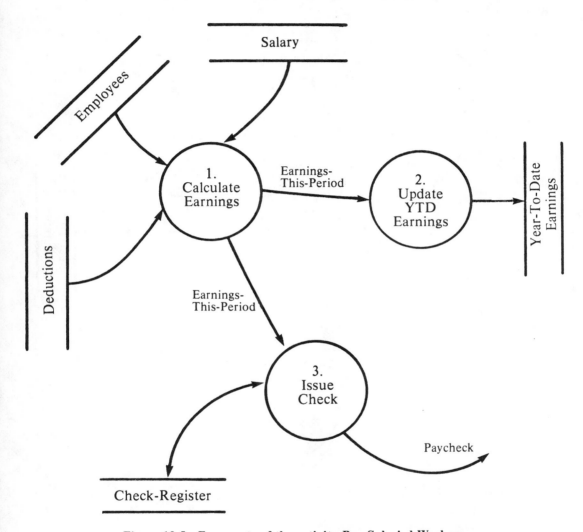

Figure 19.5. Fragments of the activity Pay Salaried Workers.

Most of the nonessential sequence in current physical models results from the use of imperfect technology to implement existing systems, although some nonessential sequence arises from the way analysts use modeling tools to depict the existing system. In general, the artificial ordering of actions in a current physical model shows up in two forms: The system performs actions in sequence that could be conducted in parallel, and the system contains time delays, or breaks, in a stream of activity.

19.1.5.2 *Artificial order for independent activities*

Figure 19.6 shows the same detailed activities as Figure 19.5. However, Update YTD Earnings and Issue Check, which were sequentially independent in Figure 19.5, are carried out in sequence in Figure 19.6. Why did this happen? Typically, either or both of two major causes are responsible: processor limitations and infrastructure limitations.

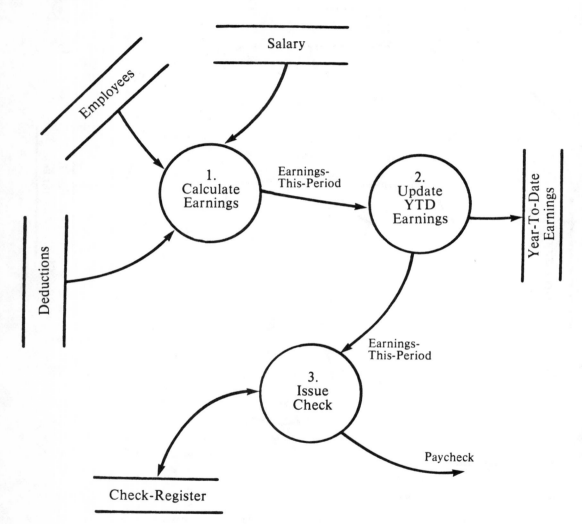

Figure 19.6. Activity fragments in artificial sequence.

Perhaps in the existing system, the three activities in Figure 19.5 are all carried out by the same processor, and that processor, like most, can't do two things at once. If the processor is human, he or she may make an arbitrary decision about whether to issue the paychecks first or to record the earnings data first. Maybe the processor will switch back and forth, some days writing the checks first and other days updating the earnings records first. It doesn't matter which order is chosen, since a correct result is produced either way. These processor limitations could be surmounted if the system had a processor that could do two things at once, or even if it had two sequential processors, one to write checks and one to record the earnings. If the existing system lacks such resources, however, instances of artificial sequence will result from the limitations of the processors.

Sometimes, the existing system does employ enough processors to perform two activities simultaneously only to be thwarted by the lack of sufficient infrastructural support. Figure 19.7 shows two processors, Bob and Jack, who are certainly capable of working independently, and the essential model of their activities in Figure 19.8 reveals that their tasks are not related. So why does the current physical DFD portray these actions sequentially? In this case, the reason is not a lack of adequate processing technology, but a lack of adequate access to the common data requirement: the sales details contained in the SR100 computer report. As it turns out, Bob and Jack share an office, and so they receive only one copy of the SR100. While this efficiency tactic might work, it does obscure the fact that Bob and Jack carry out two asynchronous and unrelated tasks. If the infrastructure that makes the sales details available to them were just a little snazzier, they both could do their work at the same time, and it would be that much easier to see the essence of their actions. In this case, the infrastructural limitation on access to data causes the physical characteristic, not the processor limitation.

Giving a sequence to what should be parallel activities is a major cause of extraneous activities and extraneous data within dataflows. Figure 19.9 shows this cause and effect relationship. There is no reason why with perfect technology you couldn't create a book receivable and package books at the same time, but these activities appear in sequence in the figure as the result of limited infrastructure. If you assume that some of the data needed for packaging a book is irrelevant for creating a book receivable, then Create Book Receivable contains an extraneous activity: It must pass through the data needed to package books whether it transforms the data or not. Since the extraneous data for each activity fragment must be brought in by dataflows, then both Book-Order and Book-Invoice contain extraneous data elements. All data extraneous to dataflows and activities should be identified when you find parallel activities that have been tied together in sequence.

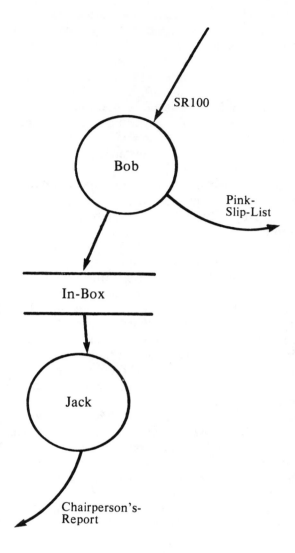

Figure 19.7. Artificial sequence caused by an infrastructural limitation.

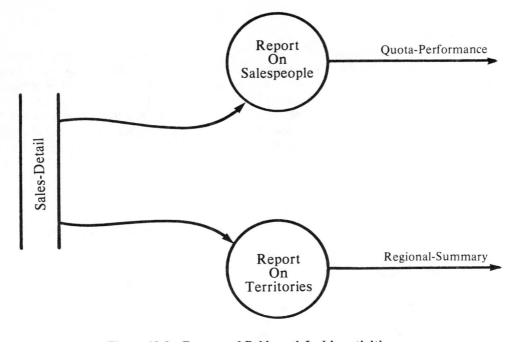

Figure 19.8. Essence of Bob's and Jack's activities.

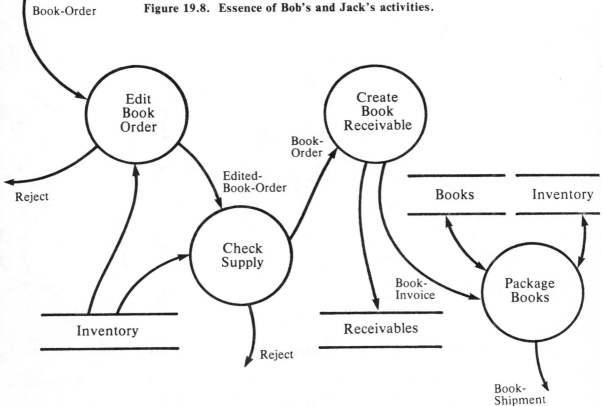

Figure 19.9. Fragments of Satisfy Book Order arranged in an artificial sequence.

19.1.5.3 Artificial time delays

Before the discovery process starts, the current physical model has data stores between fragments like those in Figure 19.10. These batch data stores allow one or more processors to synchronize their activities to make up for capability or capacity limitations. In the reduction step, you remove as many of these purely physical data stores as you can find. Yet, even if all the batch data stores were removed, the activities that used the stores have physical features: For purely physical reasons, the procedures in those activities are arranged to operate on batches of data. Some of these procedures are therefore nonessential.

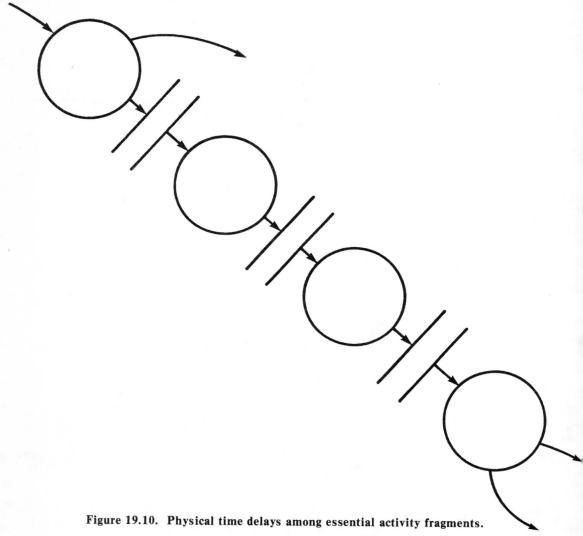

Figure 19.10. Physical time delays among essential activity fragments.

Figure 19.11 shows some of the typical activities needed to process batches of data. One activity that is not apparent in the figure is the test to see whether the batch is ready to be processed, which is part of the activity Submit Completed Batch. Often batch-processing activities are executed repeatedly until the batch has been fully processed, such as the Add Document To Batch activity in the figure. When you find batch-handling procedures in the fragments assigned to an essential activity, you have identified some subtle, but nonetheless physical, features of the fragments.

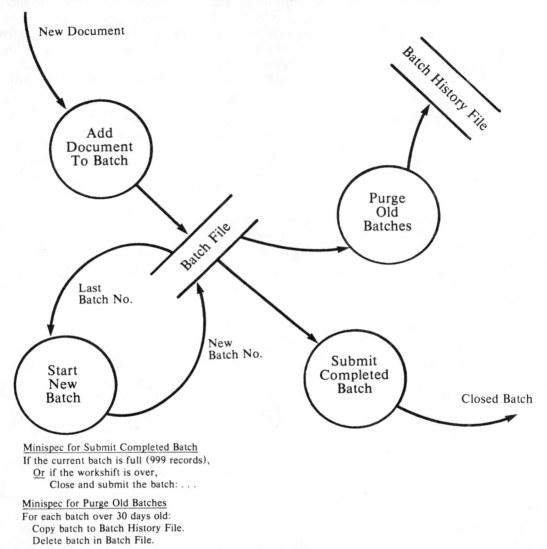

Minispec for Submit Completed Batch
If the current batch is full (999 records),
    Or if the workshift is over,
        Close and submit the batch: . . .

Minispec for Purge Old Batches
For each batch over 30 days old:
    Copy batch to Batch History File.
    Delete batch in Batch File.

Figure 19.11. Batch-handling activities.

19.2 Physical characteristics of essential memory fragments

Most of the same physical features found in activity fragments also appear in the essential memory fragments accessed by an essential activity. These features are fragmentation, extraneousness, redundancy, and convolution.

19.2.1 Fragmentation

The physical data store associated with an essential activity usually contains only a portion of the data that characterizes any one object. In fact, the typical physical data store contains data elements from many objects. This is what we call the fragmentation of essential memory. That fragmentation is present in the classified set of fragments that you must recast into an essential activity model.

The data dictionary definitions below show examples of fragmentation within physical data stores.

Telephone-Reservation-File
$$= \{ \text{Check-In-Date} + \text{No.-Of-Nights} +$$
$$\text{No.-Of-Guests} + \text{Guest-Name} +$$
$$\text{Guest-Address} + \text{Guarantee-Flag} +$$
$$\left(\begin{bmatrix} \text{Credit-Card-ID} \\ \text{Company-Name} + \text{Company-Address} \end{bmatrix} \right) +$$
$$\text{Quoted-Rate} \}$$

Front-Desk-Reservation-Board
$$= \{ \text{Guest-Name} + \text{Check-In-Date} +$$
$$\text{No.-Of-Nights} + \text{No.-Of-Guests} +$$
$$\text{Guest-Address} + \text{Guarantee-Flag} + \text{Quoted-Rate} \}$$

Front-Desk-Message-Board
$$= \{ \text{Room-No.} + \text{Guest-Name} + \text{Date-Received} +$$
$$\text{Time-Received} + \text{Guest-Message} \}$$

Room-Status-List
$$= \{ \text{Room-No.} + \begin{bmatrix} \text{``Occupied''} \\ \text{``Vacant-Clean''} \\ \text{``Vacant-Not-Ready''} \end{bmatrix} \}$$

Guest-Folio
$$= \{ \text{Room-No.} + \text{Guest-Name} + \text{Guest-Address} +$$
$$\text{Check-In-Date} + \text{Room-Rate} +$$
$$\text{No.-Of-Guests} + \text{Guarantee-Flag} +$$
$$\left(\begin{bmatrix} \text{Credit-Card-ID} \\ \text{Company-Name} + \text{Company-Address} \end{bmatrix} \right) +$$
$$\{ \text{Room-Charge} \} + \{ \text{Guest-Credit} \} \}$$

Room-Rate-Card
$$= \{ \text{Room-No.} + \{ \text{No.-Of-Guests} + \text{Room-Rate} \} \}$$

The data elements describing the guest and his or her reservation are fragmented among several physical files. Notice that the name of each file gives you a clue as to the identity of the processor for whom the file was created. The Telephone-Reservation-File serves the reservations operator, the Front-Desk-Message-Board is used by the desk clerk, and so on. The data describing the rooms is similarly fragmented. Rate information, which changes infrequently, is printed on a card that is available to the front desk clerk and to the reservations operator. The condition of an individual room, however, may change every day. So, this data is allocated to a different container, one that is shared by the front desk staff and the housekeepers.

Fragmentation of stored data results from a combination of two factors. First, the fragmentation of essential activities has the consequence of fragmenting the stored data needed by a given activity. Second, essential memory must be fragmented so that the infrastructure does not have to distribute large numbers of data elements, which would either overload the storage capacity of a central physical file or the capacity of processors that have to search through large data stores.

19.2.2 Extraneousness

The physical data stores accessed by the activity fragments contain extraneous data elements, or those that would be unnecessary if the essential activity were implemented with perfect technology. There are two types of extraneous data elements. The first type is the purely physical data element, which exists solely because of data storage and retrieval technology; examples are pointers, record offsets, and parity bits. The second type of extraneous data element includes those that appear to be needed by the essential activities in the system you are modeling. There are two kinds of essential-looking stored data elements that are actually physical features resulting from imperfect data storage technology. The first kind results from the consequence of setting up the physical data stores to handle the data needs for many essential activities. So, of all the data elements that an essential activity can access, only a subset of them will be added, modified, or deleted by the essential activity and the rest are extraneous to the model finally produced for that essential activity.

A second kind of physical stored data that appears to be essential is data that the essential activity needs but that does not need to be stored. We call this data *transient* data. The most obvious instance of transient data is the data in batch data stores. Although all batch data stores were removed in the reduction step, the remaining physical data stores are likely to contain some residual transient data. This physical feature remains in the model at this point, because you do not remove any data stores that contain at least some essential memory.

Figure 19.2 shows fragments accessing a data store that has essential transient data. With perfect internal technology, the process Print Notices could begin as soon as Record New Subscription supplied it with the data it needs. But instead, Print Notices must access Subscription-Master-File for the required information, because of the batch processing technology used in the existing system. Although other activities may need to use data from Subscription-Master-File, as far as this essential activity is concerned, the data elements in that file do not need to be stored.

19.2.3 Redundancy and convolution

Chances are good that there are multiple copies of the same data elements in the different data stores accessed by the fragments of an essential activity. For an example, look at the data dictionary definitions for the data stores in Subsection 19.2.1. Front-Desk-Message-Board and Room-Rate-Card are used by the essential activity Check In Guest, and they contain a data element in common, Room-No.

All of the reasons for keeping more than one copy of a given stored data element are physical. Perhaps duplicate copies are used as backup in case one copy of the data is destroyed, or possibly the same keys are used to implement interobject relationships. Redundant data also arise because the implementors couldn't afford to provide a central, globally accessible data store to supply stored data to all processors in the system.

The final physical characteristic to recognize in essential memory is convolution. We already discussed the convoluted activities that result from variety show data store organization. Similarly, there will be convoluted variety show data structures, since the physical stored data and the physical access activities are mirror images of one another.

19.3 Physical names

Regardless of whether a feature of the classified fragments is physical, the feature may have a physical-sounding name. Perhaps there is a dataflow called Bob's-Report or one called Monthly-Charge-Tape, or a data store called Rolodex-File. Naturally, you must change these physical names when you derive the essential activity model.

19.4 Summary

In this chapter you saw the effects of imperfect internal technology within the partially essential fragments. The partitioning among fragments of the same essential activity is one such effect. Similarly, the imperfections of data storage and retrieval technology forced the current system's developers to fragment objects among different data stores. Consequently, data stores often contain some of the same data elements as other stores, and the activities that update these redundant data elements must also be repeated. The classified fragments contain activities not needed by the essential activities they belong to, and in a similar way data stores contain data elements not used by the essential activity that accesses the stores. Moreover, the cost of the imperfect technology rewards system developers for organizing the physical data containers in a convoluted way. This forces some fragments to perform sort and extract activities to get stored data. Finally, activity fragments may be arranged in a sequence when they could be carried out at the same time.

Although by now you have removed the obvious signs of imperfect technology — the processor boundaries, the infrastructure, and the administration — all these physical characteristics are still left in the classified fragments used to construct the essential activity model. In the next two chapters, these physical features are eliminated as you go through a process of re-creating and modeling an essential activity.

Chapter 20

Deriving an
Essential Activity

To derive the essence of an essential activity from classified physical fragments, you follow an eight-step procedure:

1.   Select an essential activity.

2.   Uncover the core of the essential activity.

3.   Remove all extraneous physical features.

4.   Consolidate the remaining fragments.

5.   Partition the essential memory.

6.   Minimize the essential accesses.

7.   Establish the essential order among the activities.

8.   Establish the physical ring.

Since this derivation procedure is both long and involved, we first want to give you a sense of the overall strategy behind it: This strategy says to base the definition of an essential activity upon the responses that it must produce.

20.1 Response-based derivation

To derive an essential activity, you start with an output, or response, and follow it back through the process that creates it. You especially consider what information is needed from essential memory to create the response and what information can be acquired directly from the outside world. (This information from the outside world makes up the inputs to the activity.) You repeat this process for every essential response produced by the activity.

The procedure varies slightly according to the type of response produced by the essential activity. A response is either *external,* going to the outside world, or *internal,* going into the system's essential memory. Figure 20.1 illustrates these two main types of response, either one of which or both can be produced by an essential activity, depending upon whether the activity is wholly fundamental, wholly custodial, or a mixture of the two. Essential activity 1 in Figure 20.1, Issue Schedule Copy, is a pure fundamental activity, so its one response is external: the dataflow Student-Schedule. Issue

Schedule Copy does not have any custodial duties, so it does not produce updates to the data stores. Enroll Student In Course is a pure custodial activity, and except for the Reject dataflow, its responses are entirely internal: updates to the Student, Course, and Enrollment data stores.

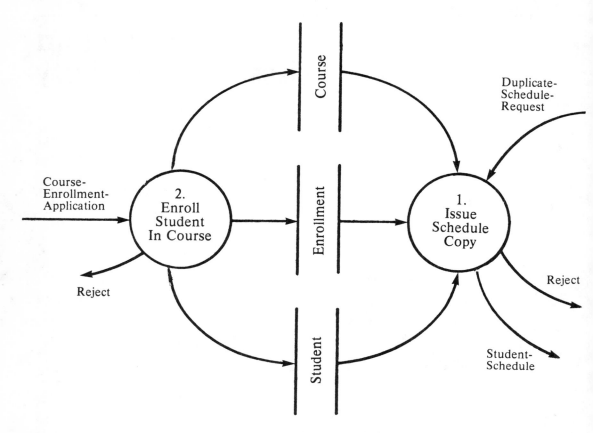

Figure 20.1. Two types of essential response.

To study an external response fully, you must discover what actions are required to create the response and what data elements are required to carry out those actions. Some of these data elements may be available directly from the outside world, while others may come from the system's essential memory. In order to produce a Student-Schedule in our example, the essential activity must have a record of the courses in which the student is enrolled, together with information about the student and the classes. These are its requirements for essential memory. Issue Schedule Copy must also know for which student the duplicate schedule is to be produced. This information must come from the outside world; therefore, there is also a requirement that the incoming dataflow Duplicate-Schedule-Request contain an identifying data element, such as Student-ID-No.

In the case of an internal response, you must define how the data to be stored is acquired from the outside world, and what other essential memory is required to validate the incoming data. For example, Enroll Student In Course in Figure 20.1 makes three updates to essential memory: It records information about the enrollment itself in the Enrollment data store (for example, date of enrollment, audit/pass-fail/grade), and it records the relationship between the student and course in the appropriate data stores. To perform these custodial duties, Enroll Student In Course needs data from the Course-Enrollment-Application, which identifies the student and the courses in which the student wants to enroll. In order to make sure the application is correct, the custodial activity has to decide if the information supplied by the outside world is consistent with the information in its memory. Thus, Enroll Student In Course must check that the student identified on the Course-Enrollment-Application is registered in the Student data store, and that the courses on the application are in the Course data store.

The derivation procedure described below is much more detailed and rigorous. The additional complexity is necessary in order to contend with such practical problems as the lingering physical characteristics of the essential fragments, and the difficulty of modeling essential memory in a fashion that is both effective — easy to look at — and free from physical bias. But don't let the details obscure the underlying approach: You are working backward from the essential responses to the requirements for activity, memory, and external inputs.

20.2 Selecting an essential activity

When you carry out the eight-step procedure presented in this chapter, you work on one essential activity at a time, as though it were completely isolated from the rest of the system. The result is a free-standing model for each activity. Thus, your first step is to select an essential activity to model from the DFD of classified essential fragments.

Since you do not worry at this time about interactions between essential activity models, you could randomly select the activity. However, because the knowledge gained from each model can obviously be applied to the next one you build, the essential activity to be worked on first is the one that will give you the most knowledge about the system's essence.

The activities that contain a large amount of essential information are pure fundamental activities, fundamental and custodial activity mixes, or custodial activities that will affect more than one object data store. Therefore, leave until last custodial activities that only update a single object data store (for example, Change Magazine Subscriber's Address). Since there are probably many essential activities in the DFD that provide much useful information about the essence, you must use your own judgment. But don't worry too much about which is the absolutely right one to select. If you cannot decide, choose the essential activity with the most dataflow and data store interfaces.

20.3 Uncovering the core of the essential activity

After selecting an activity, you want to isolate its logical core, or what is left after the external interface activities are removed. These are the activities that allow the system to communicate with the imperfect technology of the external world, as discussed in Chapter 6. As you recall, the principle of perfect internal technology states that when you derive an essential model, you can remove only the effects of the imperfect technology that is internal to the system.

Your first chance to use this principle is in reducing the physical model. As explained in Chapter 16, physical ring features are the transportation and administrative features kept in the reduced data flow diagram if they exist because of technology used by the outside world. Although these features are true requirements, they camouflage the other essential features, since the inputs and outputs between the essential activity and the outside world take the form of packed data channels. If you apply a response-based approach to a model that includes these physical interfaces, you are likely to view a physical bundle containing several distinct responses as a single response. Instead of tracing the separate path of each response, you may try to trace one terribly complicated path for the bundle of responses. As a result, you would probably fail to derive an essential activity that is as unbiased and simple as it could be. You also increase your chances of failing to recognize all the true requirements.

To avoid omitting essential features, you keep the requirements for the physical interfaces with the outside world, but you separate them from the rest of the essential activity. This separation leaves you with a logical core within the essential activity. You then apply the response-based derivation procedure to this core. In other words, you begin the procedure using the responses of an essential activity *before* they are altered to pass through the physical interfaces to the outside world.

To separate the fragments that handle the physical interfaces, trace the path for every net input and output that connects the classified fragments to the outside world. In each case, mark the point where the activity stops dealing with the physical characteristics of either the input or output and starts performing activities that are independent of these physical interfaces. In both cases, you establish the purpose of the fragments by asking if the activity would be needed even if the world outside the system had perfect technology.*

Figure 20.2 shows an example of physical ring fragments separated from logical core fragments by a dotted line. The logical core of this essential activity is made up of the processes Check Supply, Cut Invoice, and Pack Goods, although Pack Goods is at best a marginal member, since the packing process is somewhat dependent upon the nature of the external infrastructure that must carry the package. Take Order and Ship Goods, on the other hand, are entirely dependent upon the technology through which this essential activity communicates with the outside world. Take Order provides the activity with a way to use the telephone system, and Ship Goods sends outputs through the postal system.

---

*Those of you familiar with structured design will recognize this tactic as part of the transform-centered design approach, used to identify the transform center before creating an initial program hierarchy [33].

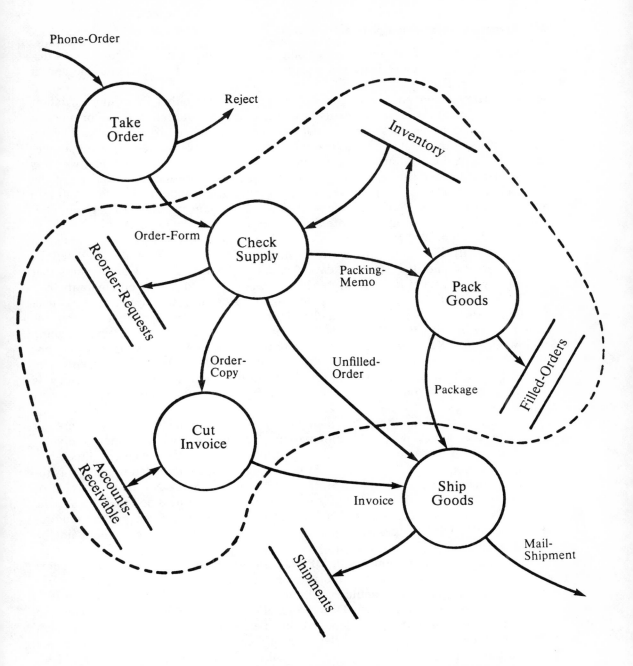

Figure 20.2. External interfaces separated from other essential fragments.

20.4 Removing extraneous physical features

The logical core from the previous step still has some physical features. As discussed in Chapter 19, these features typically are extraneous data elements that appear in data stores or dataflows.

Any data element in a response is obtained from data either in an input dataflow or in essential memory. Data elements in these sources should fall into one or more of the following categories: pass-through data, which is unchanged by the essential activity and appears in the final response; ingredient data, which is used to derive output elements but which does not appear in the final response; and identification data, which is used to identify other data elements within essential memory that are required in the response. The model at this point also includes many data elements that don't fall into any of these categories. Such data elements are found either in dataflows between physical activity fragments or in data store accesses and are logically unnecessary.

The first step to eliminate the extraneous data elements is to select an essential response to analyze. Working backward from the response, you look at each dataflow, data store access, and data store for data elements. When you find data elements that are neither pass-through, ingredient, nor identification data, ruthlessly eliminate them.

The essential activity fragments in Figure 20.3 and the accompanying data definitions below provide examples of the kinds of unnecessary data elements you encounter. The activity Find Candidate Suppliers has only one response, the outgoing dataflow Qualified-Suppliers-Report, so you begin your analysis by comparing its contents with those of the data stores and the incoming dataflow, DPR715. The only data element to pass through the essential activity from the input flow to the output flow is Stock-Description, so you cannot eliminate it from the essential activity, as it is a part of the required response.

Because none of the other data elements in DPR715 makes it all the way through the activity, they are candidates for elimination. But you must evaluate them individually. The reason becomes clear if you first consider the data element Stock-No. Although it is not part of the response, Stock-No. cannot be eliminated because it is required in order to make use of the Stock-Supplier-Index. Stock-No. is an example of identification data: It is used to identify the suppliers that are qualified to sell a specific product. Notice that not even perfect internal technology would change this requirement, since the DPR715 comes from the outside world; its contents would not be affected by changes within the system.

The remaining data elements in DPR715 are unnecessary. They do not become part of the output, they are not ingredients of the output data elements, and they are not identification data. They should be eliminated.

Just as you have pruned the DPR715 dataflow of extraneous elements, you also prune the data stores. The data elements Supplier-Name, Supplier-Address, Contact-Name, and Contact-Phone-No. are elements of Supplier-Catalog that wind up in the Qualified-Suppliers-Report, so they are essential. Supplier-Phone-No., Date-Qualified, and Date-Of-Last-Order play no role in this essential activity, so they should be eliminated.

DPR715   = * Report of stock items to be ordered *
                                = Report-Date + Calendar-Date +
                                  {Stock-No. + Stock-Description +
                                  Quantity-Required}

Qualified-Suppliers-Report   = {Stock-Description + {Supplier-Detail}}

Supplier-Detail   = {Supplier-Name + Supplier-Address +
                                  Contact-Name + Contact-Phone-No.}

Stock-Supplier-Index   = {Stock-No. + {Supplier-ID-Code}}

Supplier-Catalog   = {Supplier-ID-Code + Supplier-Name +
                                  Supplier-Address + Supplier-Phone-No. +
                                  Date-Qualified + Date-Of-Last-Order +
                                  Contact-Name + Contact-Phone-No.}

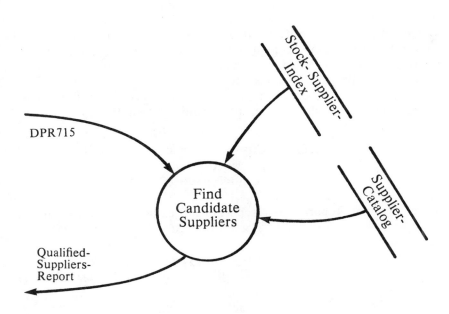

Figure 20.3. Activity containing unnecessary data elements.

This step leaves you with the element Supplier-ID-Code in the Supplier-Catalog, and the entire Stock-Supplier-Index. The essential activity Find Candidate Suppliers first looks up the Stock-No. in the Stock-Supplier-Index to obtain the Supplier-ID-Code, which it uses to access the Supplier-Catalog. Neither of these two data elements is a part of the activity's final response, and they are not the ingredients to any calculations, so their only essential role would be as identification data. We already know Stock-No. is essential identification data, but what about Supplier-ID-Code? Is it, too, identification data, or is it nonessential?

Supplier-ID-Code is a facet of the system's *internal* data storage and retrieval technology. Establishing a Supplier-ID-Code that is unique for each supplier and an index that relates each Stock-No. to a list of Supplier-ID-Codes is one way of many to implement the essential memory access capability required by this essential activity. If this system used database technology, the segment for each Stock-No. might be linked by a list of pointers to the associated Supplier segments. You don't want to keep a data element that exists only because of a particular technology; instead, you want to find the underlying requirement that transcends all possible implementations. In this case, the underlying requirement is the ability to associate a given stock item, identified by Stock-No., with the suppliers that are qualified to sell that stock item. Supplier-ID-Code is therefore unnecessary and should be eliminated.

Figure 20.4 and the data definitions below provide a slightly more complicated example of the process of identifying unnecessary elements. The two processes in Figure 20.4 are fragments of an essential activity called Forward Loss Report, which is a planned response carried out by an organization that reports lost credit cards to the issuing companies. By using this service, a person need only remember the telephone number of the reporting organization, rather than a separate telephone number for every credit card he or she carries, along with its identifying number and expiration date.

Credit-Card-Loss-Report = Customer-Account-No. + Date-Lost + Time-Lost +
$\left[\begin{array}{l} \text{``Lost-All-Cards''} \\ \text{Card-Type + (Card-No. + Card-Expiration-Date)} \end{array} \right]$

Company-Loss-Report = * report of lost cards to one credit card company *
= {Card-No. + Card-Expiration-Date + Customer-Name}

Card-Registry = {Customer-Account-No. + {Card-Type + Card-No.
+ Card-Expiration-Date + (Date-Lost)}}

Customer-File = {Customer-Account-No. + Customer-Name +
Customer-Address + Subscription-Date +
Account-Expiration-Date}

Company-Master = {Card-Type + Company-Name +
Loss-Report-Phone-No.}

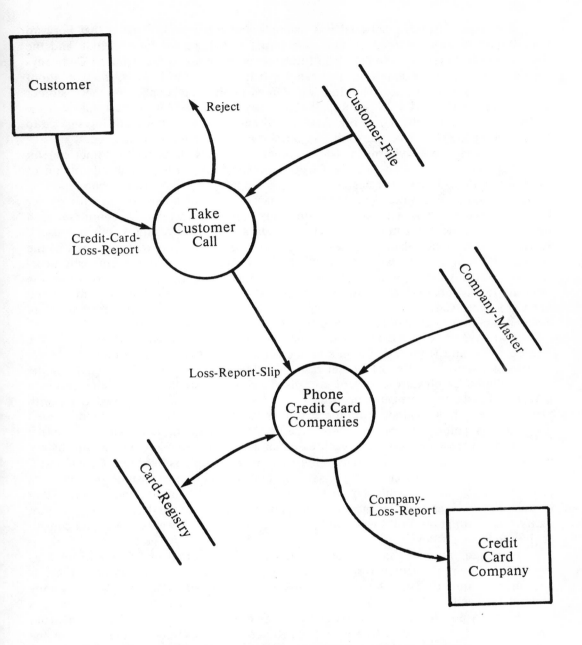

Figure 20.4. Fragments of the activity Forward Loss Report.

The main difference between this example and the previous one is that Forward Loss Report has two responses: One is external and therefore fundamental, and the other is internal and so custodial. The fundamental response is the dataflow Company-Loss-Report. From its definition, you can see that it needs Card-No., Card-Expiration-Date, and Customer-Name. The first two elements might be supplied by the customer as a part of the Credit-Card-Loss-Report, but it is more likely that the customer will say "I lost all my cards" or "I lost my Visa® card" and leave it to the reporting service to keep track of the Card-No. and Card-Expiration-Date for each card registered.

Of course, the activity needs to remember the details about the customer's credit card, but does it also need to store Customer-Name? According to the definition of Credit-Card-Loss-Report, the customer does not supply his or her name when reporting a loss. Instead, to protect against duplicate names, the customer provides his or her Customer-Account-No. with the reporting service. So, Customer-Account-No. is a necessary data element, not because it becomes a part of the output, but because it serves as identification data to retrieve other stored data elements that are a part of the output. To sum up, the data elements required in order to produce a Company-Loss-Report are the Customer-Account-No., which must come from the customer at the time the loss is reported, the Customer-Name, which is a part of essential memory, and the Card-Type, Card-No., and Card-Expiration-Date, which are usually retrieved from essential memory, but may be supplied by the customer making the report.

You have yet to account for data elements entering this essential activity from the outside world and from the system's own essential memory. To determine their role, examine Forward Loss Report's other response: the update of the Card-Registry by the process Phone Credit Card Companies. This is a simple custodial act: Whenever a card is reported lost, the corresponding entry in the Card-Registry data store is updated with Date-Lost, which is obtained from the customer via the Credit-Card-Loss-Report.

At this point, you cannot know whether this update to the system's essential memory is necessary. You study each essential activity in isolation and cannot inspect every other essential activity to determine whether Date-Lost is ever referenced, and if it is, whether that reference is truly essential. For now, you must assume that this data element is a legitimate component of the system's essential memory, and therefore that it must also be a part of the incoming dataflow Credit-Card-Loss-Report. You do not verify this assumption until the second pass of the derivation process, when you consider the model as a whole.

Although you now can account for the data elements that make up both of Forward Loss Report's essential responses, and the associated identification data, there is one more required data element: Loss-Report-Phone-No. This is the telephone number for each issuing company that the Phone Credit Card Companies process dials in order to issue a Company-Loss-Report. Loss-Report-Phone-No. is ingredient data, since it is used to produce a Company-Loss-Report while never becoming a part of the report itself.

Loss-Report-Phone-No. is also a physical data element, having plenty to do with technology, so you might question its inclusion in a logical model. The explanation, again, is in the distinction between external and internal technology. Loss-Report-Phone-No. is an essential data element simply because the technology upon which it is dependent is external to this system. If it were a phone number used by one processor to transmit the data on the Loss-Report-Slip to another processor, you would eliminate it. However, because it is used by this system to transmit a Company-Loss-Report to an entirely different system, you must keep it.

Knowing all the stored and transient data required to produce this activity's essential responses, you now can eliminate the remaining data elements. Credit-Card-Loss-Report need not contain Time-Lost. In addition, the following data elements can be removed from the data stores: Customer-Address, Subscription-Date, Account-Expiration-Date, and Company-Name. These deletions might seem disagreeable, since the data elements certainly sound essential. After all, how can the companies correspond with customers without their address? And how will they know when to demand a renewal fee if there is no record of the Account-Expiration-Date? These data elements may well prove to be essential when you study the activities that use them, but for now, since they are not essential to this activity, they should be eliminated.

20.5 Consolidating the remaining fragments

You must still resolve two major problems in the logical core. First, you have to minimize any stored data accesses set up for an awkwardly structured variety show file. Second, you must place detailed activities in essential order.

Neither of these problems can be solved unless you first eliminate the partitioning of activities and stored data that has existed up to this point. By consolidating the essential fragments, you can clearly see the essential order of the activities. Consolidation also allows you to eliminate redundant data elements and the activities that maintain them. In fact, no partitioning at all may be needed if the remaining features of the essential activity are not too complex. Combining the data stores enables you to re-create a nonredundant set of essential accesses.

The consolidation of the essential activity and memory fragments is relatively easy. You join all of the activities that are part of this essential activity as if they were to be specified in a single, imaginary minispec. Similarly, you combine all of what you know about the data elements and relationships used by the essential activity as if they were to occupy a single, imaginary data store. In the process, you eliminate any data element and relationship redundancies still remaining. At the end, you are left with a set of detailed activities and a set of data elements and the relationships between these data elements.

Figure 20.5 shows how the fragments of Forward Loss Report might look after consolidation if you were to commit them to paper at this stage. Notice that the two activity fragments from Figure 20.4, each of which was carried out by a different processor, are now a single process (with a single minispecification). By consolidating the two fragments and their minispecs, you eliminate the remnants of the infrastructure through which the two processors communicated. The only such remnant visible in the DFD in Figure 20.4 is the dataflow Loss-Report-Slip, which is a paper form in the existing system. The minispecs for Take Customer Call and Phone Credit Card Companies mention the use of this form for transmitting and receiving information about lost credit cards. Removing the dataflow and merging the minispecifications eliminates this vestige of the system's present incarnation.

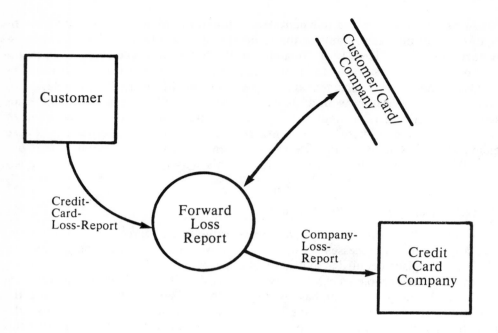

Figure 20.5. Fragments of Forward Loss Report after consolidation.

Your approach to consolidate data stores is similar. For example, the three stores in Figure 20.4, minus their nonessential components, are merged into one Customer/Card/Company file in Figure 20.5. The data dictionary definition below reveals that this new data store contains only those data elements that are absolutely essential to the activity, either because they are used by it to produce its responses, or because they must be stored by the activity:

Customer/Card/Company = {Customer-Account-No. + Customer-Name +
{Card-Type + Card-No. + Card-Expiration-Date +
Loss-Report-Phone-No. + (Date-Lost)}}

While you are still new to this technique, you may need to perform the consolidation on paper, creating a new DFD and new minispecs and data dictionary definitions to help you thoroughly understand the process. However, you will quickly learn to make the same transformation in your head, and the intermediate documentation will become unnecessary. The important point is that you cease to *think* of the fragments individually, placing the fragments in a sort of mental blender and thoroughly erasing all signs of their former organization.

20.6 Partitioning the essential memory

Now, in the first step to re-create the essential activity, you must create a new view of the essential memory used by the activity, one that avoids technological prejudice. The best way to do this is to use object partitioning, as discussed in Part Two. In this section, we explain how object partitioning enables you to view essential memory as one or more groups of data elements and relationships, and to include relationships between the chosen groups in your view.

First, however, we need to quickly review object partitioning as a means to structure essential memory. An object consists of one or more data element groups, each of which describes one of the real-world objects that the object data store represents. Such a group is called an object occurrence. In most cases, each object occurrence can be uniquely identified by specifying the value of one or more identifier data elements. A data element should be allocated to only one object. Attributing the same data element to two or more object stores makes essential memory complex and technologically biased.

To obtain an object-based partitioning of essential memory, follow these four steps:

1.   List all the data elements that must be a part of essential memory.

2.   Name the objects that are described by those data elements.

3.   Create a data store for each object name, and put into it the data elements that most closely describe that object. (This step is called data element attribution.)

4.   Resolve anomalies in the attribution of data elements to object data stores.

Of these four steps, the first is accomplished earlier when you identify the essential stored data elements in order to remove extraneous physical features. Consequently, only the remaining three steps require elaboration.

20.6.1 Identifying objects

With your list of essential data elements, think about what they tell you or what they describe. If they were adjectives in a sentence, for instance, what nouns would they modify? The answers to this question make up the list of candidate objects. Most objects come from one of three or four sources. Examining these sources is a fast way to pick out possible objects. The first place to look is in the environment that surrounds the system. You can look at the context diagram for "significant others" that interact with your system, such as entities that participate in external events. (Some obvious examples of this kind of object are Customer, Supplier, and Government.)

The essential inputs and outputs to the system are another rich source of objects. Systems often need to remember stimuli and responses for future reference. In an accounts payable system, the invoice sent in by a vendor is a likely object, and so is the payment sent out by that system. Tax-Return and Tax-Refund are likely to be objects in the essence of the Internal Revenue Service.

In business systems, objects may be contractual relationships between your system and one or more external systems. Objects are also likely to be products, services, or other resources that are part of the system or managed by it.

To illustrate the process of hunting for objects, consider the data flow diagram in Figure 20.6 and its accompanying minispec and data dictionary entries below. Clearly, you must do some work before you even have a list of data elements. To obtain one, you have to analyze the response of the essential activity Get Outstanding Orders, and determine its true need for stored data. The only response is the dataflow named Receivable-Stock, which gives the total dollar value of the orders for a single stock item, identified by Stock-No., that are found in the Open-Orders-File.

Data dictionary definitions for Get Outstanding Orders

Open-Orders-File = {Purchase-Order-No. + Supplier-Name +
Issue-Date + {Order-Detail}}

Order-Detail = Stock-No. + Stock-Description + Order-Value

Receivable-Stock = Stock-No. + Stock-Description + Total-Value-Ordered

Minispec for Get Outstanding Orders

For each open order:
   For each Order-Detail,
      If Stock-No. of Order-Detail equals Stock-No. of input,
         Add Order-Value to Total-Value-Ordered.
Issue Receivable-Stock.

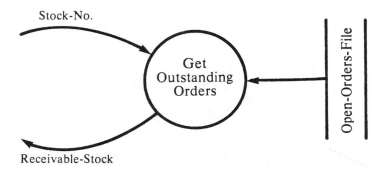

Figure 20.6. DFD for the essential activity Get Outstanding Orders.

Comparing the response to the input dataflow reveals that Stock-Description and Total-Value-Ordered must come from essential memory. Open-Orders-File does not contain the total value of the orders arranged by stock item, so Get Outstanding Orders must look through all the entries in the file, searching for those that have the same Stock-No. as the one it is interested in. Each Order-Detail entry shows the value of a single order for a given item of stock. These are added by Get Outstanding Orders to make up the Total-Value-Ordered, which is part of the desired result. There are a number of physical aspects to all of this, but for now we concentrate on identifying the required stored data elements.

To produce the Receivable-Stock dataflow, Get Outstanding Orders must find in memory the following information: Stock-No., Stock-Description, and Order-Value. Note that this process can perform its function without Purchase-Order-No., Supplier-Name, and Issue-Date. Because only the first group of data elements should be stored, your search for objects begins with them.

One object that is described by both Stock-No. and Stock-Description is an item of stock, which is the most obvious name for this object. The object described by Order-Value is not quite so obvious. Certainly, Order-Value tells you something about an item of stock — namely, the dollar value of a particular order for that item. However, Order-Value also describes an order; each order may include orders for a number of stock items and therefore include a number of corresponding Order-Values. Judgment is called for: You could lump Order-Value together with Stock-No. and Stock-Description in the Stock object, or you could choose to view it as an attribute of a distinct object, such as Order-Item.

While this decision is ultimately subjective, there are principles and guidelines to assist you, primarily from the related fields of entity-relationship analysis and information modeling. We particularly recommend the work of Flavin as a source of guidance on object identification and data element attribution [15]. In this case, the best decision is to declare Order-Value to be part of a second object, Order-Item. Having identified two objects described by the three elements of essential memory, you proceed to the next step.

20.6.2 Attributing data elements to object data stores

For each object identified in the previous step, you create a data store that bears the name of the object. Into each data store, you place the essential stored data elements that correspond most closely to the object for which the store is named. For our example, you create one data store called Stock and another called Order-Item. Since Stock-No. and Stock-Description are attributes of Stock, and Order-Value is an attribute of Order-Item, you end up with these two data stores:

Stock   = {Stock-No. + Stock-Description}

Order-Item = {Order-Value}

These object data stores replace the still physical Open-Orders-File in the data flow diagram of Figure 20.6.

Forming object data stores isn't always as easy as in this example. When you are working with many objects described by even more data elements, matching each data element to a single object data store becomes much more difficult. For a simple example of the problems that can arise, we return to the partially logicalized fragments of Forward Loss Report in Figure 20.5.

What objects are described by the elements in the data store Customer/Card/Company? That's not too difficult a question: Customer-Name says something about a customer (an external entity); Loss-Report-Phone-No. describes the company to which a loss must be reported (another external entity); and Card-No., Card-Expiration-Date, and Date-Lost all describe an individual credit card that was issued by one company to one customer (an agreement between external entities). Based on these observations, you create three data stores: Customer, Credit-Card-Company, and Credit-Card-Account.

Now, how do you allocate the data elements to these stores? It's easy enough to allocate Customer-Name to the Customer store, to put Card-No., Card-Expiration-Date, and Date-Lost into the Credit-Card-Account data store, and to make Loss-Report-Phone-No. the sole occupant of the Credit-Card-Company file. But what do you do with Card-Type? Here's where the trouble begins. On the one hand, it specifies the kind of card it is, but it also identifies the credit card company that issued the card plus all other cards of the same Card-Type. That is, "American Express" is a Card-Type that applies not only to your American Express® card, but to all cards issued by the American Express Company. One tempting compromise is to allocate data elements that seem to describe two objects to two object data stores, resulting in the following attribution:

Customer = {Customer-Name}

Credit-Card-Company = {Card-Type + Loss-Report-Phone-No.}

Credit-Card-Account = {Card-Type + Card-No. + Card-Expiration-Date + Date-Lost}

This is not a terrific solution, and redundant attribution of data elements may force you to reconsider and amend your initial attribution of elements to objects.

20.6.3 Resolving attribution anomalies

The most common problem with would-be logical stored data models is redundancy of information, particularly in data stores. Be suspicious of stored data models that allow redundancy, since they may have inherited the trait from existing systems. Although you may occasionally allow redundant stored data in order to produce a more readable model, such occasions should be few, and always suspect. Furthermore, for each stored data element, there is a single right object data store, and the urge to place an element in more than one store often stems from an inability to decide which object it most closely describes. To avoid the lazy solution (that is, when in doubt, place stored data anywhere you want), we recommend that you try, at least initially, to make your object data stores totally nonredundant.

So back to the question about Card-Type: Should it be part of the Credit-Card-Company store or the Credit-Card-Account store? Our answer is that Card-Type best describes the object Credit-Card-Company. We base this decision on our earlier observation that a single value of Card-Type is associated with a single credit card company, and that value applies to all the cards issued by that company. Even though Card-Type does indeed describe an individual card as well as the company that issued it, that attribute originates with the company that issued the card and thus is inherited by a card from the company.

Following up on this decision, you eliminate Card-Type from the Credit-Card-Account data store, leaving you with the desired product: a nonredundant model of the memory used and updated by the essential activity Forward Loss Report.

Data element inheritance can fool you into allocating a single item of essential memory to more than one object data store. Your only defense is vigilance: Check your stored data model for redundancy, and rigorously justify every attribution, using guidelines from information modeling and other data modeling disciplines whenever applicable.

20.7 Minimizing the essential accesses

After partitioning essential memory, you are ready to specify the accesses, those activities that obtain, add, modify, or delete parts of essential memory in order for the essential activity to perform its job. In specifying each access, you want it to deal with the minimum number of data elements and to find those data elements in the most direct way possible.

To specify the accesses for an essential activity, you must reverse the direction of the approach taken so far: Instead of starting from the response and working backward, you now start from the input and work forward, examining each detailed activity in turn to determine the accesses needed, always keeping in mind what the ultimate response should be. As you inspect each activity, you look first for accesses that are used to validate the incoming data. Then you look for accesses that obtain data elements from memory; the activity either uses these data elements to derive other data elements or uses them directly in the response. Finally, you look for accesses that update memory to reflect either the event or the response that just occurred.

Figure 20.7 shows these three kinds of accesses in the order just described. The first access validates incoming data: It compares the employee ID in the Time-Worked dataflow with the employee IDs in the Employees data store in order to verify that the employee indeed works for the company. The second access obtains each employee's hourly wage and deductions from memory and uses them to derive the amount on the employee's paycheck. The third access updates essential memory by recording the employee's weekly earnings in the Earnings data store. Once you know the accesses needed by an essential activity, you re-create those accesses. In the next few pages, we treat the accesses that validate incoming data and those that obtain data from memory as one kind of access, calling them simply accesses to obtain data.

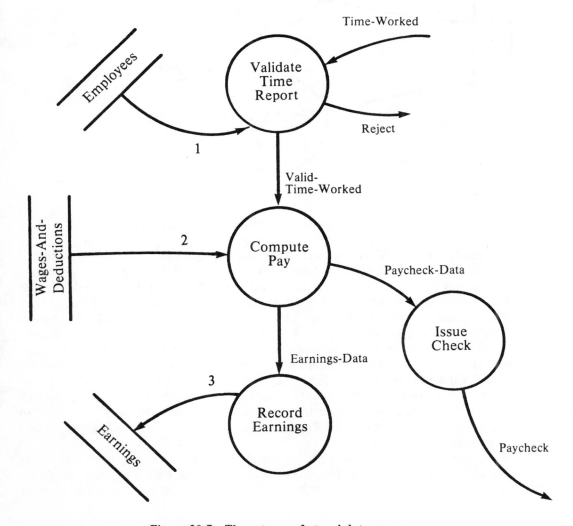

Figure 20.7. Three types of stored data access.

Because knowing how an activity identifies the elements it needs from memory is such an important part of creating an essential access, we quickly review the two basic methods used to identify data, which are shown in Figure 20.8. Get Employee Phone No. obtains data from essential memory using identification information in the input Employee-Name. In the second method, Issue Employee Phone Directory obtains data from memory without any identification help from the input. It does not need to identify a particular object occurrence, since it will obtain data from every object occurrence if every one contains a phone number (although another such process might access only a subset of stored data, such as "all employees due for twenty-five-year service awards").

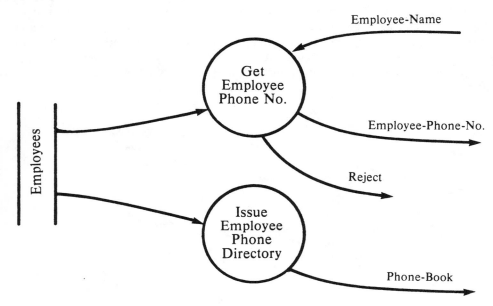

Figure 20.8. Two ways to select elements from data stores.

To specify an essential access, start by listing the data elements that the essential activity requires. You then consider how the activity will identify these data elements. The activity may be able to identify the data elements directly, using the value of the identifier of the object occurrence to which the elements belong. An example of such an access requirement is, "Find the checking account balance associated with the account number in the incoming dataflow." Another possibility is for the access activity to go through the object and use the values of other data elements to find what it seeks: for example, "Find all names and addresses for customers whose hair color is blond." In this case, Customer is an object described by the data elements Name, Address, and Hair-Color. Often, an access must identify the desired data elements by going through other objects first. "Find all credit cards associated with a given customer" or "Find all parcel values within such and such a consignment on such and such a ship" are access requirements that first access an object different from the one they seek.

Next, you choose the minimal path to the data sought. That is, you specify an access that searches through the fewest object occurrences and data elements. To determine the shortest path, you often start by examining the access paths actually taken by the essential activity fragments in the current physical model. This may be confusing at first, since you apparently discard the physical data store definitions when you create the object data stores in the previous step. In actuality, you must retain those definitions since they offer valuable clues to the minimal access paths.

To show how you use access paths from the current physical model to find the minimal access paths, return to the essential activity, Get Outstanding Orders, in Figure 20.6. Recall that Open-Orders-File did not contain the total value of orders for a given item, just the value of each individual order. So, in order for Get Outstanding Orders to obtain the information it really needed, it had to go out of its way in two respects: First, it had to leaf through all the Order-Detail entries to find those that applied to the stock item in question; second, it had to sum the individual Order-Values to produce the Total-Value-Ordered.

You re-create the essential accesses for this activity by eliminating these two detours. First, there is no reason that Get Outstanding Orders should receive stored data at the level of detail of Order-Value. Even though computing Total-Value-Ordered would pose no problem with perfect technology, it still is not essential for this activity to do so. This activity could produce its response using less-detailed stored data, specifically, Total-Value-Ordered. So, the first step is to restate the essential level of detail of stored data; Get Outstanding Orders now requires only Stock-No., Stock-Description, and Total-Value-Ordered.

We say that a piece of data is at a higher level of detail than some other piece if that data can be derived directly from the other piece without using any other data. For example, weeks is at a higher level of detail than days because the number of weeks can be found by dividing the number of days by seven. Similarly, Total-Value-Ordered is at a higher level of detail than Order-Item, since adding the Order-Items gives the Total-Value-Ordered. On the other hand, the amount of a paycheck is not the hourly wage at a higher level of detail, because another piece of data — the number of hours worked — is needed to calculate the paycheck. While minimizing accesses, make sure that each essential activity receives stored data at the exact level of detail it needs. There is no reason for an activity to perform the extra work of deriving data. To change the level of detail received by an activity, you change the level of detail of the data in the data store.

The second change in our example is so that Get Outstanding Orders does not have to browse through all the Order-Details in search of those that apply to a particular stock item, identified by Stock-No. Given perfect technology, there would be no need for Get Outstanding Orders to read, check, and perhaps discard all of the entries in the Open-Orders-File. A more sophisticated data retrieval facility would allow the essential activity to retrieve just the subset of stored data associated with a given Stock-No. Since this represents a simpler access requirement, it becomes the minimal access.

In this example, however, the first step you take — stating the requirement for stored data as Total-Value-Ordered — solves both problems. If Get Outstanding Orders can retrieve a single data element that is at the right level of detail, it does not need to access all the Order-Details so that it can add their Order-Values.

After deciding on the minimal access, check if the object data stores need to be changed. When partitioning the essential memory used by Get Outstanding Orders, you attributed Order-Value to the object Order-Item. Therefore, the Order-Item object is no longer necessary, and Total-Value-Ordered can be allocated to the Stock data store, resulting in the DFD of Figure 20.9 and a new definition for the object Stock:

Stock = {Stock-No. + Stock-Description + Total-Value-Ordered}

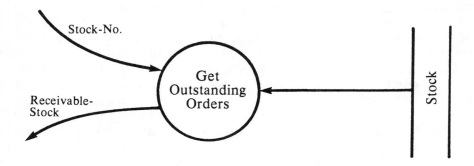

Figure 20.9. Essential memory model for the activity Get Outstanding Orders.

Developing the minimal essential access is often more difficult than this example indicates. So to guide you through more complex cases, we offer three tactics for minimizing accesses. The first tactic is called *inversion*. You first specify what you think is the minimum access, then you reverse the direction and see if the new access is simpler than your first attempt. Below are two minispec portions that show essential accesses before and after the application of this tactic.

Before

    For each Course,
        For each Student enrolled in Course,
            If SSN of Student matches SSN of input,
                (Student is found) . . .

After

    For each Student,
        If SSN of Student matches SSN of input,
            (Student is found) . . .

Notice that the inverted access is shorter and simpler.

The *compound identifier* tactic is used when more than one identifier is required to find the data. The minispec portions below show the application of this second minimizing tactic.

Before

>> For each Customer,
>>> If Name and Address of Customer match input,
>>>> For each Credit-Card registered to Customer,
>>>>> If type of Credit-Card matches input,
>>>>>> (Customer's Card is found) . . .

After

>> Find the Credit-Card associated with the Customer-Name and Address and Card-Type of the input.

Using each identifier individually to find the desired data, as in the first minispec, often results in a very complicated access procedure. To apply the compound identifier tactic, you simply use both identifiers at once. As in the second version of the access, you thereby avoid a lot of unnecessary accessing.

You apply the third tactic when the access activity uses an interobject relationship and must identify data elements without help from identifiers. In this case, you have to choose carefully the object for the activity to start its search. In the minispecs below, two essential accesses obtain the same data, but they start from different objects in essential memory.

Before

>> For each Tour,
>>> For each Customer on this Tour,
>>>> For each Passenger booked by this Customer,
>>>>> Add Passenger-Name to Passenger-List.

After

>> For each Customer,
>>> For each Passenger booked by this Customer on any Tour,
>>>> Add Passenger-Name to Passenger-List.

The second of the two accesses turns out to be the simpler. So, the last tactic is shifting the starting point of an essential access in hope of re-creating a simpler access.

These same guidelines and tactics can also be used to re-create accesses that add, modify, or delete essential memory. Below are several examples of accesses that update essential memory.

>> Create a new Invoice with a unique Invoice-No.

>> Add Invoice-Amount to Total-Billed-Amount.

>> Establish a link from Customer to Invoice.

These three examples show the different ways essential memory can be modified. The first access creates an object occurrence, the second changes the value of a data element, and the third establishes a relationship between two objects. When accesses to update interobject relationships are specified, you also give the direction and ratio of the relationship.

It is difficult to be sure at this point that you have specified these accesses correctly because many of them add, modify, or delete memory for the benefit of another essential activity. Since the scope of your work is limited during this step to the essential activity at hand, you cannot be sure that you have established the proper updating accesses for another activity. Thus, you might have to revise your choice of objects in the second pass, when you examine all the essential activities together. Naturally, if you change the object partitioning, you will have to change the essential accesses you just devised. Especially likely to be modified will be the custodial activities, the accesses that update essential memory.

20.8 Establishing the essential order among the activities

You conclude the work of deriving an activity's essence by establishing the essential order for the detailed activities. Certain detailed activities must be executed in order, since some activities can't be carried out until others are completed, yet you don't want to imply an order among actions that isn't necessary.

The basic tactic you use to establish essential order is to compare each detailed activity with every other one within an essential activity. When comparing two activities, check if one activity must be executed before the other or whether they can be executed simultaneously. If you find action A must precede action B because the results of A are needed by B, place actions A and B in order. Otherwise, the activities should not be ordered.

As you construct the detailed essential procedure, don't create breaks in the sequence of actions unless the breaks are absolutely necessary. Within a single essential activity, you should have no time delays whatsoever. The only time delays in an essential system are those that separate one essential activity from another; in this case, the two essential activities communicate only through essential memory.

If you find that you have created time delays among the tasks within a single essential activity, you have erred in one of two ways. Either you have created an unnecessary time delay or you have mistakenly lumped together two or more distinct essential activities. Ask yourself why there must be a pause between the two actions (to wait for another event, perhaps?). If you determine that the time delay really is necessary, then you must create two different essential activities.

20.9 Establishing the physical ring

With completion of your work on the logical core, you now establish the physical ring that acts as the interface between the essential activity and the imperfect technology of the system's environment. The ring activities to be added are the same as those you removed to uncover the logical core. You may have to modify some details of the original ring activities, since some data elements they dealt with may have been eliminated during the intervening steps.

To establish the physical ring, look at each input and output of the essential activity. For the input interface to the essential activity, you re-create activities that construct the logical input expected by the logical core. These activities perform two tasks:

They filter extraneous data elements from the input, and they reconstruct transactions whose dataflows have been fragmented into segments, each of which travels in a physically separate data container. Perhaps the most common example is the way transactions entering an online computer system are fragmented into separate screens, each of which holds a portion of the data necessary to the transaction. The screens are chosen for the convenience of the human operators, but because no one screen holds an entire transaction, the automated system must gather several screens of data before it can proceed with its planned response. Figure 20.10 shows two examples of physical ring activities that both remove extraneous data elements and reconstruct logical transactions.

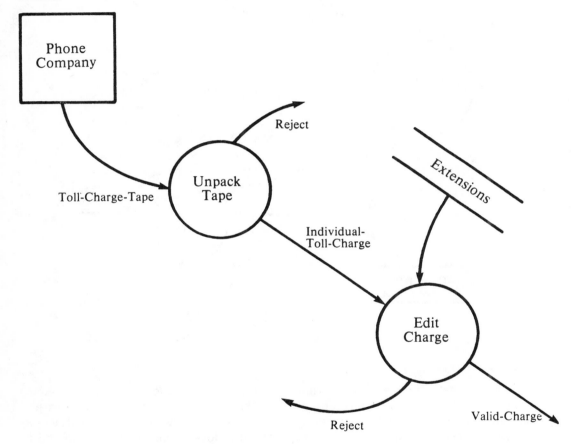

Figure 20.10. Physical ring activities for inputs.

For the output interface to the essential activity, you establish activities that translate the output from the logical core into the physical form expected by the outside world. These activities format the outgoing data onto a physical medium and direct the output to its destination. Figure 20.11 shows three examples of these physical ring activities. The two activities on the right format the outgoing data by packaging it, and they then ship it to its destination.

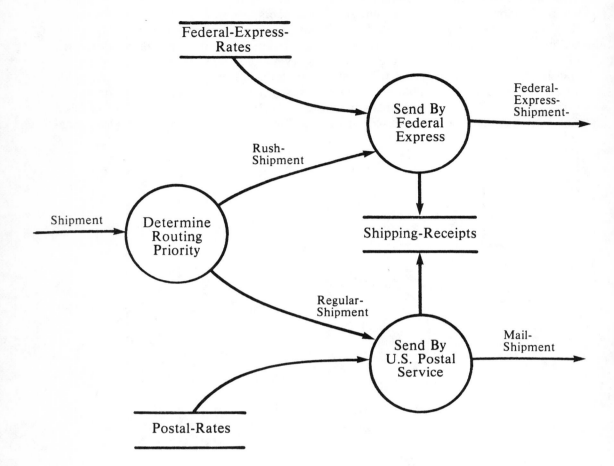

Figure 20.11. Three physical ring activities for outputs.

20.10 Summary

The reconstruction of physical ring activities completes the process of re-creating an essential activity. Many remaining physical characteristics are removed by this process, and at the same time, a new streamlined understanding of the essential activity emerges. Unnecessary data elements are removed following their identification and by carefully tracing each output backward through the process that creates it. Physical partitioning of activities also is deleted by consolidating essential fragments.

Your first step toward achieving a new understanding of the essential activity is to partition essential memory into object data stores. Next, minimizing accesses gives you a simpler model — not only are the accesses themselves simplified, but essential memory may become simpler as well. Finally, you establish the essential order among detailed activities within an essential activity. This step removes a kind of false requirement and allows you to spot any essential activities that are mistakenly lumped together. To ensure that you perform all these steps accurately, you remove the physical ring activities at the beginning of the procedure and return them at the end. The next chapter treats how to model the essential activity that you have just derived.

Chapter 21

Modeling an Essential Activity

With the derivation of the essential activity from the previous step, you use the tools of structured analysis to model it. If the essential activity is small, you need only a single DFD (with a single bubble) supported by a minispec, data dictionary definitions, and possibly a data structure diagram or an entity-relationship diagram. (We consider an essential activity to be small when its detailed activities can be modeled in a one-page minispec.) Larger essential activities require additional levels of DFDs so that the activity is partitioned into mentally bite-size pieces.

Figure 21.1 and the minispecs and definitions below make up an example of a simple essential activity model. Remember that this model represents only a single essential activity and that you must re-create and model the rest of the essential activities before bringing all of the models together.

Minispec for Produce Schedule

1. Look up Social-Security-No. in Student data store.

2. If found,
    Issue Schedule-Header.
    For each Course associated with Student,
    Issue Schedule-Detail.
    Otherwise,
    Reject Social-Security-No.

Some data dictionary definitions

Course	=	{Course-No. + Course-Title + Credit-Hours + Room + Time}
Student	=	{Social-Security-No. + Student-Name + Student-Address}
Student-Schedule	=	Schedule-Header + {Schedule-Detail}
Schedule-Header	=	Student-Name + Student-Address
Schedule-Detail	=	Course-Title + Room + Time + Credit-Hours

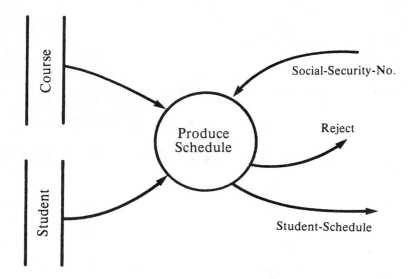

Figure 21.1. Data flow diagram for Produce Schedule.

The size of an activity is the major issue in essential activity modeling, and so two approaches are necessary because modeling large essential activities presents different problems from modeling small ones. Before we explain the two approaches, however, we first discuss the modeling decisions that must be made for all essential activities regardless of size.

21.1 Creating the high-level essential activity model

The high-level essential activity model is a single DFD, such as the one in Figure 21.2. This DFD is from an airline's frequent flyer incentive system, which gives discounts (awards) to passengers based upon how many miles they fly with that airline. Building such a high-level model is easy: To represent the response to the event, draw a single circle and name it with a concise verb-object phrase that summarizes the actions taken in response to the event. Any inputs and outputs to the activity appear as dataflows going into and out of the circle. To portray any objects accessed by the activity, draw parallel lines and place the name of the object between the lines.

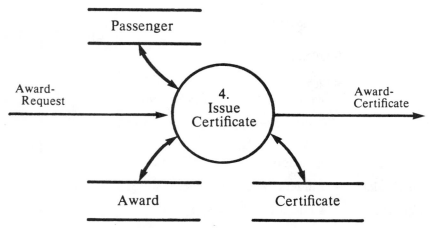

Figure 21.2. High-level model of the activity Issue Certificate.

Connect the object data stores to the activity using unnamed dataflows.* The direction of the dataflow is toward the bubble if the activity obtains data from the object or uses the object to find data from another object through an interobject relationship, and away from the bubble if the activity is adding, modifying, or deleting data in that data store.

After creating a high-level essential model, you model the detailed activities and the contents of the dataflows and object data stores. In order to choose which of the two approaches to use for the next step, estimate the size of the essential activity. In the following section, we discuss each of the modeling approaches.

21.2 Modeling the details of a small essential activity

When modeling the details of a small essential activity, you create specifically two types of documentation: A minispec to document the detailed essential activities, and data dictionary definitions for the activity's inputs and outputs and object data stores. We recommend writing the data dictionary definitions just as we proposed in Sections 8.1, 8.3, 9.1, and 9.2. Below are the data dictionary definitions and the minispec for the essential activity in Figure 21.2. Taken together with the diagram in Figure 21.2, they form the complete model of an essential activity.

This system keeps track of the mileage traveled by a passenger. When certain mileage levels are reached, the passenger becomes eligible for awards like free coach tickets and discounts on first-class fares. Each mileage level has a different award, and whenever an award level is reached, the airline notifies the passenger. If the passenger decides to accept the award, he or she sends an award request. The Issue Certificate activity transforms the award request into an award certificate. To get an award, the passenger redeems the award certificate at an authorized travel agent. The purpose of creating the last two relationships in the minispec is to keep a record of the passenger receiving the certificate and of the award level of the certificate. To follow the instruction "Issue Award-Certificate," the system looks up Award-Certificate in the data dictionary and creates a certificate that includes all the data elements listed in the definition.

Data dictionary definitions for Issue Certificate

Passenger	=	{Passenger-No. + Passenger-Name + Passenger-Address + Total-Mileage-Balance + Award-Expiration-Date}
Award	=	{Mileage-Range + Award-Level} *Mileage-Range contains the number of miles a passenger has to have flown to be eligible for the Award-Level*
Certificate	=	{Certificate-No. + Certificate-Date}
Award-Request	=	Passenger-No. + Award-Level
Award-Certificate	=	Certificate-No. + Award-Level + Passenger-Name

---

*The only controversy in creating the high-level model of an essential activity is whether the dataflows between the data stores and the processes should be named. As explained in Chapter 8, we believe naming these accesses is unnecessary, difficult, and potentially confusing. You can decide for yourself and act accordingly.

Minispec for Issue Certificate

Find Passenger associated with Passenger-No. on Award-Request.
> If found,
>> If Award-Level on Award-Request is the same as Award-Level
>> in Award associated with Passenger, and today's date
>> is before Award-Expiration-Date:
>>> Find Mileage-Range in Award associated with Passenger.
>>>
>>> Subtract Mileage-Range from Total-Mileage-Balance.
>>>
>>> Create Certificate with unique Certificate-No. and set
>>> Certificate-Date on Certificate to today's date.
>>>
>>> Create a relationship from Certificate to Award that
>>> corresponds to the Award-Level on the Award-Request.
>>>
>>> Create a relationship from Passenger to newly established Certificate.
>>> Issue Award-Certificate.
>>
>> Otherwise (expired Award or wrong level),
>>> Reject Award-Request.
>
> Otherwise (Passenger-No. not found),
>> Reject Award-Request.

Representing the detailed activities with a minispec is a little trickier than writing data dictionary definitions. First, you must select the proper tool or tools to write the minispec. Although many tools have been nominated for this duty, from graphs to plain talk, we most often choose either structured English, decision tables, or decision trees. As discussed in Chapter 8, each of these tools is attractive depending upon the task at hand, and the following guidelines can help you select the most suitable tool for writing a minispec.

Both decision tables and decision trees are useful for modeling policy when the policy decision depends upon a particular combination of several factors. If you modeled this kind of decision with structured English, the result would be a complicated group of nested "if" statements that is impossible to understand. Decision trees and decision tables are about equally suited to representing complicated decision policy. However, since we prefer graphic representation whenever possible, we recommend decision trees.

In most situations, the detailed essential activities can be nicely represented with structured English. It has enough conventions to represent all of the activities that might be part of an essential activity. However, structured English does have a drawback. The drawback is that even in small essential activities, you can still find detailed activities that do not have to be performed in any particular order. So, be careful not to impose an arbitrary order on these asynchronous activities. If you choose to use structured English, watch for parallel activities. If you find them, use the "Do in no particular order" construct that we introduced in Chapter 8.

We have one last word on representing the detailed essential activities of a small essential activity. The word is *eclectic*, which means choosing the best from diverse sources. You should be eclectic when choosing the tools to represent detailed essential activities. If you have a small essential activity that is part complicated decision logic and part complicated sequences of activities, do not hesitate to use both a decision tree and structured English to document that activity.

21.3 Modeling the details of a large essential activity

A large essential activity is one that requires a multi-page minispec. Since multi-page minispecs exceed the budget for complexity, we recommend the classic structured analysis approach to modeling a large set of detailed activities: You form the detailed activities into groups of activities, assigning each group a name, then you represent these activities and their interfaces with a data flow diagram. The result of using this tactic on the activity Offer Job Referral is shown in Figures 21.3 and 21.4.

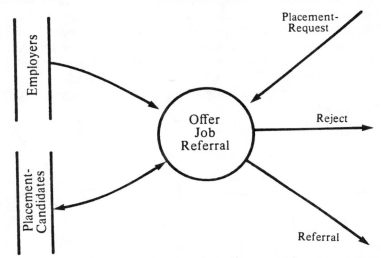

Figure 21.3. High-level model of a large essential activity, Offer Job Referral.

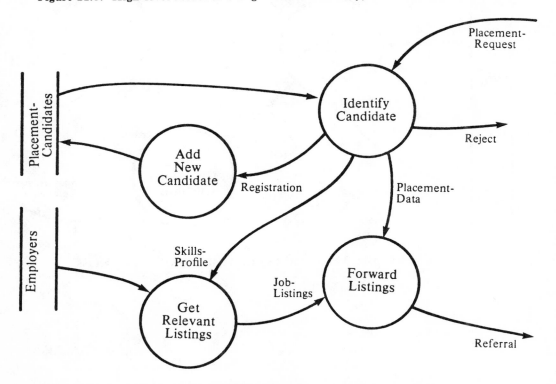

Figure 21.4. Lower-level model for the activity Offer Job Referral.

Deciding to use a lower-level DFD helps solve the problem of how to represent the details of a large essential activity in a way that is easy to understand. However, you now face a different problem: How do you partition an essential activity into separate activities for the lower-level DFD? You partition the activity by identifying intermediate data products that are produced by the detailed essential activities.

Figures 21.5 through 21.9 show portions of five lower-level DFDs. Each figure serves as an example of a type of intermediate dataflow that you can identify. The first example shows an intermediate dataflow, Available Seats, that is the product of a complicated essential access. When a complicated decision is made, the second type of intermediate dataflow results, such as Carrier-Of-Choice in the next example. In the third example, the dataflow Forecast is produced from a complicated derivation of data. The fourth example shows an intermediate dataflow, Report-Request, that results from putting a physical dataflow into logical form. New-Product-Announcement in Figure 21.9 is an example of the fifth type of intermediate dataflow, a logical dataflow just before it is converted to a physical form.

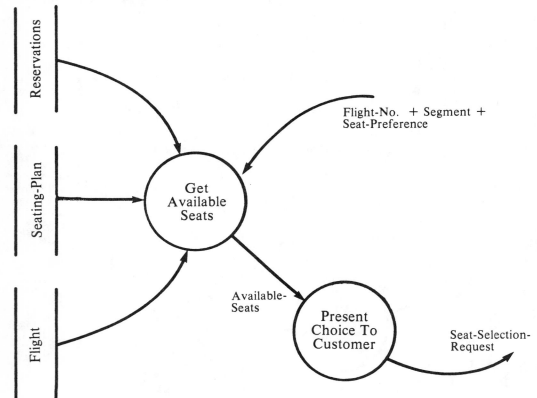

Figure 21.5. Dataflow produced by a complicated stored data access.

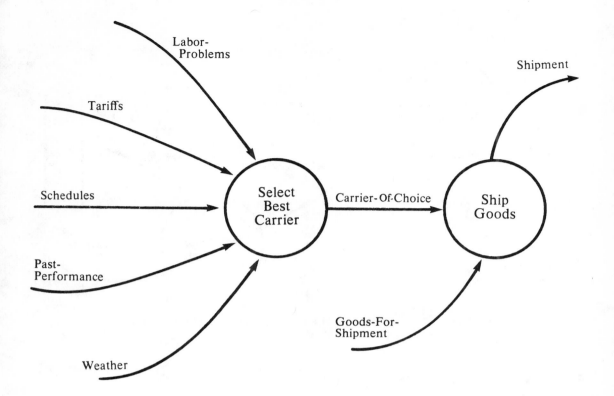

Figure 21.6. Dataflow produced by a complicated decision.

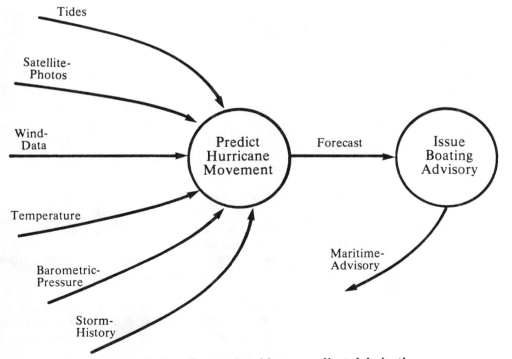

Figure 21.7. Dataflow produced by a complicated derivation.

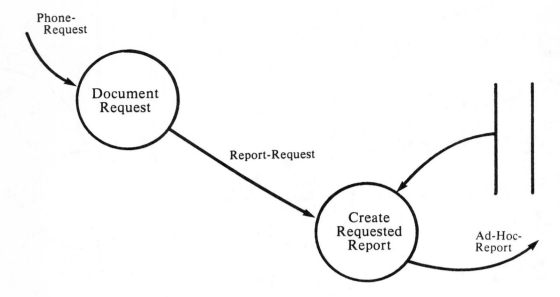

Figure 21.8. Dataflow produced by unpacking a physical dataflow.

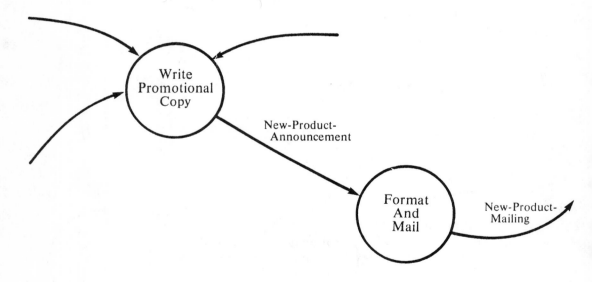

Figure 21.9. Packing a logical dataflow.

Once you have a lower-level diagram for a large essential activity, you must return to answer a previous question: Can you write a one-page minispec for each of the activities on the lower-level diagram? If any of the activities is still too big, you continue drawing lower-level DFDs under that activity. You stop when the essential activity is partitioned into one-page minispec chunks.

Be careful not to confuse the fragmentation of an essential activity, which is a physical trait, with partitioning to improve the understandability of the model. Although both fragmentation and partitioning divide essential activities into more detailed activities, there is a difference. Namely, the particular lower-level activities are chosen for a different reason. In fragmentation, the aim is to improve the efficiency of a physical system, and a choice of activities can appeal to the model reader only by coincidence. When partitioning, you can *only* justify your choice of activities because it improves the quality of the model without adding technological bias.

Once there are enough lower-level DFDs for you to cope with the complexity of the essential activity, complete the model of the essential activity with minispecs and data dictionary definitions. For each minispec, choose the appropriate modeling tool for writing it. Also, define the essential activity's net inputs, net outputs, and data stores in the data dictionary. Finally, you create additional data dictionary definitions for the intermediate dataflows used to partition the lower-level DFDs.

21.4 Modeling interobject relationships

As explained in Chapter 9, any one of a number of options are available to model interobject relationships. Although you almost always use a data structure diagram or an entity-relationship diagram to model them, we won't revoke your essential modeling license if you decide to let the essential access statements in your minispecs indirectly model the interobject relationships. If you choose the minispec alone, however, be sure to specify precisely the objects involved, the ratio of occurrences expected, and the direction of reference. That way, you'll supply all the information that data administrators need to put together their own graphic model.

21.5 Summary

Each essential activity is modeled using the tools of structured analysis. You model small activities with a one-bubble data flow diagram, a minispec, data dictionary definitions, and perhaps a data structure diagram. Large essential activities also have a one-bubble data flow diagram, but they need to be partitioned into several smaller activities, which appear on one or more lower-level data flow diagrams. Each of these smaller activities has its own minispec. No minispec should be longer than a page; otherwise, the activity should be partitioned even further.

Part Six

Integrating the Essential Activity Models

In Part Five, we presented a procedure for deriving and modeling an individual essential activity. In Part Six, we offer a method for building an understandable model of the entire essence from these individual essential activity models.

Chapter 22 discusses the integration of essential activities. Because essential activities are connected through data stores, integrating them means ensuring that all data accesses are present and correct. Although the model is completely logical after you perform this step, two problems remain: Some essential features are specified more than once, and models of large essences are too large and complicated. The two remaining chapters in this part offer tactics to solve each problem, with Chapter 23 focusing on eliminating redundant specification of essential features, and Chapter 24 advising on building concise models of large essences. This part completes our discussion of the procedure for deriving an essential model.

Chapter 22

Integrating the Essential Activities

Although the individual essential activity models are free of technological bias and easy to understand, a problem may arise when you try to combine them into a model of the whole essence. The problem is that each model's accesses to memory may be inconsistently specified from activity to activity. Assuming that you modeled the individual essential activities correctly, how could this happen?

The problem of inconsistent accesses results from using the two-pass approach to derive the essence from the reduced physical model. In the first pass, you logicalize each essential activity independently. As a result, you don't examine the relationships between the activities until the second pass, when you integrate those activities. Hence, you blind yourself until the second pass to the typical relationship between fundamental and custodial activities: Namely, the fundamental activity that requires memory is part of a different essential activity from the custodial activity that provides the memory. This relationship is depicted in Figure 22.1.

With your blinders on in the first pass, therefore, you cannot spot inconsistencies between the fundamental activities that require essential memory and the custodial activities that supply it. In the second pass, you systematically search for inconsistencies by comparing all the accesses of all the essential activities. These inconsistencies can take two forms: Either a fundamental activity requires essential memory not provided by any custodial activity, or a custodial activity provides essential memory that isn't needed by any fundamental activity.

Although the two-pass approach leaves these inconsistencies in place so long, that is no reason to abandon the approach. Indeed, the complexity of shared essential memory requirements is so great that only a two-pass approach allows you to find the inconsistencies; the first pass breaks the system into manageable pieces so that finding the inconsistencies is possible in the second pass.

22.1 Cross-checking essential accesses

The only way to resolve inconsistencies in essential activities is to examine each retrieval access performed by a fundamental activity, note the data retrieved, and examine the rest of the model to see if the proper custodial activities are specified to store and update that data. You must perform this cross-checking for both the object occurrence and the individual data elements retrieved, and you also must ensure that the use of interobject and intra-object relationships is consistent for all essential activities.

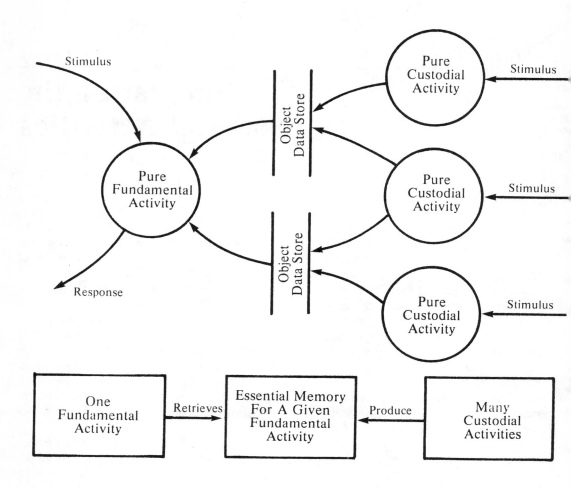

**Figure 22.1. Typical relationship between a fundamental activity
and a custodial activity.**

This sounds like a dreary task, and it is. In the worst case, you will have to check every essential activity model in order to consider the entire set of essential accesses and to search every minispecification manually to find references to each data element, object occurrence, intra-object relationship, and interobject relationship.

For example, suppose you had to cross-check the following essential access, which is required by a fundamental activity in an accounts payable system:

Find all unpaid Invoices for a Vendor associated with a given Vendor-No.

To do this, you must find one or more custodial activities that establish vendor object occurrences, that establish Invoice object occurrences, that establish the interobject relationship between a vendor and the vendor's invoices, that maintain the interobject relationship (paying the invoice, canceling the invoice, and so on), and that establish the intra-object relationship that a vendor number uniquely identifies particular object occurrences. You must also check that no custodial activities incorrectly dispose of this essential memory.

Suppose you are cross-checking this access:

Find the Loan-Balance for a Mortgage associated with a given Mortgage-No.

You must check all of the custodial activities that create, modify, and delete the Loan-Balance data element. Through this check, you insure that Loan-Balance has the proper value required by the retrieval accesses.

The use of event partitioning and individual essential activity modeling in previous steps helps to clear the way for this complicated cross-checking, since at least most physical features are gone and your model is understandable. If you are lucky enough to have an automated system that can compare the list of data elements to the text of the minispecs, the search will be much easier. However it is accomplished, this weeding out is necessary for the development of a truly logical and correct requirements specification.

In the sections below, we investigate how this detailed comparison of essential memory accesses helps you resolve the two kinds of inconsistencies in those accesses.

22.2 Resolving missing custodial activities

A custodial activity is missing when you see an essential access that retrieves memory but find no essential access to provide the memory required. Custodial activities can be missing for any given data element, object occurrence, intra-object relationship, or interobject relationship. For example, if you are checking the essential access "Find all Traffic-Tickets issued to a given Driver," you may discover there is no custodial activity to add Ticket occurrences or to create the relationship between a given Ticket and Driver. In another case, you can be checking the essential access "Find the Fine-Amount associated with a Ticket," and find no custodial activity to provide the data element Fine-Amount.

It is understandable for custodial activities to be missing at this point, since you work with too much information during the expansion and reduction processes to be able to spot these omissions and since afterward you can see only the accesses of one essential activity being modeled at a time. A custodial activity is omitted for one of three reasons. The first two are analysis mistakes: Either you set a system boundary that excludes the required custodial activity, or you inadvertently omit a custodial activity during the derivation process. Catching these analysis errors now is much better than at system integration time, for to correct them, you must either change the context of the system or backtrack through your essential modeling work to recover the wayward custodial activity.

The third reason for missing custodial activities is more complex and involves data being supplied at the wrong level of detail. For accesses to be minimal, essential activities should receive from memory data elements that are as close to the level of detail required as possible. As you examine the relationship between fundamental and custodial activities, you must ensure that the data supplied by every custodial activity is at the level needed by each fundamental activity. You will often discover that no custodial activity supplies the data element at the required higher level of detail. Instead, you will find a custodial activity that supplies the same data element at a lower level of detail. Figure 22.2 shows an example of this situation from our traffic violations system. The activity Notify Driver requires the total amount of payments made by each driver, but there is no custodial activity that stores this total. Instead, Credit Payment stores the amount of each payment, an excessively detailed form of the data.

In another case, assume that two essential activities require weekly and monthly sales totals, but the existing custodial activities provide only daily sales totals. To solve this problem, you add the appropriate custodial activity to derive the higher-level data element and store it. Because the fundamental activities often ask for data at many levels, the lower-level custodial activities will still be required if, for example, another fundamental activity requires a daily sales total. If, however, that fundamental activity doesn't exist and hence the lower level is not required, then the custodial activity for that level of detail becomes unnecessary. We show how to deal with this problem and others involving unnecessary custodial activities in the next section.

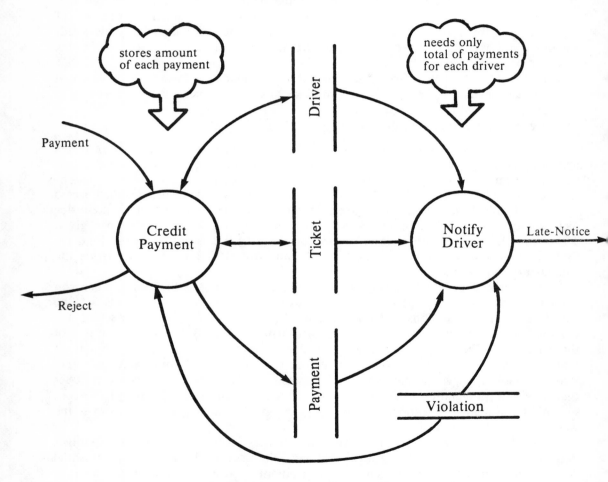

Figure 22.2. Data stored at a too-low level of detail.

In each case of a missing custodial activity, the remedy is straightforward: adding the custodial activity to the proper essential activity model that you developed in the previous step. That means changing the activity's DFDs, its minispecs, and its object data store definitions as well.

22.3 Removing unnecessary custodial activities

Despite your best efforts, the essential data stores that you defined may contain data elements and relationships that the system does not have to remember. This means that you specified unnecessary custodial activities to create this unnecessary memory. Neither these activities nor the memory they provide should be included in a specification of the current system's essence.

Some of these unnecessary activities may provide information that is completely unused by any of the system's fundamental activities, while others may store essential memory that is used by an essential activity but does not need to be remembered between occurrences of the activity — the information can be used and forgotten within one performance of the essential activity.

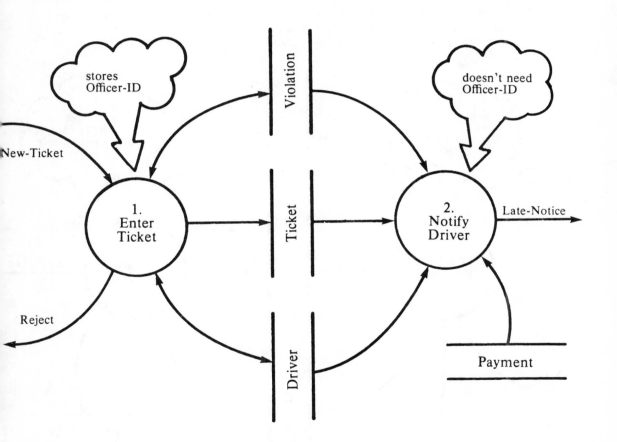

Figure 22.3. Unnecessary storage of a data element.

You missed eliminating these unnecessary custodial activities in the last step because you examined the essential activities one at a time, and thus your restricted view made it impossible for you to know whether a data element stored by one process was ever again referred to by another process. For example, Figure 22.3 shows an unnecessary custodial activity, Enter Ticket, that stores the identification of the police officer who issued a given traffic ticket. When you are studying *only* this essential activity, how can you know whether the Officer-ID is needed by any other essential activity? You can't be sure unless you check every other essential activity, including those you have yet to model, to see if this data element must be stored. So when you developed the essential activity model for Enter Ticket, you had no choice but to show that Officer-ID is a required element of essential memory.

An unnecessary custodial activity can also place extraneous intra-object and interobject relationships into essential memory. Figure 22.4 shows one such unnecessary custodial activity in Credit Payment that establishes a unique identifier for each payment. Unfortunately, the essential activity Notify Driver doesn't require unique identification of each payment. Therefore, the intra-object relationship between that payment identifier and the data elements it identifies is unnecessary, as is the activity that creates it. In a similar vein, Notify Driver does not need an interobject relationship from tickets to payments, because it will not produce a list of tickets that have been paid for. If Credit Payment created such a relationship, you would now find that particular custodial activity unnecessary as well.

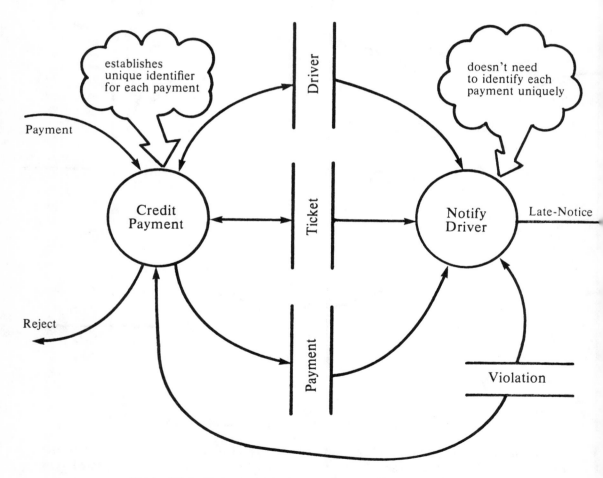

Figure 22.4. Extraneous intra-object relationship created by an unnecessary custodial activity.

Unnecessary custodial activities result from many causes. Technology is often the culprit. The major source of technology-inspired custodial activities are those activities in the existing system that keep information solely as insurance against system failure. For example, a year's worth of mortgage system activity can be reconstructed if you remember to save the individual mortgage transactions throughout the year. However, the existing backup activities are not appropriate for the new system, since it will have a different technology and therefore a different pattern of failure. You cannot be sure what data you need to back up or when you want to back it up. Since you aren't concerned with the new technology at this point in the system's development, the best thing you can do is to eliminate the backup custodial activities of the past. That way, when you plan backup requirements later on, you start with a clean slate.

Unnecessary custodial activities also result from poor maintenance of existing systems. Perhaps a system once needed a particular custodial activity, but when the need for the data store was removed, the custodial activity was not. Someone overlooked it because the system was an unruly tangle of manual procedures or computer programs. Like a soldier who doesn't know the war is over, this custodial activity continues to store data that is never used. Now is the time to remove it from the requirements statement.

A third cause of unnecessary custodial activities is specifying fundamental activities to receive data at the required level of detail, which is often higher than that supplied by the existing system. If all fundamental activities need a higher level of data than a custodial activity provides, the custodial activity may be unnecessary.

The most problematic cause of unnecessary custodial activities is the analyst's anticipating future essential requirements. In this case, the unnecessary custodial activity is currently storing data that you either suspect or know is going to be needed in the future, perhaps as part of a report that won't be needed for another couple of years. The data must be collected now because it may be impossible to get the data when the future requirements finally materialize. If the custodial activity arises for this reason, you may decide to keep it. However, you want to consider the future need for unnecessary custodial activities only. Now is not the time to consider additional future essential requirements: Don't model the activity that produces the report, for instance. You are too deep into the derivation of essence to divert your attention to future essential requirements in general.

Once you find an unnecessary essential memory requirement, you must decide just how unnecessary it is. Is it memory needed by the essential activity that stores it but not by any other essential activity? Or is it a memory component that is totally unused, even by the essential activity that captures it from the outside world?

If it's the first type, the component is simple to remove: You simply delete the custodial portion of the minispec for the essential activity that creates or modifies this memory. You also delete the component from the object data store definition. Finally, you revise the DFD, if necessary, to remove the dataflow to essential memory.

If the unused stored data component turns out to be completely irrelevant to the system, the deletions are more extensive. In addition to deleting the element from the data store and deleting the process that puts it there, you must delete any other references to it from the minispec, the DFD of the essential activity, and the definition of the incoming dataflow on which it arrived. If the unnecessary essential memory is derived within the essential activity, you also have to identify and delete data elements that are used exclusively in the derivation process. That may lead you to still more unnecessary custodial activities. In short, all traces of the utterly unnecessary essential memory components and the custodial activities that place and maintain them in memory must be removed from the model.

22.4 Summary

When there is consistency between all of their accesses, then the essential activities have been integrated. To achieve this consistency, you cross-check all the essential accesses. First, you make sure that any retrieval access used by a fundamental activity has the appropriate custodial activities. These custodial activities specify the storage of object occurrences, data elements, intra-object relationships, and interobject relationships needed by the fundamental activity. Custodial activities that store information not used by any fundamental activities are unnecessary and should be eliminated. By the end of this step, you will have added all missing custodial activities to the model and eliminated all unnecessary custodial activities, thus giving you an integrated set of essential activities.

Chapter 23

Building a Global
Essential Model

After making the accesses consistent, you could say you have a complete essential model, composed of all the separate activity models. However, your model still suffers from a significant problem: It portrays many of the same essential features more than once.

For example, Figures 23.1a, b, and c show the high-level DFD and a few data dictionary definitions for three individual essential activity models in our traffic violations system. The redundancy of these models is most obvious in the object data stores, since the same ones appear in each activity's high-level DFD, and since each of these redundant objects has its own definition. Furthermore, the same data elements appear in different data store definitions.

Redundancy is a significant problem for three reasons: It looks like a technological bias, it makes the model difficult to understand, and it makes the model unreliable. Although not really a technological bias, the redundancy of data elements in the object data store definitions looks too much like the redundant data files in the typical implementation. This can confuse someone into thinking there are still nonessential parts of the model that need to be removed.

Another form of redundancy is the appearance of the same detailed activity in different essential activities; a detailed activity shared by essential activities is called a common function. Common functions add unnecessary documentation and result in a model that is hard to read and therefore hard to verify.

Finally, redundancy in an essential model reduces its reliability by making it hard to maintain. All copies of a redundant activity or data element definition must be updated if the definition of the activity or the assignment of data elements changes. You run a great risk of overlooking some copies and thereby of making the model inconsistent.

Because of these problems, you should minimize redundancy in the essential model. Toward that end, we recommend that you integrate certain parts of the individual essential activity models:

1.   Consolidate object data store definitions.

2.   Draw a global data flow diagram.

3. Remodel the derived essential memory.

4. Factor out common functions.

We discuss each of these steps in the sections below.

23.1 Consolidating object data stores

To begin the task of combining essential activity models into a single model, you knit together the separate stored data models developed for each essential activity. The result is a set of data stores that satisfies the essential memory requirements of all the activities. Since you have already defined object data stores for each activity, now you merge any data stores whose data elements describe the same object. This is usually a simple matter of matching object data stores that have the same name and sometimes share the same data elements.

Let's look at an example of this process, using three sets of data stores from the traffic violations system. The object data stores for this process, shown in Figures 23.1a, b, and c, are divided into three groups, one for each activity that uses a data store.

Notice that even though none of the activities has the same essential memory requirements, there are objects that are common to the activities. To merge these common object data stores, you combine the data elements from data stores that describe the same object into one object definition. The result is a new, nonredundant set of data dictionary entries.

Driver   = {License-No. + Driver-Name + Driver-Address}

Ticket   = {Ticket-No. + Date-Issued + Amount-Paid}

Violation = {Violation-Type + Amount-of-Fine}

Each data store contains all of the data elements that describe a particular object, regardless of how any one activity uses those data elements.

Our example makes the process of merging data stores look trivial, and many times it is. But minor complications can arise, especially when the essential activity models are created by people working independently. What most commonly happens is the inconsistent naming of objects and data elements. If the person who built the first model in our example called the Driver object "Motorist," the consolidation would not be quite so straightforward. Not only would it not be obvious that Driver and Motorist are the same objects, but the analyst would also have to choose the better name.

A more complicated inconsistency is if one analyst chose to view the entire address as one data element, while another decided to divide it into its components: Street, City, State, and Zip-Code. You might also discover that one analyst attributed the data element Amount-of-Fine to the Ticket object, while another put it with the Violation object.

Don't wait until this stage before you try to resolve these problems. Throughout the creation of the essential activity models, someone on the project team should be responsible for the overall integrity and consistency of the model. Even if you do not manage to prevent all such inconsistencies, you should at least try to identify the most obvious ones as they happen so that you can know roughly how much work is required to resolve them.

Essential activity 1:

Driver   =   {<u>License-No.</u> + Driver-Name + Driver-Address}

Ticket   =   {<u>Ticket-No.</u> + Date-Issued}

Violation   =   {<u>Violation-Type</u>}

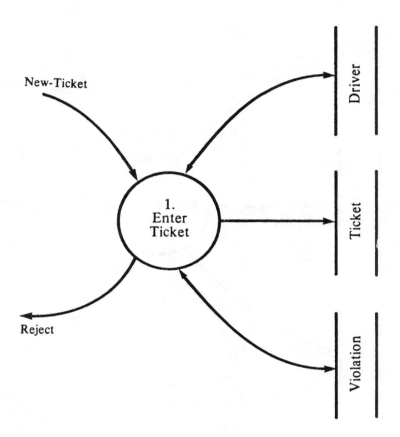

Figure 23.1a. Partial essential activity model for Enter Ticket.

Essential activity 2:

Driver	=	{License-No. + Driver-Name + Driver-Address}
Ticket	=	{Ticket-No. + Date-Issued + Amount-Paid}
Violation	=	{Violation-Type + Amount-of-Fine}

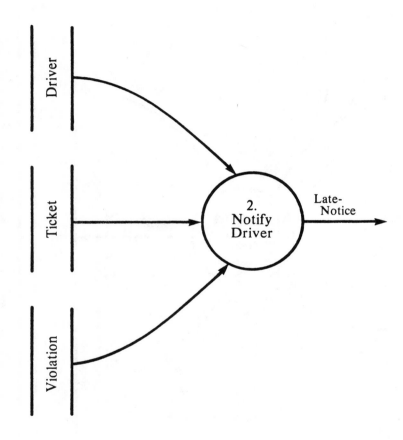

Figure 23.1b. Partial essential activity model for Notify Driver.

Essential activity 3:

Driver   =   {License-No. + Driver-Name}

Ticket   =   {Ticket-No. + Amount-Paid}

Violation =   {Violation-Type + Amount-of-Fine}

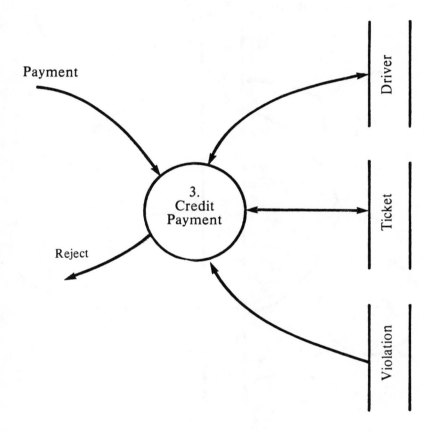

Figure 23.1c. Partial essential activity model for Credit Payment.

23.2 Drawing a global data flow diagram

In building the individual essential activity models, you create data flow diagrams showing the local object data stores that interface with the particular activity. After merging the local object data stores into global data stores, you also merge the data flow diagrams into a global diagram that shows how the activities communicate with each other through stored data.

We demonstrate this process using Figures 23.1a, b, and c, the activities in the traffic violations system. To merge these diagrams, you draw one data store for every object represented by a data store in the individual diagrams. Then, you add the essential activities to the global DFD, with the same dataflows and data store connections as before. Figure 23.2 shows the result of merging the three DFDs in this example. This technique gives you a global model that shows at least reasonably well how all the essential activities in a system interact.

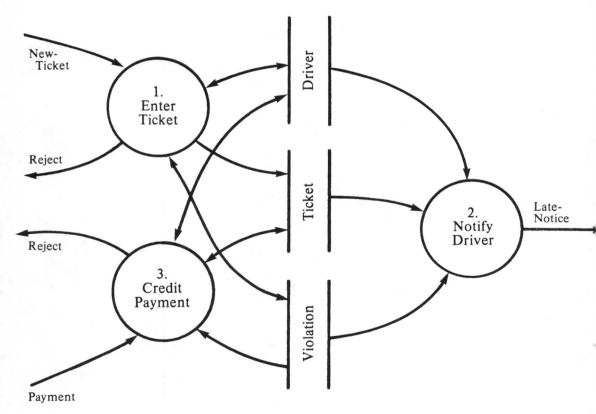

Figure 23.2. Global DFD of three integrated activities from the traffic violations system.

23.3 Remodeling derived essential memory

As stated in Chapter 22, systems sometimes need to view an element of stored data at more than one level of detail, depending upon how the data is used. For example, in the traffic violations system, one activity might require the Fine-Amount, while another activity might require the Driver's-Fine-Total. Since the Driver's-Fine-Total is nothing more than the sum of all Fine-Amounts for the violations that the driver has committed, it is the same data at a higher level. Until this point in the derivation procedure, you do not concern yourself with this issue. Both data elements would be specified as components of essential memory, each as part of a different object in this case. The Driver's-Fine-Total is attributed to the object Driver, and the Fine-Amount is attributed to the object Violation. Because Driver's-Fine-Total can be derived directly from Fine-Amounts, storing both brings about a redundancy in the essential model.

The big problem now, however, is not the redundancy of storing different levels of the same data, since a fundamental activity should ask for exactly the level of data that would minimize its work. Why should an activity that requires Driver's-Fine-Total have to piece that together from individual Fine-Amounts? To force a fundamental activity to do that would violate the minimal essential modeling principle. Instead, the problem is the way of modeling the requirement. Derived data elements should not be attributed to objects.

If you place all derived data elements in essential memory, you imply that a data element that could be derived at the time it's needed instead must be stored in memory at multiple levels of detail. That means additional custodial activities are needed to store the data at its different levels. This sounds very much like a tactic for optimizing the performance of imperfect technology. Such a tactic would be used when a processor isn't fast enough to derive the data without slowing down the performance of the system. To avoid that, the system designers decide to have the system derive the data beforehand and store it so that the slow processor does not perform the derivation. Because attributing the derived data elements to objects smacks of this tactic, it violates the principle of technological neutrality.

How do you resolve this problem, which pits one essential modeling principle against another? Our tactic is to define an essential level of detail; only those data elements at that level of detail are assigned to object data stores. All other levels of detail for that data element are derived by the fundamental activities that need them.

Figure 23.3 shows the result of remodeling essential memory that contains different levels of detail. The custodial activity on the left captures a given data item graduated in units per hour. The three fundamental activities on the right require that data item to be summarized into weekly, monthly, and quarterly totals. As a result of your applying the ideas of the last chapter, this system would now have weekly, monthly, and quarterly data in essential memory and a set of custodial activities to keep these pieces of memory up-to-date.

To produce a diagram like Figure 23.3, you start remodeling by choosing the essential level of detail for stored data. We call this essential level the *greatest common factor,* which is the highest level of detail that all the fundamental activities can use to derive the level each needs. For example, the greatest common factor among weeks, months, and quarters is the day, since weeks may not divide evenly into months and quarters. So, the essential level of detail in this example is a daily total of the hour-by-hour input.

Having determined the essential level of detail, you now must remodel the data dictionary by removing all other levels from the definitions. Then, you attribute the data element at the essential level of detail to the appropriate object.

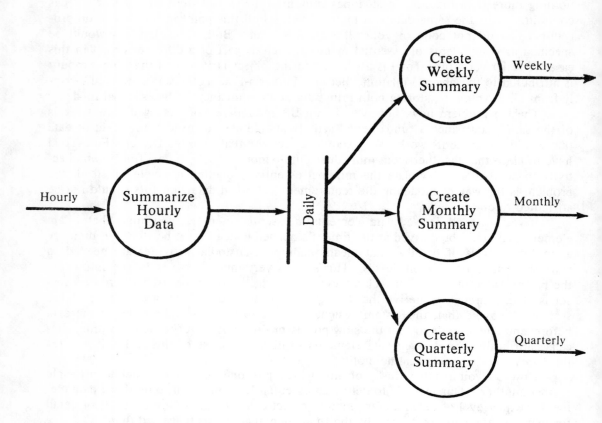

Figure 23.3. Essential memory remodeled to contain essential level of detail only.

When a removed level of detail is needed by a fundamental activity, it is derived by an essential memory derivation activity from the essential level of detail. In the example, you would specify essential memory derivation activities to derive weekly, monthly, and quarterly data from the daily data in essential memory, and then add these activities to the minispecifications for the appropriate fundamental activities.

You finish remodeling essential memory by reviewing the custodial activities that store the memory you have just worked on. When you find custodial activities that derive and store levels of data above the essential level, remove them. In a few cases, you will have to do more. If no custodial activity stores the essential level of detail, you have to redefine a custodial activity so that it does. For example, the system in Figure 23.3 didn't have a custodial activity for daily data, so one had to be provided.

Finally, you may take this opportunity to redefine the input to the custodial activity so that it reflects the system's true requirement for that input. The system in our example does not need to receive the input at an hourly level of detail, since no essential activity depends upon that level of precision. So in this case, you amend the definition of the incoming dataflow to show that the data element can be at the daily level right from the start.

23.4 Factoring out common functions

Applying event partitioning sometimes produces awkward results that must be corrected. In Chapter 17, we described the classification of essential activity fragments into essential activities based upon the events to which they respond. During that process, you sometimes encounter essential activity fragments that respond to more than one event. When such a fragment turns out to be a function that is common to more than one essential activity, you put one copy of it into each essential activity. By allowing a degree of redundancy among the essential activity models, you are able to study each of them independently. Now that the essential model is complete, you must assess whether the essential activities should continue to remain completely independent, or whether removing some of the common functions from the activities and making them separate processes would improve the overall quality of the model.

Figure 23.4 shows part of the essential data flow diagram for a blood bank. Both essential activities, Issue Blood and Discard Old Blood, reduce the amount of blood on hand, so both may appeal for new donations if the quantity in stock falls below a specified level. During the derivation of this model, the analyst identified the process of issuing an appeal for donors as a common function for both activities. So, each essential activity model got its own copy of the fragments that carry out this function.

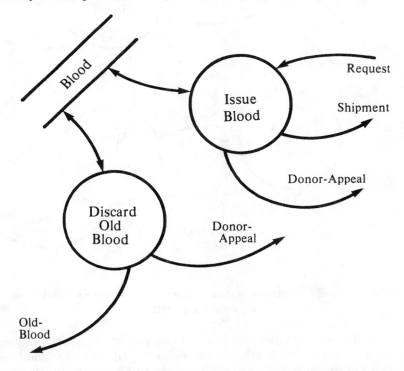

Figure 23.4. Two essential activities that contain the function Appeal For Donations.

Should you leave the diagram like this? Is this the most appropriate way to partition the essence of the blood bank? To answer these questions, you must consider how important and how substantial the common function is compared to the essential activities of which it is a part. If the common function is trivial, such as calculating a percentage, there isn't much to be gained from factoring it into a separate process. If the common function is critical but not very large, you won't make it a separate process either. For example, if the common function edits an incoming data element, drawing a separate process on the DFD to handle all incoming dataflows that contain the data element will probably complicate the diagram horrendously without adding much useful information.

You will encounter cases, however, in which a function shared by more than one essential activity is both a substantial and important planned response. If such a common function in your judgment deserves to be seen on upper-level data flow diagrams, you should create a separate process. Figure 23.5 shows the DFD of the blood bank example after factoring out the Appeal For Donations function.

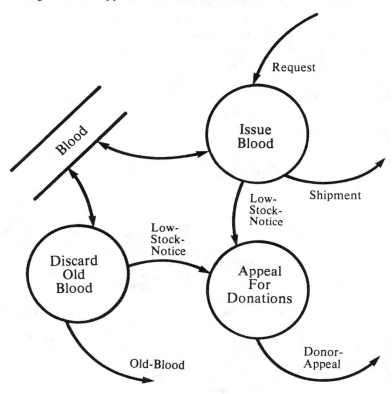

Figure 23.5. The common function Appeal For Donations factored out as a separate activity.

By revising the data flow diagram in this fashion, you are abandoning event partitioning, if only in a small way. Neither Issue Blood nor Discard Old Blood now embodies the system's entire immediate planned response to an event. But if the result is a diagram that communicates a greater amount of important information *without* becoming too complex and *without* introducing technological influences, then it is an improvement over the strictly event-partitioned version. Be true to the essential modeling goals that event partitioning is designed to achieve, not to the guideline itself.

23.5 The completely essential model

After you have integrated the individual essential activity models by making their accesses consistent, these models still contain redundancy that violates the principles of technological neutrality and concise modeling. This chapter told you how to remove that redundancy. First, you combine data store definitions for the same object in the data dictionary. Then, you produce a data flow diagram that shows individual essential activities linked through their common object data stores. Once the activities are joined on the DFD, you establish the essential level of detail and remodel essential memory accordingly. Finally, you factor out common activities.

By the powers vested in us, we now pronounce your model essential. At this point, providing you perform the procedure correctly, the resulting model depicts only the aspects of the existing system that would have to be present in any implementation, regardless of its sophistication. This model serves as an excellent foundation for the specification of new essential features and new technological requirements.

Chapter 24

Reviewing the Quality of the Model

Throughout this book, we advocate three modeling objectives: to avoid technological prejudice, to state the minimum essential requirements, and to present information effectively. Upon completing the previous step, your model achieves the first two objectives. In this, the final step of the procedure, you inspect the essential model to ensure that is also an effective one.

An effective model presents all relevant aspects of the system in an understandable form without overwhelming the reader with the complexity of any one of its pieces. Furthermore, an effective model emphasizes important information, with the key concepts most prominent. The model's organization — in this case, the organization of the data flow diagrams, data dictionary, and minispecifications — determines how well the model limits complexity and emphasizes key activities and memory.

The model produced by the foregoing steps of this procedure will be mostly effective. Many of the tactics employed during the derivation process were chosen because they result in well-organized models. For instance, event partitioning and object partitioning are examples of the best way to produce both essential and effective models. So, by the time all of the physical characteristics are removed, the model should be of reasonably high quality. There are probably a few remaining flaws whose removal would make the essential model much easier to understand and therefore more effective. These remaining defects are often the result of simple human fallibility. Throughout the derivation process, you may have been concentrating so completely on removing physical characteristics that you did not pay enough attention to the model's organization. You now have a final opportunity to judge whether the model is easily understood and whether all its pieces fit the budget for complexity.

The model may have other structural deficiencies, which result not from human error, but from applying *general* modeling principles to a *specific* system. The general principles include event partitioning and object partitioning. Although they offer a simple modeling strategy that works for most systems, these partitioning themes cause problems when you model very large systems. To partition a large essence effectively, you have to make concessions in your use of object and event partitioning. In this chapter, we discuss what concessions are necessary.

24.1 Repartitioning data stores

Using object partitioning is unsatisfactory when a large portion of the data elements are attributed to one object. For example, during a project to develop a work order system that assigns and tracks repair jobs, the analyst creates object data stores only to find that the majority of the stored data elements belong to the object Work-Order. There are other objects, but they account for only a few of the data elements.

To see why you should be unhappy with overloaded data stores, you have to recall why you use object partitioning in the first place. Object partitioning helps you avoid the complexity of showing too many data stores, but does not lump all the stored data elements into a single file either. But when object partitioning produces a data store, such as Work-Order, that contains nearly all the stored data elements, it is not meeting its objective. Instead, the model hides too much information in the data dictionary. The DFD looks like Figure 24.1: So many activities access only the one data store that the DFD does not reveal much about what the system needs to remember. A superior partitioning of stored data — one that divides the data elements more evenly among the data stores — would allow you to see more information from the data flow diagram.

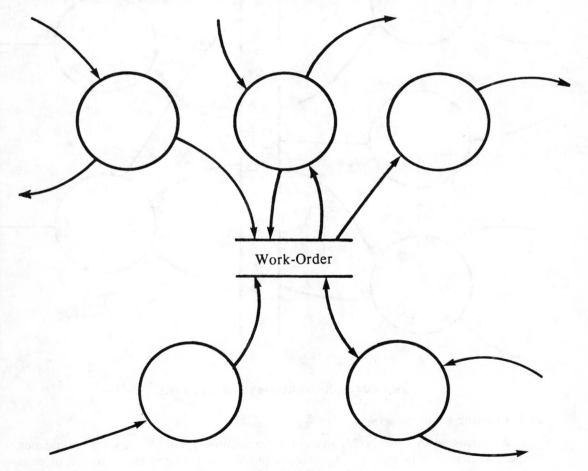

Work-Order

Figure 24.1. Object data store resulting from poor partitioning.

The true problem here is that object partitioning has not been used correctly. Perhaps the analyst has simply identified the wrong object, one that is at too high a level of abstraction. He or she could probably have identified subdivisions of the object Work-Order that also qualify as objects; these could become data stores that contain fewer data elements. Figure 24.2 shows an alternative to the Work-Order data store. Together, the data stores Work-Order-Description, Labor, and Material contain the same elements as the Work-Order store in the previous figure. Each of the resulting data stores is much smaller than the original; however, the new data stores are no less cohesive: Each contains the data elements that describe an important characteristic of the work order. These smaller data stores make the essential model more understandable.

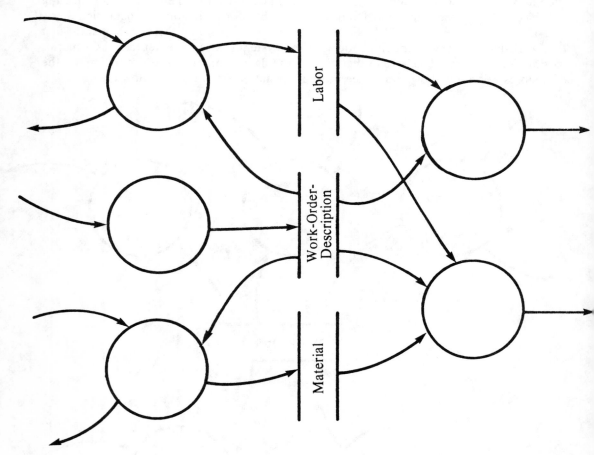

Figure 24.2. Repartitioned object data stores.

24.2 Creating subminispecs

As discussed in Chapter 23, the event partitioning approach causes a redundancy in the essential activity models: A common function is repeated in the model of each essential activity that uses that function. To eliminate the redundancy, you factor out the common functions. This means redrafting the essential DFD to show the common function as a separate activity linked by a dataflow to the essential activities that share it. That solution is fine for smaller essential models, but it causes a problem when you are modeling a large essence.

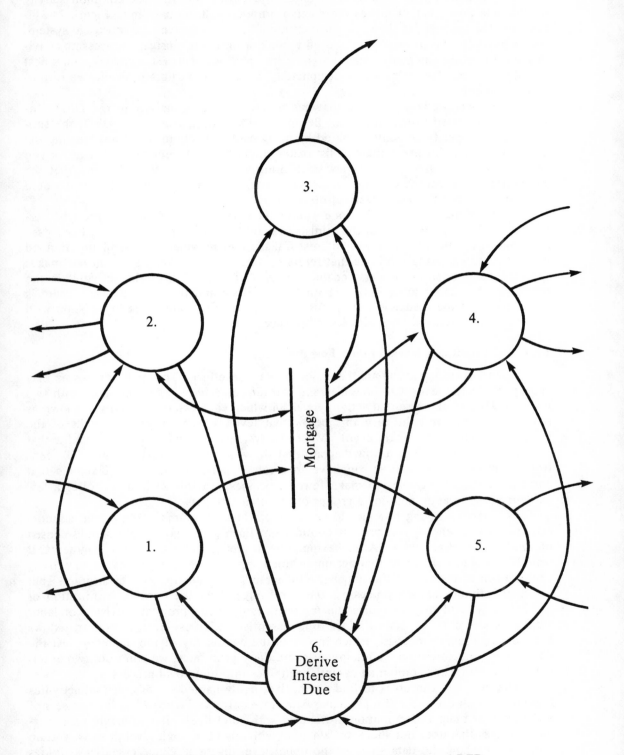

Figure 24.3. Common function that creates a poorly partitioned DFD.

If the shared activity is used by enough activities, showing the common activity on the data flow diagram makes the diagram impossible to understand. Figure 24.3 illustrates the problem; it shows an abstract model of the essence of a mortgage system. The activity Derive Interest Due is used by lots of different mortgage processing activities and therefore has many dataflows going into and out of it. This creates a rat's nest of dataflows that cannot be easily deciphered. To get around this problem, we recommend subminispecs.

A *subminispec* is a minispec that has no corresponding activity in the DFD. Instead, it is referred to in minispecs. Because it does not appear in the DFD, the subminispec can specify an activity shared by many essential activities without tangling the diagram. A subminispec contains a list of the minispecs that refer to it. Whenever you change the subminispec, you review each minispec on the list to make sure that the subminispec still carries out the required activity. Figure 24.4 shows how you would use a subminispec to specify Derive Interest Due.

Each subminispec describes one well-defined activity. This function must be performed in the same way for all the minispecs that refer to it. You would not like to see a subminispec called Generalized Interest Calculation, in which the calculation defined differs according to the minispec that refers to it. In such a case, a subminispec makes the essential model less understandable. Instead of finding all of the activities in the minispec, the reader must look at both the minispec and the subminispec. There is also a risk that the reader would confuse the part of the subminispec in question with the other unrelated activities in the subminispec.

24.3 Building upper levels of data flow diagrams

Event partitioning has another undesirable side effect: Too many processes may appear in the top-level data flow diagram. So far you have worked primarily with two levels of DFDs: the essential activity level, in which each essential activity is shown as a single process, and the diagrams below that level, which depict the details of the system's response to a single event. Most real systems respond to so many events that you can't draw a single data flow diagram that shows all the essential activities. To control the model's complexity, you have to create upper-level data flow diagrams that combine essential activities to make fewer processes. In this section, we offer a few guidelines for creating high-level groups of essential activities.

We start by saying how not to create upper-level diagrams: Don't split essential activities across upper-level process boundaries. Each group you create should consist of *whole* essential activities. After having gone to great lengths to reunite fragments, it would be a shame to undo that accomplishment by refragmenting an essential activity simply to fit a scheme for the upper-level diagrams. If your upper-level diagrams split essential activities among processes, they probably reflect the current organization of the company, and may prejudice the model toward an inappropriate division of labor among processors. In addition, splitting essential activities creates more dataflows between high-level processes, which increases the complexity of the model — just the opposite of your intention. To help you avoid these grim fates, we point out two excellent opportunities for combining essential activities into high-level processes.

The best you can do is to find a cohesive, easily named set of essential activities that contains all the users of a particular data store only. In Figure 24.5, the two sets of activities that meet this condition are circled with solid lines. If you create an upper-level process that does not share its data store with any other high-level process, you do not have to show the data store in the upper-level diagram. Since essential activities are connected through data stores, the ultimate simplification is to put not only a set of essential activities into one process, but their shared data stores as well.

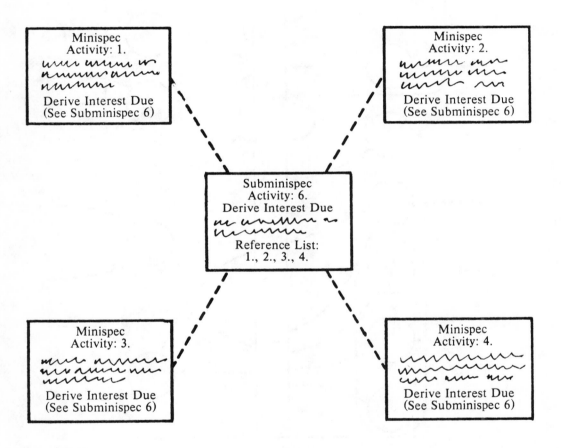

Figure 24.4. Subminispec's relationship to minispec.

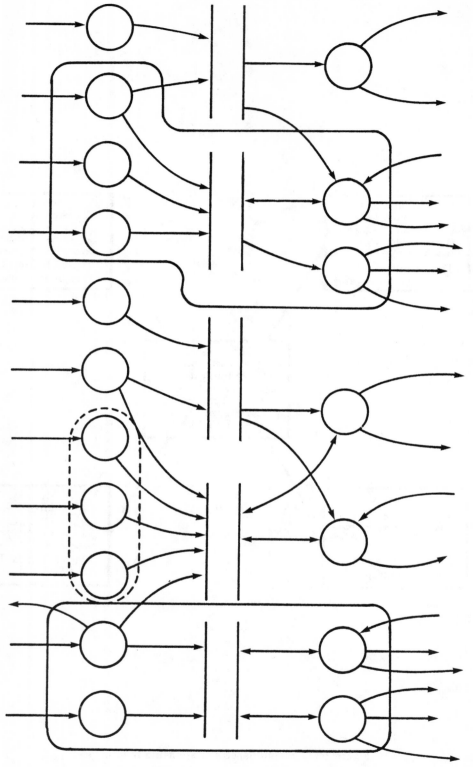

Figure 24.5. Upper-level processes identified by solid and dotted lines.

As you look for activities that share the same data store, keep in mind two warnings. First, the upper-level group must include all the essential activities that use the data store in question. If you omit even one activity that accesses the data store, the data store must be shown on the upper-level diagram. Second, you do not accomplish much if your upper-level group is not cohesive — that is, if it cannot be honestly summarized with a single verb/object name. If the best name you can give to the set of essential activities that access the Zip-Code store is "the set of activities that access the Zip-Code store," it is not a good candidate for an upper-level process. Look for groups that have a specific purpose, and name the upper-level process accordingly.

After you have exploited this first guideline to the fullest, you will probably still have far more essential activities than you care to show in a single data flow diagram. Many of these will be custodial activities, which are often trivial file maintenance functions such as Change Customer Address or Update Sales Tax Rate. Simple custodial activities like these often access only one data store: the one they update.

You can simplify the data flow diagram substantially by grouping all the custodial activities that maintain a given data store into one high-level process. Figure 24.5 shows a group of such activities circled with a dotted line. Since trivial custodial activities are so plentiful, you greatly reduce the number of processes in upper-level diagrams by following this guideline. However, you pay for the simplification by living with vague process names like Maintain Tax Data or Maintain Customer Data. Still, you rarely have a more attractive alternative, and creating a single custodial activity for a data store de-emphasizes many of the least interesting components of the system.

24.4 Summary

Just removing physical characteristics is not enough to produce an effective essential model. If the essence of a system is large, its model is likely to be unreadable unless you take steps to simplify it. Repartitioning object data stores, using the subminispec, and drawing upper-level diagrams will help you build large-scale essential models that are easy to understand.

Once you have an understandable model of the existing system's essence, it is time to define the essence of the new system. You've been working with the existing system for so long now that you may have forgotten that the system will also implement some new essential features. It is your job to define and model these new requirements, and then integrate them into the model of existing essence. These remaining tasks are discussed in Chapter 25.

Modeling
the New System

Part Seven is a brief introduction to the essential system development activities that follow the derivation of an essential model. These activities include adding new essential features to the model of existing essence, selecting the incarnation of the new system, and building the incarnation. In this part, we discuss the first two of the remaining system development activities.

Chapter 25 introduces an approach for adding new essential features to a model of the essence of an existing system, and the next two chapters explain a method for selecting the appropriate incarnation. We discuss the process of building a new physical model in Chapter 26. Then in Chapter 27, we focus on program design and show that the specification produced by essential systems analysis is easily converted into a software design blueprint.

Chapter 25

Adding New
Essential Features

The chapters through Part Six complete the description of a major activity in the system development process. By now, you know a systematic procedure for studying an existing system whose essence is similar to the one you intend to build. You also know that the procedure produces both a physical and a logical model of the existing system. This current logical model is a model of the essence of the existing system.

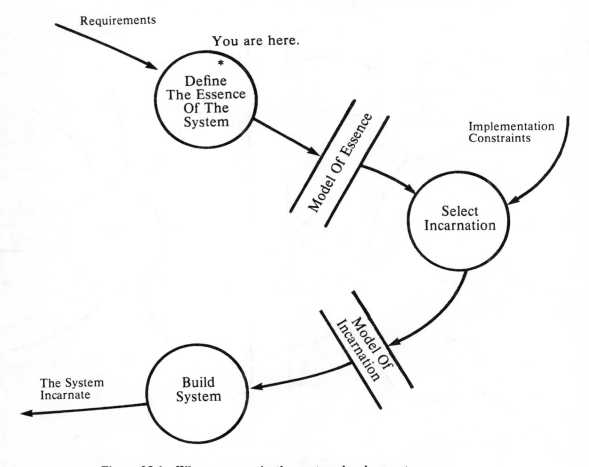

Figure 25.1. Where you are in the system development process.

Figure 25.1 shows the three activities that make up the essence of the system development process. After completing the essential model of the current system, you still are only in the first process on that diagram: to develop a model of the essence of the new system. This chapter considers how to complete that model by adding new essential features to the current model. You use a two-pass approach that resembles the two-pass procedure for deriving the essence. First, you create and model each new essential feature, and then you add these models to the current logical model.

25.1 Building mini-models

To minimize errors in the specification process, you should build a separate, small structured analysis model of each new essential feature. This *mini-model* consists of data flow diagrams, minispecifications, and data dictionary definitions.

Figure 25.2 presents the DFD portion of a mini-model for one new essential feature of the traffic violations system. The new essential feature is a "point system" for suspending the license of frequent offenders. Each violation has a point value. If a driver commits enough violations to accumulate a certain number of points, his or her license will be suspended, causing a suspension notice to be sent to both the driver and the Department of Motor Vehicles.

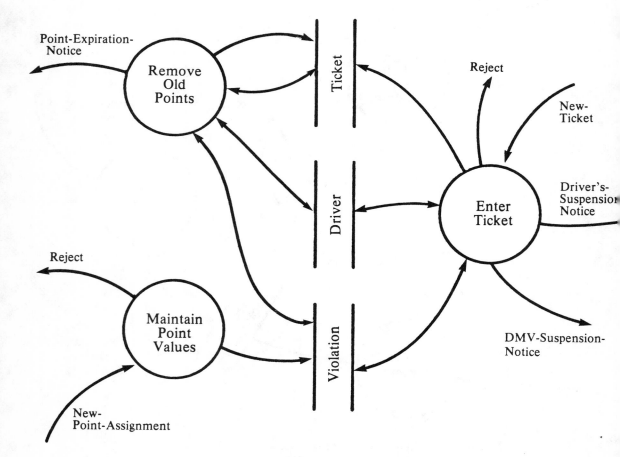

**Figure 25.2. DFD portion of a mini-model for
a new essential requirement, a point system.**

Besides minimizing errors, mini-models are useful for another reason: Mini-models offer a way to document requirements well before you finish deriving an existing system's essence. New features are often discussed in pre-project meetings, for example, and mini-models are useful to document the new features so that they can be filed away until you are ready to work with them. Also, as you work on the current model, you can build mini-models whenever you discover new requirements. Then, when you're ready to start work on the new essential features, you have the mini-models and don't need to backtrack for information.

25.2 Defining new essential features

Although representing new essential features with mini-models is not difficult, you may have trouble identifying the new essential requirements and defining them in detail. The requirements for a new system usually contain changes to both the essence and the implementation of the existing system. Because technology changes so rapidly, the majority of changes to the existing system are technological in form, with the essence remaining much the same. You must be able to distinguish the new essential features from the new implementation features in order to complete the essential model. As before, you consider the essential changes now and defer consideration of new implementation characteristics until later. How easy it is to identify the new essential requirements depends upon the skills of the people responsible for creating them. In the next two subsections, we examine the two cases of when creators understand essential systems analysis principles and when they don't.

25.2.1 When the users or analysts understand essential systems analysis

If the people creating the new essential features know the techniques explained in this book, they will have little trouble distinguishing changes to the essence from changes to the implementation. Furthermore, they can use these same techniques to create the new essential features, as if they were creating the essence from scratch.

When the user or analyst creating essence knows essential systems analysis, we recommend that he or she start by considering events, both new ones as well as existing events that aren't being handled effectively. As when creating essence from scratch, the creators make a list of these events.

In creating the new essential mini-model shown in Figure 25.2, you would produce the following events list:

- Police officer submits traffic ticket (existing event).

- Department of Motor Vehicles changes point assignment (new event).

- Points expire (new event).

You use a two-pass approach to develop the new and modified essential responses. In the first pass, you create an individual essential activity model for each response, using event and object partitioning. Model new essential activities first, then modify existing ones.

To change an existing essential activity, you first detach its model from the global model so that you can consider the changes independently. Chances are, the changes to the existing essential activities will be new object data stores and additional dataflows representing new results.

Figure 25.3 shows the DFD portion of the individual new essential activity models for the traffic violations point system. Of the three essential activities, two are completely new. Remove Old Points takes points off a driver's record after a certain amount of time from when the violation occurred. Maintain Point Values is a custodial activity that changes the point values for a particular violation. Each of these activities is the response to one of the new events on the events list above.

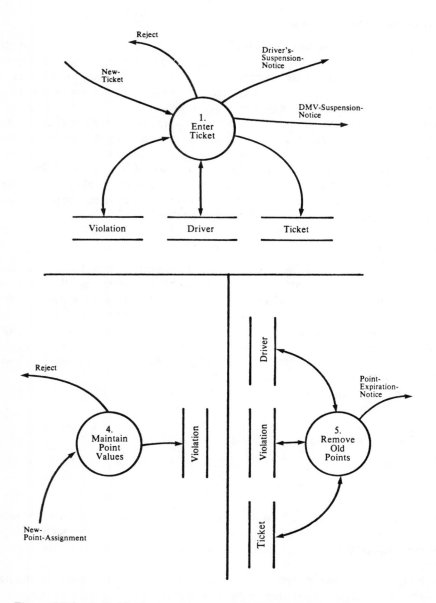

Figure 25.3. Individual new essential activity models for the point system.

Enter Ticket is an essential activity that already exists in the current essential model. It has been removed from that model and amended to account for new essential requirements. Now when a new ticket comes in, Enter Ticket assigns a point value to that ticket by adding the points for all violations on the ticket. Enter Ticket then tests the total points against the suspension limit. If the driver is over the limit, then license suspension notifications are sent to the driver and to the Department of Motor Vehicles.

In the second pass, you integrate the individual essential activity models for a new requirement into a complete mini-model for that requirement. Using the same techniques discussed in Part Six, you basically integrate the essential memory requirements of the different essential activities. You also have to review the mini-models to ensure they are easy to understand. The resulting mini-model for the point system, at least the DFD portion, would look like Figure 25.2.

25.2.2 When the users or analysts don't know essential systems analysis

Often, when the creators of the new requirements are users or user-analysts unfamiliar with the techniques of essential systems analysis, they indiscriminately mix changes to the essence with changes to the incarnation. So, before you can build a model of the new system's essence, you have to untangle the user's statements of requirements. The first step is to record all such statements in the form of mini-models.

Once you have all the mini-models, you are ready to separate the new essential requirements from the new implementation requirements. To do this separation, you examine each mini-model for the physical features given in Part Three. The physical characteristics of new requirements are exactly the same as those of existing systems, and, if anything, they should be easier to find in such a small model. Look for any requirements concerning the cost, capability, capacity, speed, and fallibility of the system's internal technology, as well as features arising from fragmentation and consolidation that are the likely result of the imperfect technology used to implement the new system. Finally, recognize those features that smack of data transportation or quality assurance that are probably requirements for the new infrastructure and administration.

If you find that an entire mini-model describes only a new way to carry out the same old essential activities, set that mini-model aside until you begin work on the new physical model. On the other hand, when you find mini-models that change the system's essence, apply the same discovery procedure presented in Parts Four, Five, and Six for deriving the essential activities of an existing system. This allows you to eliminate physical features from the mini-models and reorganize them into individual essential activity models, which are then combined into a mini-model of the complete new requirement.

25.2.3 Integrating new essential requirements

Once you have completed and reviewed the mini-models, you then integrate them with the model of the existing system's essence. This process is almost identical to the second pass of the derivation procedure, which combines the derived essential activity models into one global model. In this process, you first consider whether the new or changed fundamental activities obtain the information they require from the outside world at the time of the event, and, if not, whether they have the essential memory they need and the custodial activities to provide that essential memory. If any memory or custodial activity is missing, add it. Use the same techniques discussed in Chapter 22 to make the set of essential accesses consistent. You then consolidate object data stores, build a global data flow diagram, and factor out common functions. Finally, review the quality of the new essential model.

Figure 25.4 shows the DFD portion of the new essential model for the traffic violations system, after the activities in the new essential mini-model have been integrated with the essential activities in the current essential model.

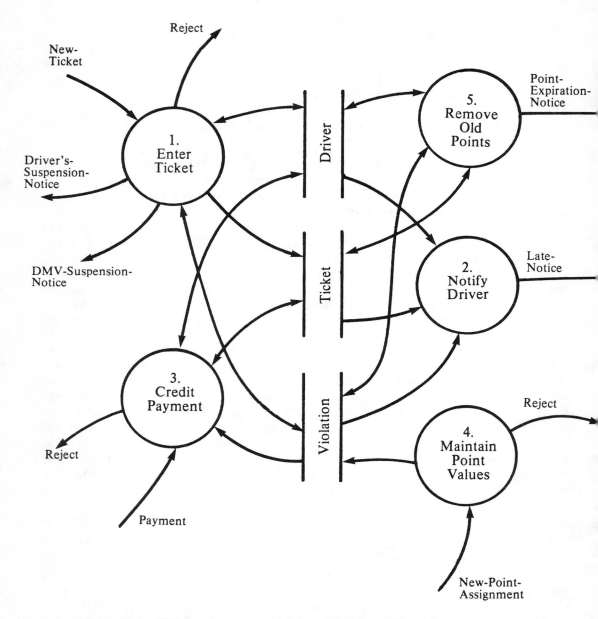

Figure 25.4. DFD for the new essential model of the traffic violations system.

25.3 Understanding the psychology of new essential requirements

The biggest problem in defining new essential requirements is psychological in origin, not technical, because of the trouble you may have in switching mental gears. After weeks or more probably months of studying the existing system, you and your project team members may find it difficult to think of new features because you are so accustomed to focusing on the old ones.

Users, too, often feel uncomfortable with the task of establishing new essential features. Throughout the modeling of the existing system, the users must record things as they are, sometimes helping the analysts derive the essence that underlies the users' jobs. They serve primarily as critics, ensuring that the emerging model accurately reflects the system as they know it.

Once you model the existing system, however, the users' role changes dramatically from recorder and critic to creator and interpreter of corporate policy. Instead of asking them "What do you do?" you are now asking, "How *should* this company do business?" Since most employees are unused to thinking in these terms, they are uncomfortable with the task of establishing even small details of corporate policy. Moreover, the question is not expressed in terms of video display terminals, database management systems, or daily file updates, thus forcing the users to think abstractly. Users may not like all that data processing jargon, but they are at least familiar with it and may have difficulty thinking in more abstract terms.

The users who work right along with the analysts to discover the essence of the existing system will probably not have much trouble stating essential requirements accurately. But in some cases, these users are not empowered to formulate corporate policy and procedures. Then you must get statements of new essential requirements from higher-ranking users who have not participated in the system development effort and who may be uncomfortable with the philosophy and process of building essential models.

These psychological blocks can slow the progress of a development project just as the effort to model new requirements gets under way. We recommend anticipating the slump by calling for new requirements long before you are ready to incorporate them into the current logical model. In this way, everyone has a chance to get used to the idea of making requirements statements assuming the new system has perfect technology. This tactic also gives the project team an opportunity to fine tune the scope of the new system and to practice developing essential models.

A word of caution is in order here: Although you try to identify all new essential features desired, you don't want to encourage unrealistic expectations of what the essence of the new system will be. A general statement of some user's grandest desire can turn into nightmarish tangles of detailed essential features, many of which are difficult to implement with the technology available. In this case, you may have to reject the user's favorite new essential requirement. No one likes to have his or her expectations encouraged only to have them dashed later on, so you must walk a fine line between encouraging ideas about new essential features and building unrealistic expectations.

25.4 Summary

Completing the new logical model is relatively easy from a technical standpoint. You have already gained extensive experience in specifying essential features and distinguishing them from implementation characteristics. You need no new tactics, just a bit of creativity. The bigger problems in adding new essential features have to do with the timing and the form of new requirements statements and the psychology of adding them. You want to model new requirements, both essential and implementation-related, when you find out about them in the form of mini-models. With this tool, you can set aside the statements of new implementation requirements until later in the development process, but beware of users' expectations for new essential features.

Once you wrangle the new requirements statements from the users, or at least get them to approve your suggestions, you simply add them to the essential model of the existing system, just as if they should have been there all along. You now have a complete model of what the new system should do, and all it would ever have to do if there were such a thing as perfect technology.

Chapter 26

Selecting
an Incarnation

Once you model the new system's essence, you must focus on how to implement that system using the available technology. Although we want you to use the concept of perfect internal technology when building models, we also remind you that no matter what technology you use to create the new system, it is less than perfect. So, you now face the task of modifying the new essential model to take into account the technological limitations that affect the new system.

If we could provide you with a procedure for transforming a new logical model into a new physical model that is just as detailed as the one for deriving the essence of an existing system, we would. Unfortunately, since no such detailed procedure yet exists, we instead state the objectives of a system incarnation, and we offer a general strategy for selecting and modeling such an incarnation.

26.1 An incarnation's general goals

You want a system that is reliable, fast, and practical. For a system to be reliable, its incarnation must be faithful to its essence, meaning the processors must possess the skills to carry out the essential activities correctly. The incarnation also must carry out its planned responses in a reasonable amount of time. Finally, the incarnation must be practical; you must choose processors, containers, and channels that are affordable, legal, and acceptable to the users.

26.2 Strategy for deriving an incarnation

Transforming an essential model into a model of a system incarnation is very much like deriving the essence in reverse. When deriving essence, you specify a physical ring for each essential activity, but focus on the essential core of those activities; when selecting the incarnation, you focus on redefining the system's physical interface with the outside world. When deriving essence, you locate fragments of a single essential activity and unite them; now, you allocate pieces of activities to processors. When deriving essence, you root out and eliminate the infrastructure and administration; now, you devise infrastructural and administrative activities that are right for the new processors. When deriving essence, you remove the many subtle effects of past optimization efforts; now, you consider ways to optimize the performance of the new system. The remainder of this chapter introduces a general strategy for building a model of the new system's incarnation.

26.2.1 Specifying the external interface

The first step is to define the physical interface between the system and the outside world. What must be inside the system depends upon its relationship with the outside world. This relationship has three parts:

- the means by which the system receives incoming dataflows

- the amount of time for the system to carry out each planned response

- the means by which the system transmits its responses to the outside world

Each part must be included in the new physical model.

Before you can define the physical means for carrying out an essential activity, you must know how it communicates with the outside world, specifically how the activity learns of the occurrence of an external event. During the creation of the essential model, you determine what incoming data elements would describe the event, but now you must also identify what medium will be used to transport those data elements. Ask, By what means will the data arrive? telephone, mail, SNA network, carrier pigeon? How will each data container be organized? Will each container hold only a single dataflow or several packed dataflows? How will the data elements themselves be coded? by handwriting, voice, ASCII, EBCDIC? Since you have assumed all along that the external technology is imperfect, the new essential model may already include some of these factors; now is the time to complete the picture.

Until now, you have assumed that the system carries out activities instantly. However, once you acknowledge the imperfection of a response mechanism, you must consider time constraints more seriously. At this stage, it is enough to record the known requirements in the minispec of the activity in question. Later in the selection process, when you allocate essential activities to processors, you can determine the impact of time constraints on the new incarnation's internal organization.

Just as you did for the incoming dataflows, you must define the format and the medium the system will use to transmit each output. Quite often, these are influenced both by the technology outside the system and the technology to be used by the new incarnation. Since you won't know everything about the new system's technology at this point, you can only roughly determine how the system will transmit its responses and return to this task when you confirm the choice of technology.

26.2.2 Allocating essential activities to processors

Systems are nested sets of processors that carry out both essential activities and the activities mandated by imperfect technology. In Chapter 12, we described the process by which fragments of essential activities are assigned to processors based upon the skill vs. cost factor and the capacity vs. cost factor. Now, we employ those same concepts to describe the allocation of essential activities to the processors in a new system.

If you had a sufficiently skilled and powerful processor, you might be able to assign all of the essential activities to it. Certainly, this is one of the simplest incarnations imaginable, since all the essential activities remain intact and few changes to the new logical model are needed. But given the size and complexity of most information systems, this is not likely to happen. If you are only slightly less fortunate, you could assign each essential activity to exactly one processor. Again, the activities remain intact, but the processors would have to share memory, thus complicating the model somewhat. But even this prospect is too optimistic, as for each essential activity, you would have to find a processor that possesses the skills, work capacity, and economy to carry out the entire activity efficiently by itself. Such a processor is rare.

Unfortunately, the best way to carry out a particular activity is most often not with only one processor but with specialized processors, each of which is suited to a portion of the activity. You then divide the detailed actions of the essential activity among the processors you select. Although the techniques required to make these decisions are beyond the scope of this book, we do give an example that describes how you transform the new logical model to show the assignment of essential activities to processors.

Suppose that you must implement a hotel system. After studying the volume of data processed by the current system and considering the response time required, you determine that the new system will need four people and a small business computer. To supply the greatest convenience to the guests and to avoid renting extra office space, you locate all of the processors in the hotel.

There are many ways to assign the essential activities of the hotel system, shown in Figure 26.1, to the five processors. One possibility is based upon two conclusions about the strengths of each type of processor: First, you decide to assign as much work as possible to the computer system so that you can minimize the processing costs. Second, you conclude you don't want to risk alienating guests by forcing them into direct contact with the computer system. Consequently, you choose to allocate everything to the computer system except the portions of essential activities that involve direct contact with hotel guests, as shown in Figure 26.2.

One of the essential activities, Cancel No Shows, requires no direct guest contact and so is allocated entirely to the automated processor. The rest are divided between the computer system and one or more of the humans. Once you assign tasks to the processors, you must give the processors a way to communicate with each other.

26.2.3 Establishing the infrastructure

Processors need to communicate with each other for two purposes: to gain access to stored data they share with other processors, and to send intermediate results from one essential activity fragment to another. Processors also require a way to access data stores that they do not share with other processors. These services are provided by the infrastructure.

In Chapter 12, we offered two basic ways to provide a processor with the essential memory it requires. You can establish a data store that is owned solely by the processor, or you can have the processor use a data store that it shares with one or more other processors. When choosing an implementation, you have the same two options.

If data is needed by only one processor, the solution is obvious: You store it within that processor and establish an access mechanism by which the processor can use and update the data store. Such an access mechanism is a part of the intra-processor infrastructure, because it helps the processor to move data into and out of a technologically imperfect container: the file.

Deciding how to implement essential memory is usually more difficult than this, because most components of memory are shared among processors. As explained previously, you must choose between the two options for providing access to memory: either establishing one container for the shared data item, or creating a private data store for each processor with redundant copies of the data element.

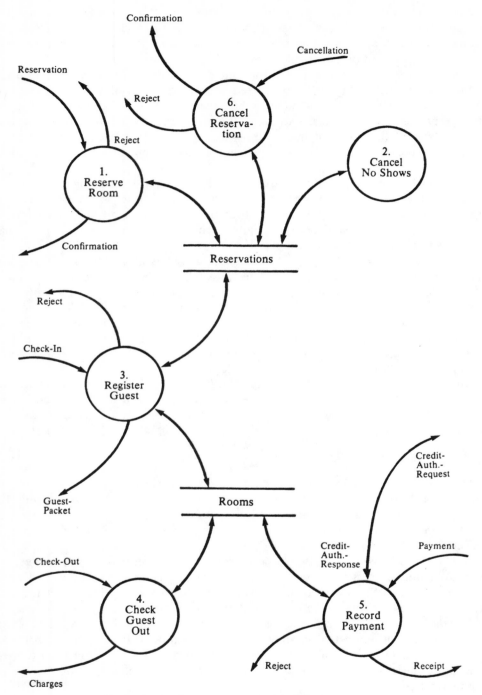

Figure 26.1. Essential model for a hotel system.

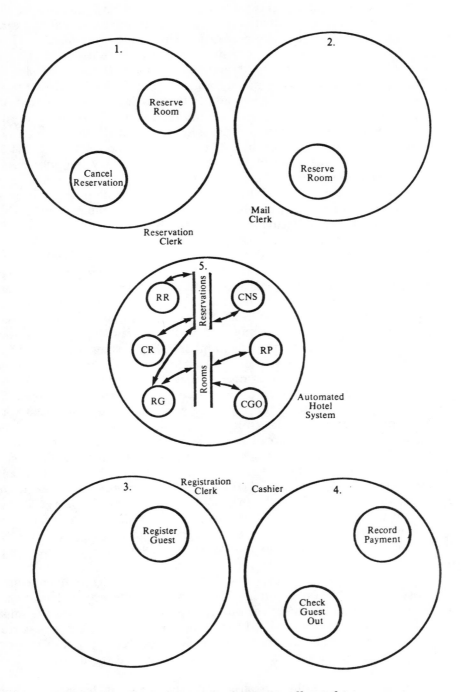

Figure 26.2. Essential activity fragments allocated to processors.

If you establish a single repository for a shared data item, you have to allow all the interested processors to access that data store. The advantage is that you can often give the custodial activities to a single processor. This reduces the number of processors updating data, and so makes the integrity of the data store easier to protect. For this reason and others, many people prefer systems that share data in this fashion.

But sharing a single data store among a group of processors makes the system more complex. First of all, you have to consider the location of the data store in relation to the processors that use it. If they are too distant from one another, you may have to establish additional transporter processes to move data from the store to a processor and back again. The format of the data presents another problem. If the processors that share the stored data are of very different types, they may not be able to understand the same stored data. For example, consider data that must be shared between a human and a computer. If the store is understandable to the human — say, a handwritten record — it may not be understandable to the computer, and vice versa. So, you have to add translation activities to convert the stored data into a meaningful form for at least one of the processors.

To avoid these problems, you may decide to store a data element in two files, each of which is privately owned by one of the processors that needs the data. This approach eliminates the need for long distance transportation between the store and the processor, because the local file can be placed close to the processor. It also eliminates the translation problem, because each file can be made compatible with its processor.

In reality, creating separate data stores doesn't eliminate these complications; it merely moves them to a different part of the model. Establishing separate, redundant data stores forces you to create clones of the custodial activity that places the data in the store. Each of the custodial clones may require the transportation and translation services previously required by the processor accessing a shared data store. So, no matter how you share stored data among processors, you create the need for both intra-processor and interprocessor infrastructural activities.

Besides providing access to stored data, the second major function of the infrastructure is to send intermediate products between processors implementing fragments of the same essential activity. First, you must select a means through which any two processors can communicate to one another. Examples are pneumatic tubes, terminals, phones, and interoffice mail. Once the means of communication is selected, you plan the activities that will allow the users to use that means. Activities like loading and unloading the pneumatic tube canisters, transferring calls, and carrying out man-machine dialogues have to be incorporated into your incarnation plan. You also provide activities that translate information from a form used by one processor into a different form acceptable to another processor. These could be activities like microfilming documents, writing down verbal information, and translating keystrokes into ASCII characters.

As you plan communication between processors, you must consider the impact of different work schedules. Ask yourself, If one processor finishes a task and is ready to communicate the results to another, is that other processor ready to receive that information? If not, you must plan batch data stores that will hold the data until the other processor is ready for it.

Figure 26.3 shows where the infrastructural activities go in the hotel system. In this example, the processors share their stored data. All of the infrastructural activities send data or intermediate products between the computer and a human processor. An important part of these activities is the translation activities that reformat the data for either humans or computer.

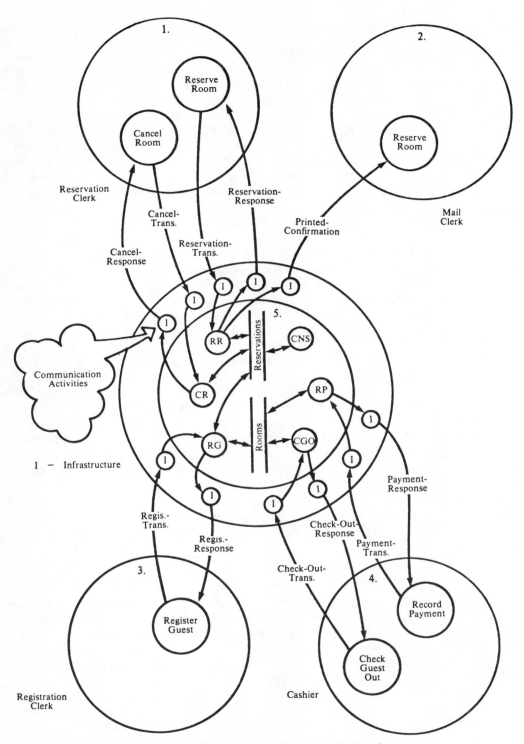

Figure 26.3. Addition of infrastructure to the hotel system.

26.2.4 Establishing the administration

Because processors make mistakes, you need to protect the new system against itself. As explained in Chapters 12 and 13, you add administrative activities to compensate for the fallibility of the processors and the fallibility of the interprocessor infrastructure. To detect errors made by each kind of component, administrative activities are of two types: intra-processor administration and interprocessor administration.

The intra-processor administration detects and possibly corrects the mistakes made by the processor itself, thus preventing the processor from sending erroneous data to another processor. To plan the intra-processor administration, you have to anticipate the errors that each processor can make. Then, you devise a way for the processor to test its own actions and to detect whether it has committed any of the anticipated errors. Finally, either you can have the processor correct its mistake, or you can direct the processor to report its malfunction to its supervisor on another administrative processor.

Where the intra-processor administration concerns itself solely with the mistakes of one processor, the interprocessor administration compensates for errors beyond the control of a single processor, such as transportation or translation failures by the infrastructure. The interprocessor administration protects each processor from erroneous data produced by another processor or from data damaged by the infrastructure. If a processor fails to detect its own errors, the interprocessor administration is a second line of defense.

To plan the interprocessor administration, you assign edits to each processor. These edits allow the processor to detect whether another processor has introduced errors into a response. You also consider designating specialized processors, like auditors, to monitor the processing. Figure 26.4 shows the effect of adding administrative activities to the hotel system. The administration is represented in that figure in two ways. First, there is a ring of edit activities (marked "E") between the infrastructure that transports the data and the essential activity fragments in the automated portion of the system that transform the data. Notice that each edit can reject incorrect data entered by the manual interface. The dataflows labeled "R" leaving the edits represent these rejected transactions. The rejects are passed back to the manual interface by the infrastructure, and then the human VDT operator corrects his or her error and re-enters the transaction. These Enter And Correct functions are also shown on Figure 26.4.

26.2.5 Optimizing the incarnation

Incarnation technology is expensive. To decrease costs, you want to make sure that you do not select an incarnation that has more capacity than you need and are willing to pay for. You look for excess capacity by examining the workload of the processors, channels, and containers built into your plans so far. If you find excess capacity, you try to consolidate activities so that they can be carried out by fewer processors, thus saving money on equipment or labor.

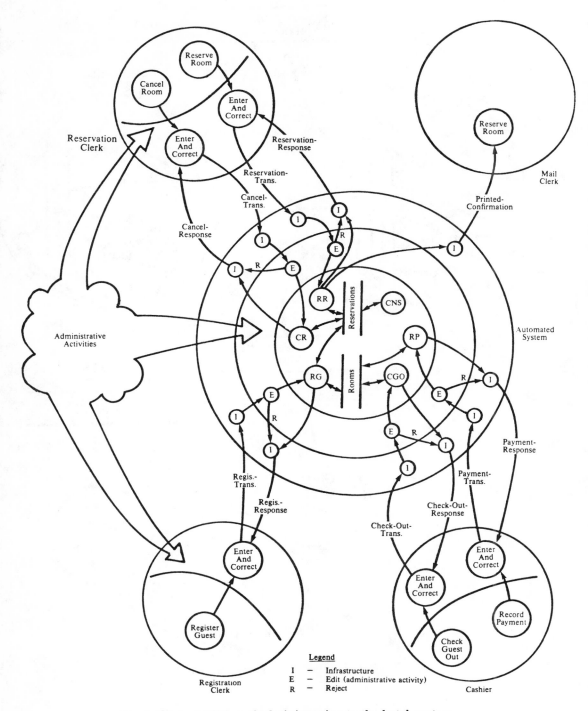

Figure 26.4. Addition of administration to the hotel system.

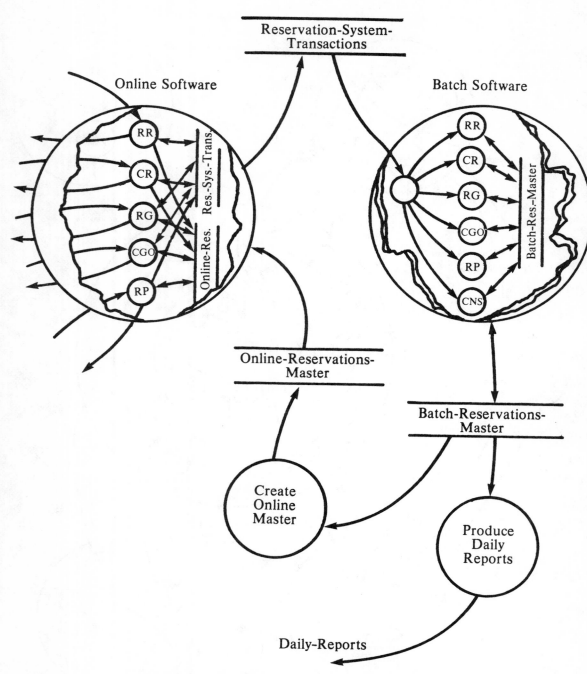

Figure 26.5. Optimization of the automated hotel system.

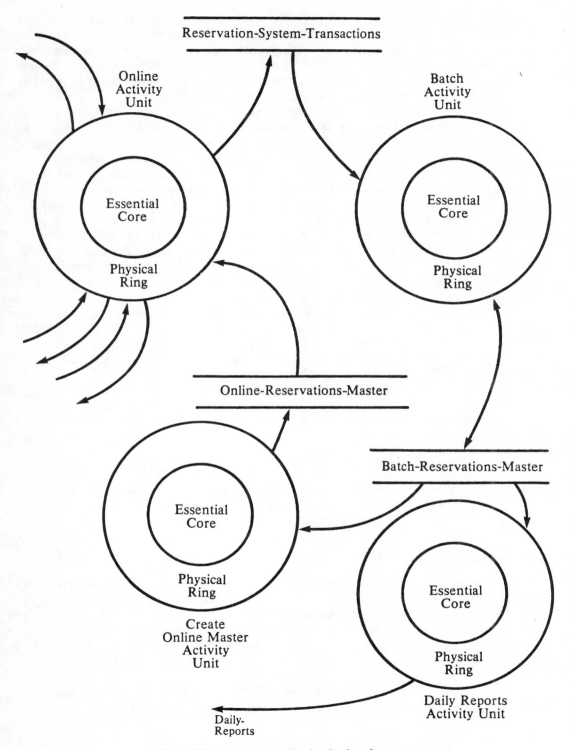

Figure 26.6. Activity units in the hotel system.

First, you examine the processors. If you find some that are underused, you can create a packed processor that carries out enough essential fragments to justify its costs. To pack a processor, you must assess both the skills and the capacity of the processor and the skills and capacity required to carry out the fragments that you'd like to assign to it. In effect, you are performing the same kind of allocation discussed in Subsection 26.2.2, but then each processor carried out only one fragment.

Once you assign the fragments to a given processor, you can optimize the processor's incoming and outgoing data channels. If all the stored data required by that processor are similar, you may choose to establish a variety show data store that consolidates the separate stores belonging to each of the fragments. You may also spot an opportunity to combine several separate input dataflows into a single packed channel.

Figure 26.5 shows the effects of optimizing the automated portion of the hotel system. The major tactic used is batch processing. Given the volume of work and the allowable response times, it was decided that online updating of the master files, Reservations and Rooms, would be too inefficient. Hence, the updating portions of each essential activity fragment are collected into a unit that will become the batch software of the system; such a unit is called an activity unit. This software will run once a day, updating the batch master file, which contains a duplicate of the essential memory originally allocated to the automated portion of the system.

Another activity unit creates the online master file from the batch master file; the online activity unit passes transactions it receives to a batch data store called Reservation-System-Transactions. All of the data stores have become variety show data stores; the dataflow from Reservation-System-Transactions to the batch software activity unit and the Daily-Reports flow are variety show dataflows.

As a result of the optimization process, your model takes on a typical appearance, as shown in Figure 26.6. In the figure, each processor is divided into several activity units; these are separated from one another by data stores. Each activity unit, in turn, can be divided into two regions:

- a core of essential activity fragments and essential memory fragments

- a surrounding ring of physical processes, dataflows, and data stores that carry out the administrative duties and communicate with the infrastructure

26.3 Summary

Your model of the system incarnation is nearly complete. After specifying the external interface of the system, you fragment its essence and allocate the fragments to processors. You also establish communication and administration facilities among and within processors. Finally, you improve the system's efficiency by packing processors, data stores, and data channels. The resulting system model consists of activity units, each of which has an essential core surrounded by administrative and infrastructural activities.

Now that we know how to build a model of the system to be implemented, we briefly look at how to use this model to create a blueprint for software design.

Chapter 27

Creating a Software Design Blueprint from the Incarnation Model

How much more work must you do on the incarnation model before you can start coding? Unfortunately, in most cases, you can't begin coding immediately because you still need to produce a blueprint or design of each program.

Since you already have the new physical model, could this serve as the software design blueprint? There appear to be three good reasons for using the new physical model for the blueprint. First, the new physical model specifies everything that the system should do. Second, it imposes no constraints on the essence of the system other than those inherent in the processors, containers, and channels that you have selected. Third, the new physical model is effective: It conveys a lot of information about the system without confusing you with details.

Given these virtues, why should you waste your most precious resource — time — by taking the modeling process any further? The reason you must continue to model the physical details of the new system is that the model at this point does not take into account several important physical features of most software environments: the details of the stored data access mechanism, the telecommunications protocol, and the programming language, among others. But one characteristic has the greatest impact on the incarnation model and must be present in the model before coding can begin; that feature is the organization of the system into hierarchical structures called programs. Organizing the system into programs is the main task of software design. In this chapter, we first describe hierarchical software and then explain how to derive a hierarchical model using a technique called transform analysis.

27.1 Designing hierarchical software

Most contemporary software environments require that individual pieces of software behave as serial hierarchies. That is, if you wish to divide a single program into modules, they must be organized in a "boss/worker" hierarchy whereby the system activates one main or boss module, which calls upon the worker subordinates to carry out activities. The boss module is inactive between the time that it calls a worker and the time that control is returned to it. A subordinate module may activate a module even lower in the hierarchy, acting as its boss. But only one module at a time can have control: A boss may not activate a second worker before the first finishes its task.

Each program consists of one large hierarchy, composed of several smaller hierarchies. System developers structure most of their applications systems in this fashion because programming languages and operating systems traditionally support only this type of software structure. Of course, there are now, and always have been, computing environments that support networks of parallel activities connected by data channels. The activities in these networks interact in much the same way as do bubbles in a data flow diagram. Co-routines are one popular version of this type of software; two co-routines can run at the same time and send data back and forth to each other. Nevertheless, boss/worker hierarchies remain the primary software structure for most systems.

This is a problem, because so far you have produced only a *network-style* dataflow model of the new system. No matter that it is a complete, unbiased, and effective model — it isn't a hierarchical model, so you can't start coding yet.

27.2 Transforming the network model into a hierarchical model

To make a software design blueprint, you must transform the new physical DFDs into a hierarchical model. Since the entire automated portion of the new physical model does not necessarily need to become hierarchical, your first step is to determine how much of the new physical model must be transformed. This determination depends upon the technological facilities, such as teleprocessors or database management systems, that are used to implement each part of the model. You may find that for some parts of the automated system, a network organization of software is possible. On the other hand, you may find that every program in the entire automated portion of the system needs a hierarchical structure. Even within the part that must be organized hierarchically, you may have to choose how many separate hierarchies will exist, and what functions will be assigned to each.

There are many ways to arrive at a hierarchical software structure, among them the design methods of Constantine, Myers, Stevens, Yourdon, Jackson, and Orr. But because you have a dataflow model of the new system, the most straightforward approach is to use a design method that derives the software hierarchy from the pattern of data flowing through the system. This leads us to structured design as defined by Yourdon and Constantine and by Page-Jones [54, 33], or to composite design as defined by Myers [31].

27.3 Applying structured design

The primary technique offered by structured design for deriving a hierarchical software structure is called transform analysis. It requires as input a very physical diagram of the software system. This diagram should be drawn so that the essential activities occur near the center, and as you move out from the center, the activities become more and more devoted to the implementation technology of the interfaces with other systems and data stores.

As discussed in Chapter 26, a diagram fitting this description is the final result of essential systems analysis. In that chapter, this arrangement of essential and physical activities is called an activity unit. A system generally has several activity units, each of which is made into a separate hierarchy.

You begin transform analysis by choosing an activity unit and identifying its transform center, which is the core of essential activities. You then create the program hierarchy by assigning parts of the essential core to be control modules, logical input modules, logical output modules, and transform modules. One control module is assigned to manage all tasks of the essential core and the physical ring. Transform modules carry out the activities in the essential core, and logical input and output modules interface the hierarchy with the physical ring of activities.

Figures 27.1 and 27.2 show an example of the results of these steps. Figure 27.1 is a physical DFD, containing an activity unit from a new payroll system; the bubble in the center is the essential core. Figure 27.2 shows the final product of transform analysis: a structure chart, the primary tool of structured design. In Figure 27.2, Pay Hourly Workers is a control module in charge of all activities in the activity unit. The long arrows connecting Pay Hourly Workers to the line of modules below it represent the flow of control between the boss module and its subordinate modules. The open tailed arrows represent flows of data between modules.

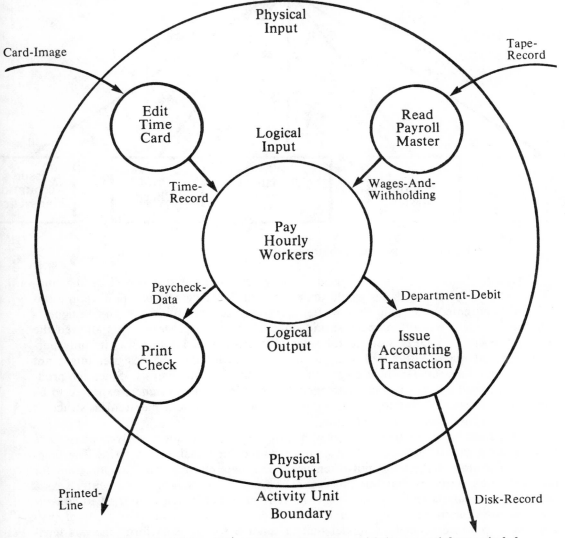

Figure 27.1. Activity unit from a payroll system, with its essential core circled.

Get Time Record and Get Payroll Master are two logical input modules, and Print Check and Issue Accounting Transaction are logical output modules. Calculate Pay is a transform module; it implements the essential core of the activity unit. Figure 27.2 shows only a portion of the final design blueprint, for it would also have data dictionary definitions and module specifications similar to minispecs.

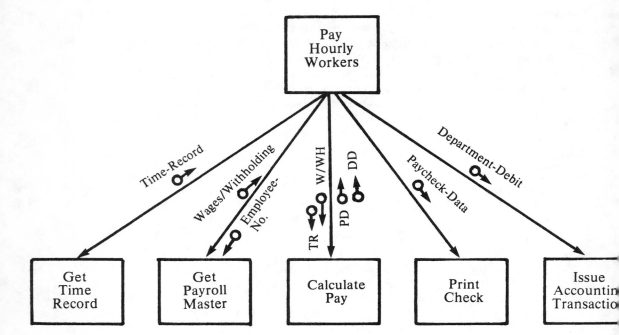

Figure 27.2. Structure chart resulting from transform analysis.

Transform analysis has been used to design hierarchical software since the mid-1970s. Originally, the designer would draw the data flow diagram, basing it upon a traditional requirements specification written primarily in narrative. So, the designer's data flow diagram was often the first and only DFD in the process. This allowed the designer to create a DFD suitable for transform analysis, since that was its only purpose. The advent of structured analysis, however, meant that the designer might not have to build a special data flow diagram just to use transform analysis. Since the product of structured analysis is a dataflow model of requirements, designers expected to be able to use it as the input to transform analysis, and to derive a hierarchical structure directly from the structured specification.

This method didn't work out so well, primarily because most analysis-phase data flow diagrams didn't look like the diagrams required for transform analysis. The practice of structured analysis did not often produce a new physical data flow diagram for the automated activities that was physical at the edges and logical at the center. As a result, many analysts and designers experienced frustration when they tried to convert the new physical model into a hierarchical model. But now you have a set of procedures to produce a dataflow model that is suitable for the transform analysis technique. The model is divided into activity units, each of which can be converted easily into a separate hierarchy of modules.

Therefore, if you develop new physical models using essential systems analysis, you should have no trouble making the transition from modeling requirements to designing the system's new incarnation. Moreover, since the need to distinguish between the logical and physical aspects of a software system is not unique to structured design, the new physical model makes an excellent starting point for other design methods as well.

27.4 Summary

Since most software environments won't let you code directly from the new physical model, you need to build a software design blueprint first. This blueprint organizes the model into programs, each of which is a hierarchy of modules that pass control up and down to each other.

There are many ways to derive a hierarchical software structure, but we recommend transform analysis from structured design. Transform analysis converts a data flow diagram into a hierarchical model called a structure chart. To use transform analysis, you need a data flow diagram made up of sections that consist of a logical core surrounded by physical interface activities. This description matches the activity units produced by our derivation process. In other words, your new physical model is easily converted into a hierarchical structure.

Part Eight

Managing Essential Systems Analysis

In this final part, we take the manager's perspective as we consider the practicality and implications of applying our techniques in the real world of system development.

Chapter 28 focuses on time as the project manager's key resource, and discusses why so much time is wasted in current development projects. Since a manager's first priority is to expedite the conduct of systems analysis, design, and implementation, while preserving the quality of the products, our goal in Part Eight is to show you how to use time efficiently. Specifically, we show you how to survive real-life projects by developing system models quickly and efficiently. Chapter 29 presents an expediting technique called blitzing, whose goal is to develop very rapidly an initial model of the existing system. Although incomplete, this model provides a useful overview of the system under study. We provide a procedure for blitzing an essential model by rearranging the steps of the discovery procedure presented in Parts Four, Five, and Six.

Chapter 30 offers another tactic for making the best use of your project's time, called the leveled approach. By performing the essential system development activities at a general level first, you learn which portions of a system deserve detailed analysis before investing the time required to model the entire system. Chapter 31 shows how to use both blitzing and the leveled approach to protect your project from threatening external forces, such as company politics and management impatience. By planning to deliver at least a working portion of the new system within two years, you minimize your project's exposure to these forces and thereby improve your chances for success.

Chapter 28

Time:
The Critical
Development Resource

Although modeling the new system's essence and modeling its incarnation are two essential system development activities, in discussing activities we did not describe only their essence. That is, we did not completely ignore the effects of the system development technology. We assumed that the activities are carried out by human beings and, therefore, we took into account whatever shortcomings are shared by all human system developers. So, our procedures are designed for use by creatures who have limited conceptual capacity and limited, short-term memory, and who make mistakes. In short, we tried to account for those project traits that transcend any one system development effort.

Many additional constraints for different projects can have a significant effect on development activities, including the size of the development team, the distribution of skills among the developers, the size and complexity of the system to be developed, and the amount of time allocated to the development effort. How well you deal with the constraints affecting your development effort determines to a great extent the project's success and even its survival.

Many issues are vital to managing a project to a successful completion: organizing the development team, planning and controlling quality assurance activities, estimating the resources necessary for development tasks, and many others. However, one project constraint is more significant than any of these issues, and that is calendar time. Only so much time is given you to finish a project, and you don't dare go too far beyond that deadline. In this chapter, we discuss why time is the most critical resource and how time is wasted in many projects.

28.1 Why time is critical

As Frederick Brooks wrote in *The Mythical Man-Month* [2], "More software projects have gone awry for lack of calendar time than for all other causes combined." Calendar time is one of two measures of time as it relates to a system development effort; it is the simple count of work days, months, or years from the beginning of the project to its deadline. The other major measure is project time: the people-hours, -days, -months, or -years needed to complete the project. High levels of project management are usually concerned with calendar time. Those managers feel that within reason the number of people-hours allocated to the project can be changed, but the date of delivery is fixed. In the next three chapters, we suggest an approach that will enable you to complete your project on time when management believes that a fixed deadline can be preserved while the number of people-hours is varied.

Brooks made his statement long before structured analysis came into existence, but it is no less true today. From our experience as consultants, we conclude that lack of calendar time is the *only* reason that structured analysis efforts fail. Unfortunately, not every structured analysis project has been a smashing success, and the failures have all been remarkably similar: Sometime during the study of the existing system, particularly late in the current physical phase, the sponsors of the project run out of patience. They complain that no real progress is being made, even though they've been applauding the intermediate products all along, often declaring that for the first time they have a useful understanding of their own business policies. Quite often it is only a single group of users who feel this sudden, vehement dissatisfaction, possibly because the group was recently approached by a rival data processing group, who offer to develop a system "just for them" in far less time than the single, corporate megasystem development effort is taking. But whether the discontent comes from user management or data processing management, the project is either canceled or disintegrated into smaller, autonomous efforts. These efforts cease to operate under a single executive and shift from serving multiple users to serving a single user organization.

The failure of structured analysis efforts rarely if ever has anything to do with the quality of the specification products. Project managers aren't fired because their data flow diagrams are badly out of balance, or because the data dictionary definitions are incomplete, or because they cannot produce one minispecification for each bottom-level process. More important, projects aren't canceled because the specification does not reflect the users' true policies and system requirements. No, the single most effective way to lose your development project is simply to take longer to produce a working system than your sponsors are willing to wait.

How should this phenomenon affect the conduct of structured analysis? Quite simply, you should make every effort to complete the analysis phase as quickly as possible without sacrificing the quality of the finished product. You must build your specification quickly or risk creating political pressure that will crush the project. So that you may eliminate the sources of wasted time, we next consider the factors that contribute to the length of many structured analysis efforts.

28.2 How project time is wasted

Obviously, project time goes to talking with users, reviewing existing documentation, drawing DFDs, conducting walkthroughs, redrawing DFDs, and especially creating and maintaining data dictionary definitions and minispecs. Knowing that these activities are performed, however, does not tell you how to reduce the amount of time spent on them. You need to know how practitioners of structured analysis waste time, time that might easily be reclaimed through better management of the process itself.

Analysts may waste time by doing the right things slowly, as is usually the result of inexperience with structured analysis. Any time you develop new skills, you must pay the price of lost time. Since there is not much you can do about minimizing the effects of this learning curve, you have to expect the first project that uses structured analysis to progress more slowly.

Analysts may also waste time by doing the right things with the wrong tools. For example, you could keep all ten thousand data dictionary definitions on index cards. This might not be wrong, but it would almost certainly waste time. There are readily available tools, everything from simple text editors to nuclear-powered data dictionary packages, that can do everything index cards can do plus a lot more, and do it much more efficiently. Clearly, you want to use the most efficient tool for the task at hand.

Neither the use of manual documentation techniques nor the learning curve accounts for the bulk of time lost during structured analysis efforts. The most time is lost on performing activities that do not help to achieve the objectives of the project. These activities should not be done at all, or they should be done to a much lesser extent. There are two main reasons that analysts squander time on thoroughly unnecessary activities: They fail to understand the structured analysis process, or they fail to grasp the purpose of the system under study.

When project team members do not truly understand the method they are using, they tend to do an activity or not do it because of what they perceive is a rigid rule, regardless of whether the activity offers any benefit to their project. The most common gap in their understanding of structured analysis is in knowing how to build the intermediate products — the current physical, current logical, new logical, and new physical models. Because earlier books on structured analysis recommend these products, project teams almost always try to build all of them in great detail. But for some projects, it is silly to build a current physical model or a new logical model, for example. Even when the intermediate product is useful, the project teams perform many unnecessary activities to develop it.

Projects also suffer when team members do not fully understand the system they are modeling, as when there is a gap between when they begin to work on a system and when they realize its full purpose. During this period (which in the worst case can last well beyond the analysis phase), team members can end up studying portions of the system that are not important to the success of the project. At the very least, they may find themselves producing overly detailed documentation for parts of the existing system that will not help them specify the new system.

28.2.1 The current physical tarpit

The single most common and most deadly result of failing to understand the method or the system is what we call the current physical tarpit. It is the reason that unsuccessful structured analysis projects usually fail sometime during the study of the existing system: Members of these projects nearly always take an unacceptable length of time to study the system, because they put far more physical detail into the current physical model than is needed to develop a new system.

You always need to build some form of current physical model, no matter how scarce project time is. Without one, you cannot obtain and verify an understanding of the existing system. You also need some information about the current physical system in order to develop an essential model of the system. The trick is to spend as little time as possible developing the current physical model, in order to avoid the current physical tarpit. You need to know what information to include in that model and, more important, what information to leave out.

When developing the current physical model, any time you spend modeling physical features could be wasted. After all, the physical characteristics built into the model will just be removed later, when you derive the current logical model. Although some of these physical characteristics are of value in finding the essence of the system, most are not, and the greater part of the current technology is probably about to become obsolete. You don't want to add to the already lengthy project time by researching, modeling, and verifying features that you don't need in the search for true requirements. For these reasons, we recommend that in your initial models of the existing system you include only physical information that is useful toward finding the essence of the system.

Figure 28.1 indicates the contents of the typical current physical model. The lowest segment of the bar represents essential information, which you definitely want in a current physical model, and the middle two segments represent physical information that is acceptable though not desirable in a current physical model. Some physical content is needed to provide landmarks for you and your user, and other physical characteristics sneak in because you can't see they are physical. The extraneous physical information in the top segment of the bar is unacceptable in a current physical model, because it isn't needed to orient you or your users. You should be able to spot this extraneous information using the framework for physical features provided in Part Three.

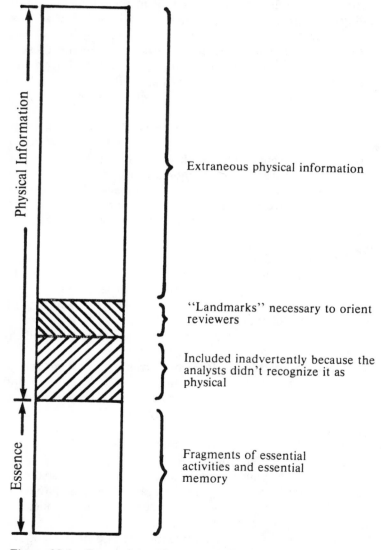

Figure 28.1. Composition of the typical current physical model.

Extraneous physical information makes up more than half of the typical current physical model. Logicalization usually yields models that are two thirds to three fourths smaller than the current physical models they were derived from. In other words, up to 75 percent of the information in a typical current physical model is deleted during the derivation of essence. In one or two extreme cases that we have encountered, the proportion was in excess of 80 percent.

If you have to delete a large portion of an intermediate product, developing much of that product was a waste of time. Suppose you have just finished deriving a current logical model of a large commercial application from a current physical model that contained 500 bottom-level processes (minispecifications). Your logical model contains only 150 minispecs: You have, in effect, discarded 350 minispecs, together with the data flow diagrams and data dictionary definitions that supported them. Some of the physical information they contained may have helped you communicate with your users, and a few physical characteristics escaped detection through no fault of your own. Nevertheless, it was certainly not necessary to spend several thousand hours developing *all* of those 350 minispecs in order to arrive at a good logical model.

All in all, the development of some form of current physical model is a regrettably necessary detour in the production of the current logical model, and the more you put into the current physical model, the longer the detour. Figure 28.2 illustrates the effect of adding even more detail to the current physical model. The two paths drawn in Figure 28.2 represent two approaches to building a current physical model: The top path shows what happens when a great amount of physical detail is modeled in the current physical model, and the bottom path shows that the same essential model can be derived using a current physical model that contains much less physical detail. In each case, the inclusion of any physical detail forces a detour away from the shortest path to the current essential model, which is represented by the horizontal scale in Figure 28.2.

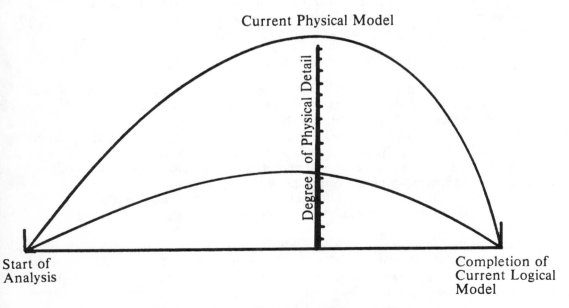

Figure 28.2. Detour resulting from the current physical tarpit.

Why do so many projects end up in the current physical tarpit? One reason is that it is easier to document the physical features of a system than it is to abstract the essential features of the existing system. After all, you have hard evidence of a physical feature. The essential features are often intangible.

The second reason is that until now analysts have had few procedures to help them derive the essence of an existing system. Because the project members do not understand how to transform the current physical model into the new physical model and eventually into the new system itself, they are not sure what kind and how much of all possible information ought to be included in the current physical model. So, in a spirit of caution, they include anything and everything, even ad hoc responses and irrelevant technological features. They may spend yet more time trying to develop their own techniques for deriving essential models. Because they lack the guidelines of a formal procedure, they spend precious hours debating whether an aspect of the current system is logical or physical or wondering whether they have obtained enough current physical information to support the logicalization process. Team members may become so confused that they even waste time debating the usefulness of logicalizing the system at all.

Another reason projects take so long is that team members do not grasp the purpose of the new system, and therefore don't know which portions contain the most important information. As a result, they spend as much effort modeling trivial portions as they do important ones. The result is a gigantic current physical model that takes much longer than it should to build, to review, to maintain, and to reduce to its logical equivalent.

Many of these projects eventually escape from the tarpit. Some do so in fine form, the project team members emerging with a good concept of a logical model and the procedure to get one. Other teams can only keep their projects alive by arbitrarily ending the current physical work and rushing directly to the new physical work, without any search for true requirements. In some unfortunate cases, the tremendous amount of time spent developing the current physical and logical models exasperates management so much that it cancels or reorganizes the project.

28.3 Summary

Different constraints affect each system development project, and how you deal with the constraints can determine whether your project survives or not. Time is the most critical constraint and so accounts for the majority of project failures.

The most deadly way to waste time is to spend too long modeling the current physical system. Although some form of current physical model is needed, most projects include far more physical detail than necessary in that model. This extraneous detail is simply eliminated during the derivation of the logical model and it isn't needed to orient reviewers. You can avoid the current physical tarpit by following procedures that tell you what information to include in the model and what parts of the system to concentrate on.

The rest of Part Eight describes a few techniques for improving the efficiency of your structured analysis efforts, and thereby avoiding disaster. Most of these techniques reduce the time wasted performing useless activities, since you naturally don't want to spend time doing something that does not move you closer to your ultimate goal.

Chapter 29

Optimizing the Process of Defining Essence

Although scarce project time must be used effectively on all system development activities, it is most critical to use it effectively at the beginning of the project. Most projects begin with a study of some existing system. If you don't carry out this study efficiently, you risk not having time to finish the rest of the system development tasks properly. You may also create the perception that your project is foundering.

In this chapter, we describe a technique that eliminates the need for building most of the current physical model. Like the procedure for deriving essence, this technique uses the same modeling principles and conventions, since it, too, is based upon the framework described in Parts One through Three. Before presenting the technique, though, we first want to review the components of the framework and show how your understanding of them can contribute to keeping a project on schedule, even without using special techniques.

29.1 How the essential modeling framework saves time

As the first component of the framework, the definition of a planned response system provides the project team with an understanding of what kind of system it can build. The project members are able to separate the ad hoc responses from the planned responses of the system being studied; they can then concentrate on the planned responses. Since time is not spent trying to describe something that cannot be built into the system anyway, the project can be much more efficient.

Equipped with the description of incarnation characteristics given in Part Three, the project team members are able to quickly understand the purpose of components in the existing system. They can identify the purely physical aspects of the system, and know not to waste time modeling them. The project avoids spending an inappropriate amount of time studying either the infrastructure or the administration of the system.

Project members can use the concept of perfect internal technology as a kind of divining rod that points out fragments of essence in the existing system. By asking themselves what parts of the current system would still exist if the system had perfect technology and by using their knowledge of how imperfect technology camouflages these essential fragments, project members are able to direct the study toward areas of the system where more of the essence is to be found.

Project members know what kinds of components make up the system essence. This knowledge helps them locate the essence when they are building the essential model of the existing system. They know to look for fundamental activities that serve to accomplish the system's purpose, the essential memory that supports the fundamental activities, and the custodial activities that establish and maintain essential memory. Because they know how to recognize the essential components, these project members avoid time-consuming dead ends.

Finally, some of the modeling principles and conventions discussed in Part Two help project teams save time. By applying the modeling principles discussed in Chapter 6, team members are able to make fast decisions about how to make their models effective. The essential model conventions discussed in Chapters 7, 8, and 9 eliminate the need for project members to take time to establish their own conventions.

29.2 Deviating from the idealized derivation approach

The idealized approach to deriving essence given in Parts Four, Five, and Six ignores most managerial constraints, including the scarcity of project time. Paradoxically, even though this approach ignores time constraints, it still helps you when you are working against the clock to derive essence. The activities and their sequencing in the idealized approach keep teams moving steadily toward the completion of their project. Moreover, the detailed procedure for discovering the existing system's essence ends a lot of confusion about what the project should do to produce the current essential model. Projects are therefore able to avoid wasting time performing unnecessary tasks or performing necessary tasks incorrectly.

Despite its improvement over classical structured analysis, the detailed discovery process has two flaws that hinder you from using time effectively. First, the process is designed to pull projects out of the current physical tarpit, not to help them avoid it altogether. To apply the discovery process, you already have to have built a current physical model. The discovery process gives you no way to focus on what's important when modeling the current physical system, so you still risk wasting time on trivial details. Second, the discovery process is designed for clarity of presentation, not for speed. In fact, the time problem is eliminated by our assuming a mostly ideal system development technology.

In the procedure, the steps are organized in a logical sequence; one step is performed at a time, and it is completed before another is begun. While the order of steps is inefficient, it does make sense. You not only learn the procedure, but you also can understand why things are done as they are. Because the idealized discovery procedure proceeds so carefully and methodically, we also call it the conservative approach.

The approach we use to expedite the current system study may seem reckless by comparison, yet it follows naturally from all we have said so far about essential models and their derivation. There are no new objectives: You are still committed to deriving the true requirements for the new system. There are no new tools, conventions, or principles. The approach doesn't really have any new activities either. It uses the same activities discussed in Parts Four, Five, and Six, but carries out certain of them in a different order and usually to a different level of detail.

29.3 The optimized procedure: blitzing an essential model

Like the conservative derivation process, the optimized procedure requires two passes to produce an essential model. In the first pass, you "blitz" an essential model. By *blitzing,* we mean quickly producing an incomplete, unverified, high-level essential model. The blitzed essential model is for the most part a set of event-partitioned DFDs and object-partitioned entity-relationship diagrams. The model does not address the entire scope or depth of the system. You make no serious attempt to identify all of the system's interactions with the outside world, and you do not model any one interaction in great detail. The blitzed data flow diagrams and entity-relationship diagrams are usually not more than three levels deep, and you rarely produce data dictionary definitions or minispecifications. A blitz usually takes a week or less. In the second pass of the optimized procedure, you obtain details about the essence and verify them. The final result is the same fully detailed essential model you would achieve by using the conservative derivation process.

Blitzing saves time by providing a preview of the final essential model. Because you have a picture of the whole, you can far more easily pick out the sections of the system that will require the most attention when modeling the details. The most obvious time-saving benefit of being able to focus your attention is that you can skip over most of the current physical modeling; then, to finish the essential model, you can devote most of your efforts to important areas and ignore the irrelevant ones.

There are two ways to perform a blitz. Using the conservative discovery process, you could build a physical model, expand that model, then reduce and classify the remains; this would give you a rough model of each activity that you could use as a high-level logical model. To achieve a rough model of the system's essential memory, you could do all of the necessary steps to obtain the rough activity models, to refine the local models for each essential activity, and to combine the local views of essential memory into one essential memory model, but you would do all these steps as fast as possible.

A second possible approach, and the one we prefer, is to produce a blitzed essential model by applying only the techniques of event partitioning and object partitioning. Then, you return to do the preliminary steps once the blitz is over.

Does this strategy make you worry that doing the partitioning first means postponing many steps and possibly causing you to make a serious mistake? Fortunately, there are two reasons that it is safe to do event and object partitioning first in order to get quick results. First, event and object partitioning rely extensively on your understanding the environment around the system, but not on your understanding the essential fragments inside the system. So, you don't have to study the current system itself before performing event and object partitioning. Second, since information presented in Parts One through Three allows you to distinguish the essential fragments in the existing system from its physical characteristics, your blitzed model is unlikely to contain the wholesale physical content that the early stages of the discovery process are intended to eliminate.

It is safe to ignore preliminary steps only when you are producing a blitzed essential model, since such a model is by definition high-level and incomplete. It is decidedly unsafe to produce only a blitzed model and no detailed model afterward because the detailed essential features will not have been properly established. Thus, most of the same tasks must be carried out whether you use a blitzing approach or the conservative discovery approach.

The procedure for blitzing an essential model consists of four steps:

1. Establish the system's purpose.

2. Create an essential context diagram.

3. Create lists of objects and events.

4. Draft the preliminary essential model.

In the following subsections, we discuss each of these steps in detail.

29.3.1 Establishing the system's purpose

The first step in the process of blitzing an essential model is to establish a simple statement of the new system's purpose or, in analyst's terms, to summarize the fundamental activities of the system. Although this statement is often difficult to express formally, understanding the purpose or at least making the attempt to state it is crucial to pinpointing the activities that the system will carry out.

Since you are using the current system to derive the essence of the new one, the purpose of both must be the same. The developers of the current system probably had only an informal concept of its purpose in mind that has been passed informally to each group of users. Now, you must create a formal statement of this purpose through interviews with this user group.

The materials supply system in Chapter 17 is a good example for illustrating the steps of the blitzing process. In the first step, we determine the system's purpose is to maintain an inventory of raw materials sufficient to support a certain level of manufacturing. This statement of the purpose provides us with a good overall understanding of the system at a low cost in time and resources, and it also serves a crucial role in the next step.

29.3.2 Creating an essential context diagram

Using this statement of the system's purpose, you now can create an essential context diagram. This second step of the blitzing procedure requires you to look at the incarnation of the system for the first time. You are interested only in establishing a boundary around the parts of the incarnation that serve the purpose of the system, and thus the boundary is drawn so that the external events are outside, while all of the activities that respond to those events are inside. You model the boundary using the context diagram.

This context diagram must be of a special sort. Because it represents the essence, its dataflows should not reflect the technology that presents them to the system or transports them to the receivers of responses. So, input dataflows such as Mail-Bag, Inter-Office-Envelope, and Pneumatic-Tube are inappropriate, as are the output dataflows Computer-Report, Device-Bus, and Inquiry-Screen. You want input and output names that reflect the essence of either the event that gives rise to the input flow or the response that gives rise to the output flow.

The essential context diagram for the materials supply system is shown in Figure 29.1. The presence of only three external entities indicates that we are using a very simple system to illustrate blitzing, although it is a technique that is most useful when applied to large and complex systems.

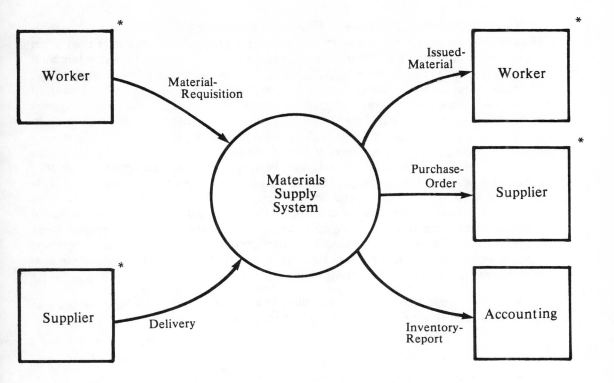

Figure 29.1. Essential context diagram for materials supply system.

29.3.3 Creating lists of objects and events

Your objective in the third step is to derive the raw material for constructing the blitzed essential model. That raw material consists of the events that the system responds to and the objects that make up the system's essential memory.

You will probably identify the external events while creating the essential context diagram. This process is very similar to the one used to identify external events and discussed in Chapter 17 as part of the discovery process. When you want to find the temporal events, however, you must break with the procedure we used earlier. The discovery procedure recommended that you find temporal events by examining the minispecs of essential activity fragments that "hang" directly off data stores and that are not triggered by external dataflows. You can't do that now simply because you have no reduced data flow diagram that shows these fragments. You have to use a more indirect approach.

The approach depends upon the assistance of a user who knows about the system and understands the guideline of perfect internal technology. In an informal discussion of the essential context diagram with the user, you examine each essential system output to see if it results from an external event on the list. Once you have found all such outputs, you assume that the rest of the essential outputs are produced by temporal events. The user identifies these temporal events, and you add them to your list.

The biggest difficulty in matching essential outputs to external events is that imperfect technology may force the system to produce an output at a time later than when the external event occurs. This may cause you to suspect mistakenly that a temporal event is responsible for the output. A batch file in some cases may hold partial output from many occurrences of the external event so that the final output can be produced efficiently all in one run. For example, in an electric utility's billing system, the system may batch meter readings from many customers and then generate all the bills at once. You expect a separate response to each occurrence of an external event, but this system produces an output that is a response to many occurrences of the event "meter reader submits reading." Confused by the system's behavior, you may be unable to identify the external event and incorrectly conclude that the output is the response to a temporal event, such as "time to bill customers."

To avoid this mistake, you ask the user if each output could be many separate outputs, each one the result of a single external event if the system had perfect technology. Now, you may have to use different terminology to avoid confusing the user, but we've done it and so can you. For example, you find the following facts about system responses in the materials supply system:

- The issued material can come directly from the response to any material requisition if the system has enough inventory.

- The purchase order also can come directly from the response to any material requisition.

- The inventory report cannot come from either a material requisition or a delivery. (At least in this system, the recipients of the report want it to cover a period of time during which inventory adjustments occur. They don't want a report that documents each change in inventory.)

From the results of your interview, you connect the outputs Issued-Material and Purchase-Order to the event "worker submits material requisition." Any responses that are based upon the occurrence of many different events, like the Inventory-Report, should be outputs from temporal events. If you can't find the event that generates a given response, you have to do some more work. You must either find a connection to one of the events you've already identified, or you must find an event that you missed before.

You may be worried that you will make a mistake in trying to identify these temporal events, and you almost certainly will. However, you suspend for just a bit your primary emphasis on correctness for the sake of efficiency. Once you have blitzed the essential model and taken stock of how you will proceed, you may then decide to do the detailed analysis necessary to validate your choice of events.

You use a different approach to identify objects during a blitz from that used in the conservative approach. In the conservative approach, you first identify the data elements needed by an essential activity. You then identify objects by forming groups of data elements from this list. That approach is theoretically appropriate: Since objects are a means to group data elements, it makes sense that you should know the data elements before choosing the groupings. However, you may also recall that producing the list of data elements involved a lengthy procedure; therefore, you need a different approach.

The approach taken to identify objects during blitzing is really the reverse of how you identify objects when deriving an essential activity. When blitzing, you identify the object first and then spend a short time looking for stored data elements to attribute to that object. If you can't find any such data elements, you eliminate the object from your list. In general, potential objects are selected from the vocabulary of nouns that everyone uses to describe the system's technologically neutral features.

Since that vocabulary can be pretty extensive, to help you focus your object identification efforts, we recommend you examine the sources of objects given in Chapter 20. These are external entities that interact with your system, contractual relationships between your system and another, inputs and outputs to your system, and products, services, or other resources that are part of the system or managed by it.

29.3.4 Drafting the preliminary essential model

After you complete the above steps, the drafting of the blitzed essential model should be easy. Be sure to apply the essential modeling principles and conventions presented in Chapters 6 and 21 and especially in Part Two and your model should turn out fine. Here's a brief review of the basic steps:

1. Create a DFD that shows the essential activities. To do this, you draft an activity for each event by drawing a bubble for the activity and showing the activity's inputs and outputs, which should match those that you identified on the context diagram.

2. Place object data stores in the DFD. Each object should correspond to one data store.

3. Connect activities to data stores. (Use common sense to decide what stored data an essential activity needs.)

4. Draft objects onto a preliminary entity-relationship diagram.

5. Connect objects on the entity-relationship diagram by applying the information gleaned in user interviews and your own judgment.

Figures 29.2 and 29.3 show the results of these steps when applied to the materials supply example.

Using scarce project time effectively requires the quick creation of high-level models. By taking the shortcut process just described, you avoid the current physical tarpit and provide useful information for project planning. However, you must remember that the resulting essential model is both high level and incomplete and that sooner or later you almost certainly will have to establish its supporting details. In the next section, we discuss one possible approach to find the details of the current system essence and still make effective use of project time.

29.4 Mopping up

Although the blitzed model has many beneficial uses, it also has the fundamental problems of being incomplete and possibly inaccurate. If we stay with the military terminology used to describe effective project management, we can call the activity of completing the model mopping up.

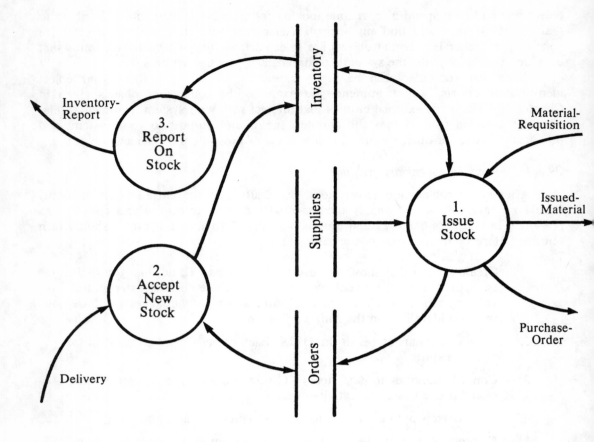

Figure 29.2. Blitzed essential data flow diagram for the materials supply system.

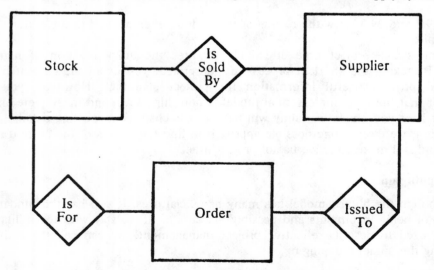

Figure 29.3. Blitzed essential entity-relationship diagram for
the materials supply system.

Like blitzing the high-level current logical model, the mopping up process is based on concepts used in the conservative approach. Not only must the final model be complete and correct, but it also must abide by the modeling principles of technological neutrality, budget of complexity, and minimal modeling. You also use many of the same techniques to expand and reduce the model, classify fragments, model individual essential activities, and integrate essential activities. An important difference from the conservative approach is that in mopping up you want to finish the detailed levels of the current essential model quickly.

Mopping up consists of these major steps:

1. Blitz a high-level current physical model.

2. Create a modeling plan telling the order in which you will model the essential activities.

3. Build individual essential activity models.

4. Integrate the essential activity models.

We discuss each of these steps in detail in the following subsections.

29.4.1 Blitzing a current physical model

Now that you are interested in the details of the existing system's essence, you find yourself in a familiar bind: You want to avoid current physical modeling, yet you will need to do some of it in order to build the essence of the system and verify your understanding of it. In blitzing the current physical model, your objective is to survey the availability of detailed information about essential features and to build an index that will guide you to the essential details.

Your first step is to create a high-level current physical DFD, which you base on the partitioning of processor boundaries by drawing a bubble for each process or group of processors. In making that high-level DFD, your major task is to connect the processors or superprocessors on the diagram. Figure 29.4 shows a typical blitzed current physical DFD.

Instead of producing minispecs, data dictionary definitions, and lower-level DFDs to support this physical model as before, you gather any existing documentation and key it to the appropriate bubble or dataflow in the high-level physical model. To replace minispecs, you can use policy manuals, run books, project request forms, procedure manuals, and even dreaded source code. Photocopies of forms, automated file definitions, and database data dictionary output can serve as data dictionary definitions. Notes identifying people to interview about particular areas of the system can be linked to the high-level DFDs. When you are ready to model the essence of a certain part of the system, you look at the high-level, current physical model to find out what documentation would contain the essence of that part. The model will also be useful later for reviewing essential features with physically minded users, analysts, or managers.

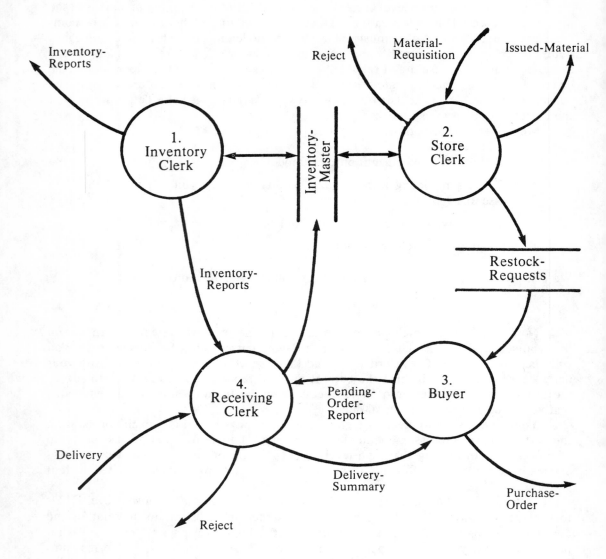

Figure 29.4. Blitzed current physical data flow diagram.

29.4.2 *Establishing an essential modeling plan*

Once you have a high-level current physical model, you develop a plan for researching the details that you need in order to write the minispecs and data dictionary entries for the detailed *essential* model. In effect, this plan tells you in what order to model the individual essential activities. To construct this plan, you obtain the essential activities from the high-level essential model, and you employ one of the basic concepts of system essence to choose the order in which you will model them. This concept is the chain of consequences that links the components of essence: The system's purpose determines the fundamental activities of the system, the fundamental activities determine the essential memory of the system, and the essential memory determines the custodial activities of the system. Once you know the fundamental activities, it should be relatively simple to establish essential memory custodial activity details.

Therefore, you want to model the details of the fundamental activities first. The planning of the detailed work of specifying the essence of the system begins with discovering which activities declared on the high-level essential DFD are fundamental activities. Either of two characteristics identifies a fundamental activity. Any activity that is triggered by a temporal event is a fundamental activity.* An activity is also fundamental if it is triggered by an external event and it produces a response that does more than update essential memory.

Once you have isolated the fundamental activities, you perform a rough analysis to establish the complexity of the essential memory accesses in the activity. You want to list the essential activities according to the complexity of these essential accesses. Temporally triggered activities often head this list; their essential activities tend to be complicated, because they usually have to find the essential memory they need without the help of identifiers provided by external input. To obtain a detailed estimate of the essential access complexity of any activity, first determine the number of data elements in the response that must come from essential memory, and then count the number of object data stores those data elements would have to be obtained from. Naturally, the more data stores accessed, the more complicated the activity.

Once you have a list of fundamental activities and the estimate of the complexity of their essential accesses, you create the plan for further detailed work. The plan guides you in performing the next step, which is to build a model of each individual essential activity. It is most effective to start by modeling the fundamental activity that has the most complex essential memory accesses and work your way to those with simpler accesses. By working this way, you can usually verify large parts of the initial stored data model early. Having much of the stored data model will help you create the details of the remaining essential activities.

29.4.3 Building individual essential activity models

Building a local essential activity model now is very different from the steps in the earlier discovery procedure, primarily because you do not have a global reduced model. Therefore, you do not have for each event a collection of essential fragments that constitutes the response to that event. Finding that set of essential fragments is the first step in the fast approach to building an essential activity model.

Starting with the activity at the top of your list, you prepare a reduced physical model for this activity alone. To do this, you trace the activity's inputs and outputs through the existing system. The high-level current physical model will tell you what documentation gives the procedures for carrying out the activity, so that you can identify the activity's components as you trace the dataflows. The result of this trace is a lower-level DFD showing the essential fragments that carry out the activity in the existing system. As you identify the components of the activity, be careful to avoid documenting the infrastructure and administration. Use what you previously learned about these basic components of the incarnation to recognize and avoid them. By so doing, you will end up with a set of activity fragments that is mostly essential. You will have

---

*In the two years since we first made this statement, we have found only two exceptions to this rule. Both of the exceptions involved custodial activities that were triggered by time.

to verify that your essential activity fragments are accurate and complete. Do this by presenting them to your users and soliciting their comments. You may have to separate comments on *physical* inaccuracies from those that uncover your errors in defining the system's essence.

Once you have produced and verified the reduced physical model for this essential activity, you apply the local essential modeling techniques discussed in Chapters 20 and 21. This activity proceeds almost as it does in the conservative discovery procedure. However, as you produce any one essential activity model, you will be noticeably less strict about ignoring what you know about other local essential models. For instance, once you know the objects for one essential activity, you would check immediately to see if they might be useful in modeling another essential activity. This will reduce the effort required to integrate the essential activity models, and may therefore speed up the modeling process.

29.4.4 Integrating the essential activity models and performing the final pass

Once you have a collection of essential activity models for the fundamental activities, you must integrate them in the same way as in the conservative discovery process, discussed in Chapters 22 and 23.

Having produced an integrated current essential model of the fundamental activities of the system, you now make another pass to incorporate the custodial activities into the model. Again, you create a ranked list of the custodial activities, but this time you are most interested in looking at activities that create object occurrences, as opposed to those that merely update existing fields. For example, a custodial activity that creates a new subscriber in a magazine subscriber file is more interesting to study than one that changes the subscriber's mailing address.

Once you have an ordered list, you do all of the same steps that you performed to model the fundamental activities: Create a set of essential fragments for each activity, build a local essential model based on those fragments, and integrate the individual custodial activity models with the already integrated fundamental activities.

29.5 Summary

When you apply our techniques to a project, you will probably have to develop some variations to suit your own environment. Blitzing high-level and detailed current essential models is one such variation to derive the essence of a current system quickly. Both blitzing and the subsequent mopping up are nothing more than a rearrangement of the steps in our discovery procedure. Although the rearranged procedures lack the precision and rigor of the conservative discovery approach and although completeness is sacrificed for efficiency, they show that valid variations of that discovery procedure can be developed. Having seen how we varied our own approach should help you to design you own variations.

To finish our discussion of the managerial side of our techniques, we leave modeling the current essence and return to the project as a whole in the final two chapters. The blitzing technique illustrates the potential value of doing a less-than-complete job in a much-less-than-outrageous amount of time. In the next chapter, we exploit this concept in order to build a leveled approach to the management of the entire system development process.

Chapter 30

Effective Project Management: A Leveled Approach

Scarcity of project time is a problem not only for the essential modeling process, but for the whole system development process as well. Fortunately, blitzing as described in Chapter 29 is a time saver that can be applied to any system development activity. The result of blitzing any activity is a high-level product that will quickly show you the parts that are most important. Then, when you perform the detailed work to complete that activity, you won't waste time working on unimportant sections.

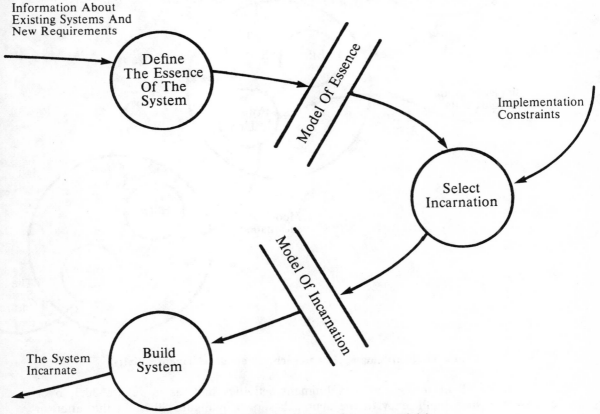

Information About
Existing Systems And
New Requirements

Define
The Essence
Of The
System

Model Of Essence

Implementation
Constraints

Select
Incarnation

Model Of Incarnation

Build
System

The System
Incarnate

Figure 30.1. Essential system development activities.

To see how to apply the blitz and follow-up approach to the entire development process, compare Figures 30.1 and 30.2. Figure 30.1 shows a diagram of the three essential system development activities. If you blitz and then complete in detail each of those activities, you will be following the plan in Figure 30.2. Using the blitz and follow-up approach in each of the major activities will help you speed the development process. However, you can work still faster if you do a blitzed, high-level version of all three activities before doing the detailed work for any of them. This results in the arrangement of system development activities shown in Figure 30.3.

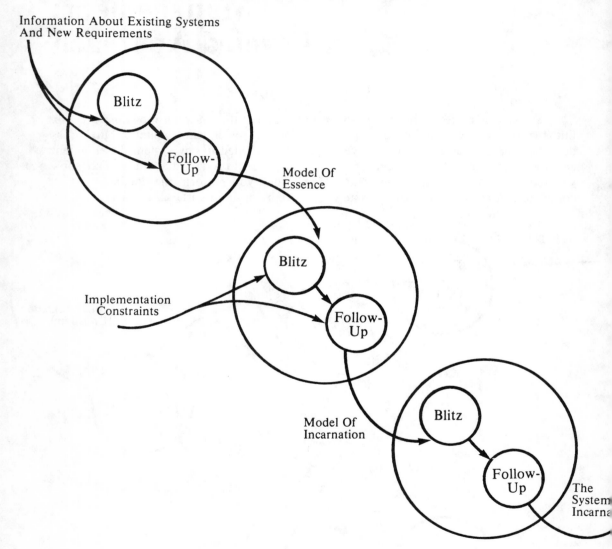

Figure 30.2. Blitzing applied to each system development activity.

Blitzing all of the system development activities first, as in Figure 30.3, overcomes a possible inefficiency in the approach shown in Figure 30.2. In the latter approach, you still complete each of the major steps before moving on to the next one.

That means you spend a lot of time completing the detailed essential model before you see a high-level picture of the new incarnation. You may discover that some of that time was wasted if your blitzed model of the new incarnation indicates that you emphasized less important areas of the system's essence.

With the approach shown in Figure 30.3, you overcome this limitation. The high-level models produced in the first activity of this approach will tell you where to focus your efforts for the entire project before you begin any detailed work.

In this chapter, we discuss how blitzing can be applied within the system development life cycle to produce distinct advantages over the traditional approach. Then, we offer some practical tactics for using the process more effectively.

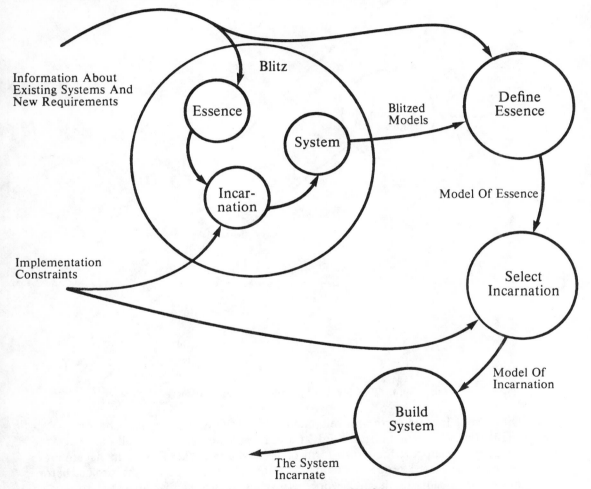

Figure 30.3. Blitzing the entire system development process.

30.1 Leveled system development

If you blitz the first two system development activities — Define Essence and Select Incarnation — one right after the other, that part of the system development process can be reorganized to consist of the two activities shown in Figure 30.4. The objective of the first activity is to build high-level models of the essence and the incarnation of the new system, and the objective of the second activity is to build the de-

tailed portions of these models. We call this method of organizing the system development process *the leveled approach* because it produces two levels of each product — one
an overview, the other detailed. The leveled approach combines our modeling techniques with the system development life cycle. That is, we apply our techniques of leveling as used in drawing DFDs to the life cycle activities that are necessary for the
development of a high-quality system.

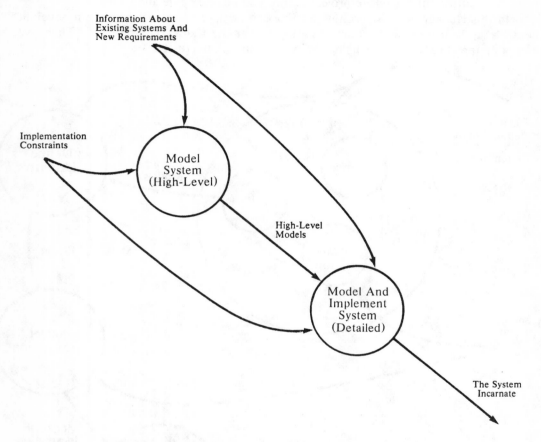

Figure 30.4. Two levels of system development activities.

The idea of leveled system development isn't unique to our approach. Brian Dickinson [13], Larry Proctor [38], and Ed Yourdon [53], among others, all use much the
same argument to support their own form of leveled approach: Project management is
more effective when the proper focus or direction is established for the project before
the project team begins its detailed work. However, our particular approach to leveled
system development is unique. Unlike the others, it supplies a purely technical method
for logical modeling and new physical modeling, and it suggests how the technical
methods can be adapted to managerial constraints.

30.2 Problems with the unleveled approach

The traditional, unleveled approach to organizing system development activities has the following characteristics:

- Each task is performed in sequence so that all the information necessary for completing a task will be attained by completing previous tasks.

- Project members complete the product of one task to the lowest level before moving on to the next task.

- Project members make only one pass through the sequence of tasks.

These characteristics cause two major problems to threaten projects using that approach. The first problem is that there is no practical way for teams using a traditional effort to focus their attention on the more vital parts of the system, and consequently they are forced to devote as much time to trivial areas as to important ones. For example, during current physical modeling, these project teams try to analyze the entire existing system because they can't properly focus their efforts. This wastes a lot of precious project time and sooner or later jeopardizes the entire effort.

When project management fails to focus the development effort, the project may end up proposing the wrong system to solve the right problem or the right system to solve the wrong problem. Since the traditional approach provides no opportunity for taking a high-level view of the project, it may be some time before management realizes that it is on the wrong track, if in fact it ever does.

The second problem with the traditional approach is that the project team has to complete work at the lowest level of detail before moving to the next step. This can slow progress considerably. For example, sometimes project management refuses to allow the project to move on to the current logical phase because the current physical system has not been documented down to the last detail. So, the project is held up even though the final details of the current physical model are almost entirely irrelevant.

The two problems with the traditional approach reinforce each other, and together, they conspire to land many projects in one tarpit or another. If you fail to focus your efforts, you will end up with a lot of detailed work to do. If you can't move to the next step of development until you work through all the details of the step you are on, your lack of focus may leave you with so much work that you may never finish the step.

30.3 The benefits of the leveled approach

The benefits of the leveled approach stem from project members taking the time to get a basic idea of the essence and incarnation of the new system before doing the detailed work. With an overview of the system available, project managers are less likely to choose the wrong system development strategy. For example, they will know when to derive the essence of a new system from that of an existing system, rather than create the essence from scratch. The managers will also be able to change the project's direction more easily, since there is a more gradual commitment to any given direction. This includes being able to change the emphasis on certain portions, the work procedure, the work allocation to team members, and the definition of deliverables.

Another advantage of the leveled approach is that it provides more information than the traditional approach for making estimates and allocating resources. Measurements made during the high-level modeling can help the project manager make resource projections and requests more accurately. In addition, the leveled approach exposes the project team members to the complexities of the work only gradually, thus avoiding their being possibly overwhelmed by the project's complexity. Finally, the approach enhances the image of the project with management, since it can perceive at an early stage that the project is designing the solution, rather than merely researching the problem.

In the remainder of this chapter, we present three basic ideas that will help you put the leveled approach into practice: plumbing the amount of detailed work in a given system development activity, assigning special workers to build high-level products, and doing system development activities to an intermediate level of completion.

30.4 Plumbing the depths of a detailed activity

Before plunging into the detailed phase of an activity, you should find out the amount of time and effort needed to finish the detailed work. Since blitzing doesn't provide you with this information, we recommend that you use a technique called plumbing.

Plumbing is a nautical term for the act of measuring water depth with a weight suspended from a line. In leveled system development, it refers to measuring the amount of detailed work needed to flesh out a given high-level model. You start plumbing by selecting one or more representative areas of the high-level model. For each area selected, you work out the detailed features of the model. When plumbing a blitzed essential model, for example, you would choose one or more representative essential activities and model them to the lowest level of detail. This process is shown in Figure 30.5.

While plumbing selected areas, you measure the time and resources taken to complete the model. From these measurements, you can accurately estimate the resources that will be required to complete the high-level model in all areas. You may also find reason to expand or contract the scope of your modeling effort.

30.5 Project scouts

Directing all project team members to participate in all levels of project activities may not be the most effective way to manage project resources. This is especially true of building high-level products, an activity that may not provide enough work to keep everyone busy. Our alternative is to establish a *scouting* function, which can be filled either by a single person or by a whole group. The scouts are responsible for performing the high-level work. The rest of the project team follows the trail of the scouts, by building the detailed levels in the areas identified by the scouts.

For project scouts to be effective, they must have a head start on their work. The remaining members of the project only start to build detailed products after some interval has elapsed. In this way, the other team members won't be idle while the scouts are working, and the project manager will have some lead time to alter the direction of the project according to what the scouts find.

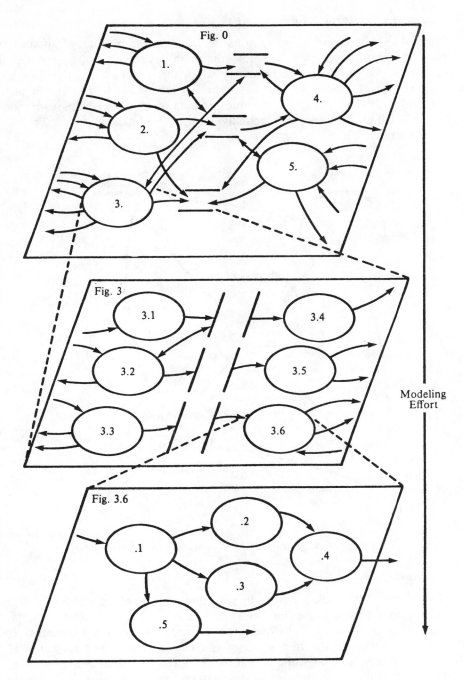

Figure 30.5. Plumbing a blitzed essential model.

30.6 Intermediate levels of development

There are sometimes good reasons for building the product of an activity at some intermediate third level that is below the high level, yet above the detailed level. Intermediate-level products are necessary when the system being developed is so large that the high-level models do not have enough information to help you plan your approach to the detailed work. So the intermediate-level model helps you to locate areas requiring detailed work without actually doing the detailed work. This approach is displayed in Figure 30.6.

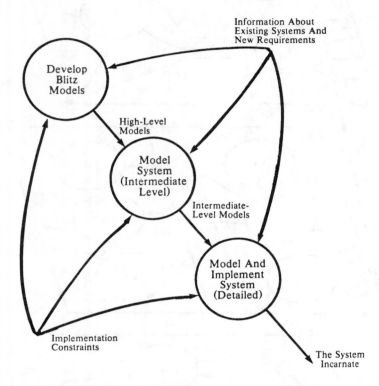

Figure 30.6. Three levels of system development activities.

30.7 Summary

To practice the leveled approach, you blitz high-level models for both the current and new system before modeling any system details. This approach is helpful for just about any project: It shows developers the important parts of the system to focus on, it helps managers choose the best strategy and direction for their project, and it makes more accurate estimates possible. We provided a few tactics to help you use the leveled approach (plumbing the depths of an activity, using project scouts, and building intermediate-level models), but many more tactics need to be invented. Still, no matter how many tactics are added and how precise the leveled approach becomes, it will always be some rearrangement of the activities that would be performed in an ideal system development environment. Therefore, an understanding of the ideal approach to system development is crucial to the proper application of the leveled approach.

Chapter 31

Planning Long Projects: The Project as Moving Target

Blitzing and the leveled approach to project management are techniques that reduce the time lost during structured analysis projects. By applying these techniques and using automated development aids such as data dictionary packages, you should be able to produce high-quality structured specifications quickly.

Unfortunately, producing the specification on schedule may not be enough to keep your project alive. Many projects are canceled even though the quality of their work is good and they meet the deadlines for their intermediate products. These projects are canceled because of factors that are beyond project management's control. Some projects are sitting ducks for the destructive effects of these external factors. In this chapter, we investigate this most frustrating and potentially deadly predicament: Assuming that you have optimized your project's performance, how can you insure your project against destruction from factors outside your control? Our objective now is to give you some tactics that will turn your project into a moving target: A project reduces its vulnerability to these external forces by producing a series of working systems that each take only a short time to build.

31.1 Destructive external factors

These six factors beyond the control of a project team can destroy a project:

- *politics:* Someone somewhere above you in management may want to do you in. He or she could be out for vengeance, interested in self-preservation, or trying to destroy his or her competition. This factor rarely surfaces as such. Instead, political motives lie behind other destructive external factors [1].

- *staffing changes:* Key members of your user management or key members of the technical staff may be transferred or promoted out of your project or the total project staff may be cut. In both cases, manpower needs in other areas of the organization disrupt your project. The departure of these project participants often results in an irreversible loss of momentum and know-how.

- *budget reallocation:* Many projects are longer than the organization's budgetary process; management will be making a new budget well before such a project is finished. If management priorities have changed by the time the new budget is decided, money may be taken from your project and used for some other purpose. That cut in funding may be the beginning of the end. Even a well-planned and executed project needs the means to accomplish its work.

- *system environment changes:* Customer preferences may change. New scientific or engineering approaches may be developed. The economy may boom or go into a recession. Any of these changes may cause management to reevaluate the need for your project.

- *management impatience:* Well into a project, management may decide it can't put up with the system as it is, and it can't wait for you to finish your project to replace the system. Even though your project is running according to schedule and even though management endorsed that schedule, management now insists on canceling your project and starting another project to produce a quick fix.

- *major changes in technology:* If a new generation of hardware or software technology emerges during your project, you may find that the need for the project disappears. For example, you no longer need to build a huge system to support a centralized, mainframe-supported database if distributed databases appear that can maintain data integrity as well as the centralized database can.

Any one of these factors could destroy a smoothly running project, at any time from project initiation on. Often, a project is scuttled by several of these factors occurring at once. However, all projects are not equally likely to fall victim to problems beyond their control; certain projects are more exposed to these problems. These are sitting duck projects.

31.2 Sitting duck projects

Since the above problems are beyond your control, about all you can do is limit the length of time your project is exposed to them. Short projects often avoid these problems because they are usually completed before the effects of external factors can be felt. Projects that are scheduled to turn over a system only after a long period of time are much more likely to be affected by these external factors. Lengthy projects are like sitting ducks waiting for the pressure of external factors to build up and overwhelm them.

In our experience, projects that produce a complete product in two years or less are far less likely to be damaged by external factors. Projects that won't produce anything for more than two years are far more likely to fail because of external factors, especially because of management's patience running out. Any project that takes significantly longer than two years to produce a working system is a sitting duck.

Whether or not you produce your intermediate models on schedule has little effect on whether your project becomes a sitting duck. Producing intermediate models during your two-year safety window may keep management happy during that time, but probably will not save a project that has just passed the two-year mark without producing a working system. There is no substitute for an acceptable system as proof of success. Users and upper DP management will love your structured analysis documentation, but don't be fooled. As time moseys on, they will still want hard proof that the project is accomplishing something real. Structured analysis models prove that work is being done, but the psychology of system development success demands that you produce an actual working system within two years.

31.3 Turning a sitting duck into a moving target

The obvious way to prevent your project from becoming a sitting duck is to make sure that you produce a working system within 18 to 24 months. However, this obvious solution has an obvious flaw: Some projects are chartered to produce systems that can't possibly be completed in two years' time. In that case, the solution is to divide the end product of the project in such a way that you can get at least part of the final system working every 18 to 24 months for as long as it takes to complete the whole effort. If you take this approach, the project is always reasonably close to implementing something useful for the organization. It therefore becomes harder to destroy or easier to redirect if there should be a dramatic change in the external factors.

You probably just felt a sense of déjà vu. "Versioned implementation" of a software system — producing several versions or releases — is not a new idea. Many other authors have advocated similar strategies [13, 21, 31, 33, 34, 53]. However, versioned implementation is a little different from what we have in mind. Versioned implementation is designed to overcome the impossibility of testing and installing a large system all at once. Versions are usually not created until the construction phase of development. The purpose of our strategy — to protect the project from external factors — is different. You need to build versions because the entire project is not viable if you attempt to build all the system at once. Therefore, the two-year efforts are established at the very beginning of the project. As the project progresses, you will be able to redefine the two-year efforts as necessary.

To plan the strategy for turning your project into a moving target, you have to answer a number of questions. First, you must ask whether you have a project that must be divided into two-year increments. If you decide that you have to partition the project, then you must answer an additional question: How far along in the project do you plan the different versions? The leveled approach to project management can provide you with answers to these questions. It can be used to produce project versions that will keep your project a moving target.

31.4 Model-based project planning

When using the leveled approach, you produce a high-level version of the products of the various phases of system development. You can use these high-level models to estimate the scope of the effort needed to carry out the project charter. In many cases, these estimates will indicate that your project can be successfully completed within the two-year window of safety.

Figure 31.1 shows how the leveled approach is applied if the project fits the 18-to-24-month time frame. The high-level or blitzing activities occur first. The high-level models are used both to estimate the length of the project and to determine which portions of the system the developers will concentrate on. The blitzed models are then fleshed out by the detailed activities. In other words, since the project doesn't need to produce separate versions, it proceeds normally.

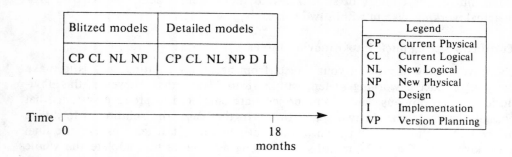

Figure 31.1. Simple project plan.

In the typical case, however, high-level blitzing reveals that the project will take far more than two years to complete. These days, huge projects (any project measured in people-centuries is huge) are chartered because of the powerful technologies available. High-capacity, superfast computers; database and teleprocessing software; and sophisticated networks of minicomputers or microcomputers make it possible to implement vast systems. An outgrowth of the incarnation technology explosion is the MIS concept, which has launched hundreds of projects for the purpose of automating enough of the company's information processing activities to place the organizational pulse at the fingertips of upper management.

An even more recent holy grail is the integrated corporate data resource. Supporters of this notion make impressive cases for the development of large software systems that consist of several traditional application systems and that are organized around centralized databases. The question is whether the system development technology can support development on such a large scale.

System developers can attempt large-scale efforts by using the leveled system development approach and the moving target project concept. Figure 31.2 shows how you would approach this problem. Just as before, you start with a blitz of the major models. This blitz confirms what you already suspected: You have hooked a whale of a project.

Next, you must decide at what point in the project you will plan the versions. This depends on how long the project is, based on the estimates from the blitzed models. The chart in Figure 31.2 shows that in this case, developers estimated that they could produce the detailed current physical, current logical, and new logical models for all versions and still have time to plan versions and select and construct the first version all within 18 to 24 months. Figure 31.3 shows a similar picture, except in this case the project developers felt they could finish the detailed design before planning the versions. Notice that by the end of the first 18 to 24 months for both projects, work on the second version has begun.

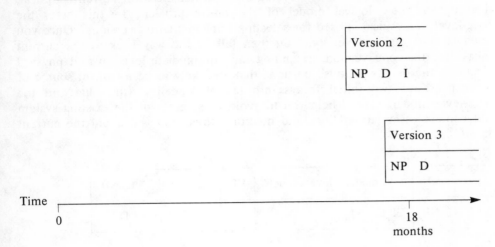

Figure 31.2. Project plan that calls for selecting versions after detailed new logical model is completed.

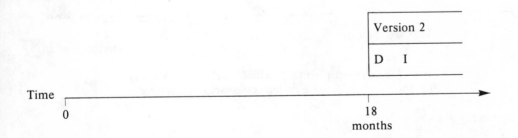

Figure 31.3. Project plan that calls for selecting versions after detailed preliminary design.

So far, we have assumed that you can complete a detailed new logical model or perhaps a detailed design, plan future versions, and build the first release within 24 months. However, some projects are so large that just producing the detailed new logical model could take 18 to 24 months. Your major option in this case is to expand the number of levels that you use in the leveled approach to project management. Instead of having only high-level and detailed models, you should plan on building an intermediate-level model. You would use this intermediate-level model to do version planning.

A project plan using the leveled approach is displayed in Figure 31.4. Notice that after the initial blitz, you proceed to develop an intermediate-level current physical, current logical, and new logical model of the entire project. In this case, the intermediate-level new logical is used for selecting implementation versions. Once you have defined at least the first version, you then fall all the way back to the current physical phase for that one version, extending the intermediate-level current physical model to capture the details of those portions that are likely to be a fruitful source of system essence. Your two initial forays into logical modeling (the blitz and the intermediate-level model) should help you to avoid those areas of the existing system that do not contain essential activities and memory; thus, you will avoid the current physical tarpit.

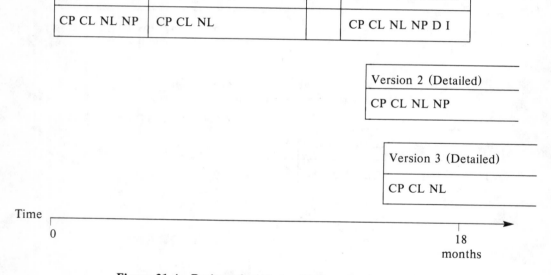

Blitzed models	Intermediate-level models	VP	Version 1 (Detailed)
CP CL NL NP	CP CL NL		CP CL NL NP D I

Version 2 (Detailed)
CP CL NL NP

Version 3 (Detailed)
CP CL NL

Time

0   18
                                                              months

Figure 31.4. Project plan that calls for selecting versions after intermediate-level new logical model is completed.

31.5 Summary

All of the best technical and management efforts can come to nothing if a project lasts longer than 18 to 24 months without producing a working system. We have seen projects torpedoed that were working effectively; these projects had become sitting ducks because they went more than two years without producing a system. This won't happen to your project if you take the planning precautions recommended in this chapter. Projects that deliver systems every 18 to 24 months maximize their chance of survival. To plan and execute your 18-to-24-month efforts, use the leveled project management approach and turn your projects into moving targets.

Afterword

... **from Stephen M. McMenamin**

At the beginning of *Essential Systems Analysis,* we cited the identification of true requirements as the central problem of system development. We illustrated the types and causes of false requirements that sneak into requirements specifications, and their consequences. Our alarmist rhetoric, together with your own experiences, should have sharpened your desire to identify the true requirements — the essence — of every system you study.

Essential Systems Analysis has equipped you to achieve this objective. The concepts introduced in Part One heightened your awareness of the distinction between essential and nonessential system features. The principle of perfect internal technology alone will help you to distinguish essential system requirements from technological constraints. The anatomical patterns found in existing system incarnations, as set forth in Part Three, will allow you to spot many nonessential activities, data channels, and data storage containers.

The modeling strategies introduced in Chapters 10 and 14 prepare you to derive the essence of a system from scratch or from a model of its present incarnation. The latter strategy, in particular, is described in great detail in Parts Four through Six. You will experience little difficulty transforming current physical models into their essential equivalents using this procedure. Once you have identified the essential components of a system, the modeling guidelines in Part Two will show you how to depict essential activities and essential memory using the tools of structured analysis. You will find event partitioning and object partitioning to be among the most useful system modeling tools you have used.

Part Seven introduced the system development activities that follow the definition of essential requirements: new requirements modeling and the transition from analysis to design. These chapters placed essential systems analysis in the larger context of the entire system development process. In the same way, Part Eight elaborates on the technical method presented in most of the book by addressing some of the managerial and political issues that affect system development efforts. Most important, Part Eight shows you how to use essential systems analysis principles to prevent or overcome such nontechnical problems as company politics and managerial impatience, and how to use these techniques to develop high-level system models for project planning and control.

The identification of true system requirements remains an important yet rarely achieved goal of systems analysis. We hope that *Essential Systems Analysis* will help you to develop effective essential requirements models that contribute to the success of your system development efforts.

. . . from John F. Palmer

Essential Systems Analysis was originally designed to solve the very specific problem of specifying the true requirements, the essence of the system you are developing. As this book showed, specifying essence correctly is not a trivial matter. Not only did we present two new technical strategies (creating essence from scratch and deriving essence), but we also introduced a strategy for blitzing an essential model, which enables you to cope with the hazards of the real world as you specify a system's essence. To support these strategies, we had to introduce new definitions of systems, system features, and modeling principles.

Despite the amount of material already covered, we had the audacity to deal with two more problems that you will face in a system development effort: selecting an incarnation and planning large-scale system development efforts. Our brief treatment of these problems revealed an important truth: Techniques that help you specify true requirements are central to solving these other system development problems.

When we discussed the anatomy of an existing incarnation in Part Three, our purpose was to show you how it camouflages the essence of a system. However, the components of system anatomy (processors, containers, infrastructure, and administration) were just as important to our discussion of selecting an incarnation. The imperfect technology that drives the shape of an existing system similarly drives the shape of your new system. The process of deriving essence also helped us address the problem of selecting an incarnation. As we mentioned in Chapter 26, selecting the incarnation is very much like deriving the essence in reverse.

We applied our concept of system and the principle of perfect technology, both designed to specify essence, to system development as a whole. The result, presented in Chapter 5, was the dual strategies approach. In establishing the blitzing approach for essential requirements definition, we found another tactic that would prove useful for other system development activities. Blitzing is especially valuable for managing a large-scale system development effort as a whole.

Having shown how our approach leads to methods for both selecting an incarnation and managing projects, we offer a final conclusion: The identification of essence, the distinction between logical and physical, is *the* central issue in system development.

System development efforts that aren't able to specify essence soon begin to unravel into an expensive chaos of people, paper, and equipment. We sincerely hope that the ideas of *Essential Systems Analysis* will bring order to this process, helping you to develop a correct specification, and ultimately a correct system, in the shortest amount of time.

Bibliography

[1]   Block, R. *The Politics of Projects.* New York: Yourdon Press, 1983.

[2]   Brooks, F.P., Jr. *The Mythical Man-Month.* Reading, Mass.: Addison-Wesley, 1975.

[3]   *Central Billing System Functional Requirements Model.* SofTech, Inc. Waltham, Mass.: 1976.

[4]   Chen, P. "The Entity-Relationship Model — Toward a Unified View of Data." *ACM Transactions on Database Systems,* Vol. 1, No. 1 (March 1976), pp. 9-36.

[5]   Conway, R., and D. Gries. *Introduction to Programming: A Structured Approach Using PL/I and PL/C-7,* 2nd ed. Cambridge, Mass.: Winthrop Publishers, 1975.

[6]   Couger, J.D. "Evolution of Business System Analysis Techniques." *ACM Computing Surveys,* Vol. 5, No. 3 (September 1973), pp. 167-98.

[7]   Date, C.J. *An Introduction to Database Systems,* 2nd ed. Reading, Mass.: Addison-Wesley, 1977.

[8]   Davenport, R.A. "Data Analysis for Database Design." *The Australian Computer Journal,* Vol. 10, No. 4 (November 1978).

[9]   De Bono, E. *The Mechanism of Mind.* New York: Penguin, 1976.

[10]   —————. *Practical Thinking.* New York: Penguin, 1976.

[11]   DeMarco, T. *Structured Analysis and System Specification.* New York: Yourdon Press, 1978.

[12]   —————. *Controlling Software Projects: Management, Measurement & Estimation.* New York: Yourdon Press, 1982.

[13]   Dickinson, B. *Developing Structured Systems: A Methodology Using Structured Techniques.* New York: Yourdon Press, 1981.

[14]   *Distributed Systems Handbook.* Digital Equipment Corp. Maynard, Mass.: 1978.

[15]   Flavin, M. *Fundamental Concepts of Information Modeling.* New York: Yourdon Press, 1981.

[16]   Galbraith, J. *Designing Complex Organizations.* Reading, Mass.: Addison-Wesley, 1973.

[17] Gane, C., and T. Sarson. *Structured Systems Analysis: Tools & Techniques.* New York: Improved System Technologies, 1977.

[18] *An Introduction to SADT® Structured Analysis and Design Technique.* SofTech, Inc. Waltham, Mass.: 1976.

[19] Jackson, M.A. *Principles of Program Design.* London: Academic Press, 1975.

[20] Kent, W. *Data and Reality: Basic Assumptions in Data Processing Reconsidered.* Amsterdam: North Holland, 1978.

[21] Lister, T.R. "Designing Efficient Systems." *The Yourdon Report,* Vol. 2, No. 1 (January 1977), pp. 5-7.

[22] Martin, J. *Computer Data-base Organization,* 2nd ed. Englewood Cliffs, N.J.: Prentice-Hall, 1977.

[23] _____. "The New DP Environment and How to Design for It." Course offered by Technology Transfer, Inc., 1979.

[24] McMenamin, S.M. "Seven Questions about Structured Analysis." Paper presented at SHARE 53, August 28, 1979.

[25] _____, and J.F. Palmer. "Advanced Structured Analysis Workshop." 4.1. ed. Course offered by Yourdon, Inc., 1981.

[26] _____. "The Transition Between Analysis and Design: An Annotated Case Study." Paper presented at GUIDE 53, November 1981.

[27] _____. "Structured Analysis and Design Workshop." 8th ed. Course offered by Yourdon, Inc., 1982.

[28] _____. "The Transition Between Analysis and Design." Paper presented at GUIDE 54, May 1982.

[29] Miller, G.A. "The Magical Number Seven, Plus or Minus Two: Some Limits on Our Capacity for Processing Information." *The Psychological Review,* Vol. 63, No. 2 (March 1956), pp. 81-97.

[30] Myers, G.J. *Advances in Computer Architecture.* New York: John Wiley & Sons, 1978.

[31] _____. *Composite/Structured Design.* New York: Van Nostrand Reinhold, 1978.

[32] Orr, K.T. *Structured Systems Development.* New York: Yourdon Press, 1977.

[33] Page-Jones, M. *The Practical Guide to Structured Systems Design.* New York: Yourdon Press, 1980.

[34] _____, and T.R. Lister. "Principles of Packaging." *The Yourdon Report,* Vol. 4, No. 4 (September-October 1979), pp. 5-7.

[35] Palmer, J.F. "Blitzing a Logical Requirements Model Workshop." Course offered by J.F. Palmer, 1983.

[36] _____. "Policy on Policy: A Structured Analysis Parable." *The Yourdon Report,* Vol. 5, No. 2 (March-April 1980), pp. 6-7.

[37] _____, I. Morrow, and G. Schuldt. "Structured Analysis Meets Database." Yourdon Users Group II Panel Discussion, 1979.

[38] Proctor, L.L. "A Cycling Approach to Successful Project Management." *The Yourdon Report,* Vol. 7, No. 2 (May-June 1982), pp. 5-7.

[39] Rose, L.A. "Packaging." *The Yourdon Report,* Vol. 3, No. 6 (December 1978-January 1979), pp. 2-3, 7.

[40] Ross, D.T., and K.E. Schoman, Jr. "Structured Analysis for Requirements Definition." *IEEE Transactions on Software Engineering,* Vol. SE-3, No. 1 (January 1977), pp. 6-15.

[41] Scherr, A.L. "Distributed Data Processing." *IBM Systems Journal,* Vol. 17, No. 4 (1978), pp. 324-43.

[42] Schuldt, G. "Information Modeling." Paper presented at GUIDE 54, May 1982.

[43] Simon, H.A. *Sciences of the Artificial,* 2nd ed. Cambridge, Mass.: MIT Press, 1981.

[44] Smith, J.M., and D.C.P. Smith. "Database Abstractions: Aggregation and Generalization." *ACM Transactions on Database Systems,* Vol. 2, No. 2 (June 1977), pp. 105-33.

[45] Stelmach, E. *Introduction to Mini Computer Networks.* Maynard, Mass.: Digital Equipment Corp., 1974.

[46] Stevens, W.P., G.J. Myers, and L.L. Constantine. "Structured Design." *IBM Systems Journal,* Vol. 13, No. 2 (May 1974), pp. 115-39.

[47] *Structured Analysis and Design Technique Overview.* SofTech, Inc. Waltham, Mass.: 1976.

[48] Sundgren, B. *Theory of Databases.* New York: Van Nostrand Reinhold, 1975.

[49] Tsichritzis, D.C, and F.H. Lochovsky. *Data Models.* Englewood Cliffs, N.J.: Prentice-Hall, 1982.

[50] Weinberg, G.M. *An Introduction to General Systems Thinking.* New York: John Wiley & Sons, 1975.

[51] _____. *The Psychology of Computer Programming.* New York: Van Nostrand Reinhold, 1971.

[52] Weinberg, V. *Structured Analysis.* New York: Yourdon Press, 1978.

[53] Yourdon, E. *Managing the Structured Techniques,* 2nd ed. New York: Yourdon Press, 1979.

[54] _____, and L.L. Constantine. *Structured Design: Fundamentals of a Discipline of Computer Program and Systems Design,* 2nd ed. New York: Yourdon Press, 1978.

Index

Accesses, essential, 41, 78-84, 97, 331
  consistent, 291-93
  defining, 79-80
  minimizing, 270-76
  modeling, 80-84, 97
  notation, 81-83, 94
  of shared data container, 138, 334
Activity:
  unit, 339-40, 342-44
  see also Custodial, Fundamental
Administrative facility, 30-31, 125
  consolidation of, 158-59
  interprocessor, 145, 148-49, 165, 336
  intra-processor, 123-25, 127, 148, 158, 188, 197, 336
  physical ring and, 200-201, 257-59
  removal of, 197-99
Arrow notation:
  in DFD, 81-82
  in DSD, 94-95
  in ERD, 99-100
  in structure chart, 343-44

Batch data stores, 194-95, 250-51, 334
Batching, 125-26, 144, 216-17, 340
Blitzing, 357-62
  models for planning, 377-80
  of system development, 367-69
Brooks, F., 349, 350, 383
Budget for complexity principle, 42-43, 363

Channel, data, 154-56, 180-84, 340
Chen, P., 99, 383
Cohesion, 184-86
Communication facility, 133-47
  for data translation, 141-43

for intermediate products, 145-47, 334
Complexity:
  budget for, 42-43, 363
  of data dictionary, 83, 87-88
  of DFDs, 66-68, 78, 83, 228, 314-17
  of essential model, 233-37, 310ff.
  of minispecs, 74
  of reduced physical model, 204ff.
Conglomeration, 32, 151-67
  in book selling system, 151ff.
  of essential activity fragments, 151-52, 265-66
Constantine, L., 5, 342, 385
Construct:
  parallel, 75, 282
  sequence, 74-75
Container, 16, 125
  private data in, 140-45
  shared data in, 133-39, 331, 334
Context diagram, 69-70, 76, 358-59
  essential, 358-59
  of materials supply system, 206, 207, 222, 359
  of payroll system, 70
Convolution, 32
  of essential activity fragments, 241-44
  of essential memory fragments, 253
Core, essential, 257-60, 276, 329-40, 342-43
Custodial activity, 20-22, 50, 54, 215
  essential memory update and, 76-78
  finding, 293-97
  identifying, 105-106
  minispec and, 78
  payroll example of, 20-21

Data dictionary, 41, 69
  data elements in, 85-86
  essential fragments and, 252
  expanded, 181-84
  external response and, 77
  frequent flyer example of, 281
  modeling essential accesses and, 81-84
  notation, 19, 82, 86-87, 92-93, 107, 279
  objects and, 86-93
  payroll system example of, 69
  physical model and, 168
Data elements:
  attribution, 268-70
  extraneous, 260-65
  in data dictionary, 69, 77
  modeling, 85-86
  object partitioning and, 59-61
Dataflow, 180, 188
  double-headed, 80
  naming, 78, 81
  notation, 281
  unbundling, 181-83
Data flow diagram, 41
  blitzed current physical, 364
  context, 69-70, 76, 358-59
  documenting essential activities, 279-87
  event-partitioned, 51ff., 69ff., 107
  expanded, 176-81
  global, 228, 304
  interobject relationships and, 94
  leveling of, 42, 66-68, 176ff., 314-17
  mini-model and, 322
  modeling accesses in, 80-81
  modeling objects in, 86-93
  new physical, 342
  object-partitioned, 60, 107
  of payroll system, 20, 21, 69-72
  of traffic violations system, 50-53, 322
  physical model and, 168
  temporal event and, 71-73
Data store, 19, 188
  consolidation of, 157-58, 266, 299-303
  extraneous, 260
  private, 140-45, 331, 334
  object, 269-70, 300-303, 311-12
  removal of batch, 194-95, 250-51

  repartitioning, 311-12
  sharing, 133-39, 331, 334
Data structure diagram, 93-100
Data translation, 141-42
Definition, data element, 85-87
DeMarco, T., 5, 6, 67, 85-86, 93, 99, 157
  expansion of DFD, 177
  object partitioning, 57
  specification strategy, 7
Design, software, 341-45
Dickinson, B., 370, 383
Documentation:
  mini-model and, 322-23
  of blitzed model, 363
  of essential activity, 279-87
  of physical model, 168, 363

Entity-relationship analysis, 57, 269
Entity-relationship diagram, 41, 93, 95-100, 279
  blitzed, 361-62
  notation, 95, 99
Essence, system, 3-9, 16
  components of, 17-22
  incarnation of, 25
  logical requirements, 5-7, 16-22
  modeling, 16, 41, 101-102, 106-109
  of Good Skate Co., 107
  of system development, 38, 41, 321-22, 367
Essential activity, 17-22
  accesses and, 78-84
  consolidation of, 151-67
  core of, 257-59, 329-40
  deriving, 255-78
  documenting, 279-87
  external responses and, 78-84
  fragmentation among processors, 29-30, 128-33, 330-333
  minispec and, 41
  modeling, 69-84
  partitioning by, 49-57
  results of, 76-78
  single processor and, 121-28
  two-pass strategy and, 234-37, 323-26
Essential activity model, 69, 279-87
  examples of, 279-87
  global, 97, 323
  high-level, 280-81

individual, 255-78, 324, 365-66
    of large activity, 283-87
    of small activity, 281-82
Essential fragment, 199-200
    classifying, 202-29
    consolidating, 265-66
    external event and, 205-20
    physical characteristics of, 238-53
    temporal event and, 221-25
Essential memory, 17-20, 22, 56, 120-21
    accesses, 78-84, 94, 97, 270-76, 291-93
    documentation of, 41
    fragment, 251-53
    modeling, 85-100
    partitioning of, 57-61, 266-70
    payroll system example, 18-20
    processors and, 120-21
    shared data and, 133-39
    skating system example, 105
    updates to, 76-78, 217, 275-76
Essential model, 41
    blitzing, 357-62
    current, 321-22
    first-cut, 226-29, 361
    for hotel system, 332
    for traffic violations system, 322, 325
    global, 97, 299-307, 323
    minimal, 45-46, 78
    overview for deriving, 168-71, 233-37
    partitioning themes for, 49-68
    plan for, 364-65
    principles for, 42-46, 356
    strategies, 40-42, 101-15
Essential order, 74-75, 244-51
Essential requirement, 3-9, 34, 322-28
Event, 12-14
    blitzing and, 359-60
    external, 14, 49, 50, 69-71, 205-20
    identifying, 49-56
    in payroll system example, 18
    in reduced physical model, 204-20
    in skating system example, 103-106
    in traffic violations system, 50-53
    partitioning, 49-57, 69, 204-26, 312, 314
    processors and, 120-21
    temporal, 14, 50-51, 71, 221-25
Examples:
    blood bank system, 45, 307-308
    Bob and Jack system, 23-24, 247-49
    book selling system, 151-52, 166, 239-40
    course and student system, 60-63, 80-87, 93-96, 255-57, 279
    credit card system, 262-66, 269-70
    frequent flyer system, 280-82
    Get Outstanding Orders activity, 267-69, 273-74
    hotel system, 331-40
    Lou Pole system, 122-26
    magazine subscription system, 240-41
    materials supply system, 205-29, 358-364
    mortgage system, 181-84, 241-42, 313-14
    payroll system, 17-18, 64, 69-72, 77-78, 244-46, 271, 343-44
    safe deposit system, 64-65
    Satisfy Book Order activity, 147, 239, 249
    Satisfy Material Request activity, 52, 54
    ship, consignment, and parcel system, 81-82
    shuttle ticketing system, 129-33
    skating system, 103-108
    subway token system, 10-14
    tax system, 162
    traffic violations system, 50-53, 293-97, 300-305, 322-26
    worker system, 160-61
Extraneousness, 30-31, 240, 247
    of data elements and stores, 188, 260-65
    of essential fragments, 253

False requirement, 3-4, 66, 74-75, 93
Flavin, M., 85, 86, 95, 99, 269, 383
Fragmentation, 29-30, 32, 238-40
    of essential activities, 128-33, 330-33
    of essential memory, 252
    processors and, 129-32
Fundamental activity, 17-22, 50, 54
    examples of, 17-18, 103-104
    identifying, 103-104
    response to, 76-78

Gane, C., 5, 6, 67, 99, 384

Implementation technology, 3-7, 16ff., 43-45
  external, 45
  limitations of, 29
  new features of, 323
  perfect internal, principle, 44-45
Incarnation, 23-33
  consolidation and, 152-66
  expansion of, 175-87
  in Lou Pole system, 122
  model of, 25-29, 41-42, 109-10, 168ff., 329-45
  multiple processor, 128-50, 161
  of nested processors, 163-64
  planned response and, 25-29
  single processor, 119-27, 162
  strategy to choose, 329-40
  superprocessor, 165-67
Information modeling, 8, 41, 85, 269
Infrastructure:
  consolidation of, 153-57
  external interface and, 96, 257-59, 330
  interprocessor, 133-47, 188-96
  intra-processor, 124-25, 127, 331-35
  removing, 188-96
Interactive system, 10-12, 35
Interobject relationships, 93-96
  extraneous, 296
  modeling, 287
  semantic, 98
Interprocessor administration, 145, 148-49, 165, 336
  of superprocessors, 165
Interprocessor infrastructure, 133-47
  private data and, 140-45
  removing, 188-96
  sharing data and, 133-39
Intra-object relationships, 92-93
  extraneous, 296
Intra-processor administration, 124-25, 127, 148, 158, 188, 197, 336
Intra-processor infrastructure, 124-25, 127, 331-35

Jackson, M., 93, 342, 384

Leveling:
  for project management, 367-74
  of DFDs, 42, 66-68, 176ff., 314-17
  of processors, 163-64

of system development, 369-74, 377-80
Logical:
  defined, 5-8, 16
  model, new, 102, 110
  system requirement, 5-7, 16-22, 43-44

Minimal modeling principle, 45-46, 70, 93, 310, 363
Mini-model, 322-23, 325
Minispec, 41, 69
  complexity of, 74
  example of, 108, 275, 279, 282
  expanded, 184-86
  false requirement and, 74-75
  in physical model, 168
  interobject relationship modeling, 96
  modeling accesses, 80
  one-page rule, 279, 283, 287
  updates to memory and, 78
Model:
  blitzed, 357-62, 367-69, 377-80
  classified, 207-29
  current logical, 321-22
  current physical, 109-12, 168, 170, 175-87, 351-54
  essential, 16, 41-46, 69ff.
  expanded physical, 175-87
  global, 97, 299-307, 323
  hierarchical, 341-42
  incarnation, 25-29, 41-42, 109-10, 168ff., 329-45
  local, 97
  logical, 5
  minimal, 45-46, 70
  new logical, 102, 110, 322-27
  new physical, 341-42
  planned response, 73-76
  principles of, 42-46
  reduced physical, 188-201
  stimulus, 69-76
  system, 39-46
Modeling:
  essential activity, 69-84
  essential memory, 85-100
  principles, 42-46, 310, 355-56, 363
  strategy, 40-42, 112-15
  tools, 41, 69, 75ff.
Mopping up, 363-64
Myers, G., 342, 384

Object, defined, 57, 267
  blitzing and, 360-61
  data stores, 269-70, 300-303, 311-12
  identifying, 267-69
  instance, 57
  interobject relationship, 93-96
  intra-object relationship, 92-93
  modeling, 86-93
  notation, 280-81
  occurrence, 57, 60, 79, 87, 267
  partitioning, 57-61, 266-70, 311-12, 323
Orr, K., 342, 384

Packed channel, 154ff.
  eliminating, 180, 258
  example of, 150, 182
Packed processor, 151-56, 160-61, 340
Page-Jones, M., 342, 384
Partitioning themes, 49-68
  event partitioning, 49-57, 69, 204-26, 312, 314, 323
  leveling and, 66-68
  object partitioning, 57-61
  private component file, 64-65
  subtopic partitioning, 89-92
  technical bias and, 88-91
Perfect internal technology principle, 44-45, 194, 201, 305, 310, 329ff., 355
Physical:
  defined, 5-8, 16
  names, 180, 186, 254
  ring, 200-201, 257-59, 276-78, 329, 339-40, 343
  system characteristics, 23-32
Physical model, 351-54, 356
  abstract expanded, 189, 193, 199
  abstract reduced, 190, 203
  blitzing, 363-64
  current, 109-12, 168, 170, 175-87, 257ff., 351-54
  essential model and, 168-71
  event in, 204-20
  expanded, 175-87
  mopping up, 365-66
  reduced, 188-201, 202ff.
Planned response, 12, 14, 25-29
  in payroll system example, 18
  modeling, 76-78
  system, 12-14

Planning, project, 36-37, 375-80
Plumbing, 372-73
Principles, modeling, 42-46, 310, 355-56, 363
Processor, 16, 188
  anatomy of, 119ff.
  boundary, 190ff.
  conglomerate of, 151, 160-64
  fragmentation and, 29-30, 129-32, 330-31, 333, 341
  imperfect, 35-36, 39
  multiple, 36, 128-50, 160-61
  nested, 160-65, 330
  optimizing performance, 125-26
  packed, 151-55, 160-61
  single, 119-27, 160, 162
  super, 165-67
  system development, 35-36
Proctor, L., 370, 385
Project management:
  constraints, 349-56, 375-76
  leveling, 367-74
  planning, 36-37, 375-80
  quality control activities, 37

Redundancy, 30, 270
  of essential activity fragments, 241, 307-308
  of essential activity models, 299-300, 307, 312-14
  of essential memory fragments, 253, 304-306
Requirements specification, 7, 344
Requirement, system:
  defining, 34
  essential, 3-9, 34
  false, 3-4, 66, 74-75, 93
  logical, 5-7, 16-22, 43-44
  new, 102, 322-28
  physical, 23ff.
  true, 3-9, 11, 49, 56, 66
Response, 12-14, 22
  ad hoc, 12, 25-27
  derivation of essential activity, 255-59
  event partitioning and, 49-57, 69, 204-26, 312, 314, 323
  external, 76-78, 215-17, 255-57
  in payroll system, 18
  in skating system, 104
  internal, 76-78, 215-17, 255-57

planned, 12-14, 25-29, 51ff., 73-76, 120-21
Ross, D., 5, 99, 385

SADT, 5
Sarson, T., 5, 6, 67, 99, 384
Schoman, K., 5, 385
Scout, project, 372-73
Simon, H., 10, 385
Stevens, W., 342, 385
Stimulus, 18, 22, 49
    external event and, 69-71
    in payroll system, 18, 69-70
    modeling, 69-76
    temporal event and, 71-73
Stored data: see Essential memory
Structure chart, 343-44
Structured analysis, 5-8, 56-57, 344
    project failure, 350-51
    tools, 8, 41, 69-72, 85, 101, 106, 235, 279, 350-53
Structured design, 258, 342-45
Structured English, 41, 72, 74-75, 282
    parallel construct, 75
Subminispec, 312-15
Subtopic partitioning, 89-92
Superprocessors, 165-67
Synchronization, 144, 146
System:
    characteristics, 29-32
    interactive, 10-12, 25, 35
    modeling, 39-46
    multiple processor, 128-50
    planned response, 12-14, 25-29, 51ff.
    single processor, 119-27
System development process:
    adding essential features, 321-28
    blitzing of, 367-69
    essential activities of, 38, 41, 169-70, 321-22, 367
    models and, 39-46
    resources, 349-54
    strategies, 34, 101-15

Technological neutrality principle, 43-44, 88-91, 93, 194, 238ff., 310
Technology:
    bias in reduced physical model, 238-54
    external, 197, 264
    imperfect, 29-32, 35-37, 98, 121ff., 330
    perfect internal principle, 16, 44-45, 194, 201, 305, 310, 329ff., 355, 363
    system anatomy and, 119ff.
Tools, analysis: see Data flow diagram, Data dictionary, Minispec
Transaction center, 154-56
Transform analysis, 342-44
Translator activity, 190-92, 194
Transportation facility, 30-31, 133
    internal, 124-25, 128
    on DFD, 190-91, 194
    physical ring and, 200-201, 257-59, 276-78
    stored data and, 140-43, 190-91, 334
True requirements, 3-9, 11, 49, 56, 66

Updates, modeling, 76-78

Variety show data store, 158, 164

Weinberg, V., 5, 99, 385

Yourdon, E., 5, 342, 370, 385